MW00760545

Lecture Notes in Electrical Engineering

Volume 707

The book series *Lecture Notes in Electrical Engineering* (LNEE) publishes the latest developments in Electrical Engineering—quickly, informally and in high quality. While original research reported in proceedings and monographs has traditionally formed the core of LNEE, we also encourage authors to submit books devoted to supporting student education and professional training in the various fields and applications areas of electrical engineering. The series cover classical and emerging topics concerning:

- Communication Engineering, Information Theory and Networks
- Electronics Engineering and Microelectronics
- Signal, Image and Speech Processing
- Wireless and Mobile Communication
- Circuits and Systems
- Energy Systems, Power Electronics and Electrical Machines
- Electro-optical Engineering
- Instrumentation Engineering
- Avionics Engineering
- Control Systems
- Internet-of-Things and Cybersecurity
- Biomedical Devices, MEMS and NEMS

For general information about this book series, comments or suggestions, please contact leontina.dicecco@springer.com.

To submit a proposal or request further information, please contact the Publishing Editor in your country:

China

Jasmine Dou, Associate Editor (jasmine.dou@springer.com)

India, Japan, Rest of Asia

Swati Meherishi, Executive Editor (Swati.Meherishi@springer.com)

Southeast Asia, Australia, New Zealand

Ramesh Nath Premnath, Editor (ramesh.premnath@springernature.com)

USA, Canada:

Michael Luby, Senior Editor (michael.luby@springer.com)

All other Countries:

Leontina Di Cecco, Senior Editor (leontina.dicecco@springer.com)

** Indexing: Indexed by Scopus. **

More information about this series at http://www.springer.com/series/7818

Jitendra Kumar · Premalata Jena
Editors

Recent Advances in Power Electronics and Drives

Select Proceedings of EPREC 2020

 Springer

Editors
Jitendra Kumar
National Institute of Technology
Jamshedpur
Jamshedpur, India

Premalata Jena
Indian Institute of Technology Roorkee
Roorkee, India

ISSN 1876-1100   ISSN 1876-1119   (electronic)
Lecture Notes in Electrical Engineering
ISBN 978-981-15-8585-2   ISBN 978-981-15-8586-9   (eBook)
https://doi.org/10.1007/978-981-15-8586-9

This Springer imprint is published by the registered company Springer Nature Singapore Pte Ltd.
The registered company address is: 152 Beach Road, #21-01/04 Gateway East, Singapore 189721,
Singapore

Preface

This book related to the Electric Power and Renewable Energy Conference 'EPREC-2020' constitutes the high-quality papers which provide knowledge of advancement in power electronics. EPREC-2020 was organized by National Institute of Technology, Jamshedpur, India, in May 2020. It covers the issues related to application of power electronics in power system, renewable energy, and green energy. The selected papers also contain optimization techniques used for various applications, inverter-based topologies, Clamped Multilevel and High-Frequency Resonant Inverters, harmonic elimination, induction heating, and power quality issues. These diverse topics provide an appealing collection for academicians, students, engineers, and researchers.

We appreciate the effort of authors and reviewers for their enormous contributions. We convey heartiest thanks to the entire team of Program committee, chairs, co-chairs, and student coordinators. We also appreciate the role of organizers. We are very much thankful to the entire team of Lecture Notes in Electrical Engineering (LNEE) of Springer. We hope that this volume helps to enhance the knowledge and provide the information about the application and advancement in power electronics, and this can be more informative, inspirational, and educational.

We also hope to see you at a future EPREC event.

Jamshedpur, India   Jitendra Kumar
Roorkee, India   Premalata Jena

Contents

About the Editors

Dr. Jitendra Kumar is currently Assistant Professor at the Department of Electrical Engineering, National Institute of Technology Jamshedpur, India. He received his B.Tech. from IMS Engineering college, Ghaziabad, India, in 2009, M.Tech from the National Institute of Technology Kurukshetra, Kurukshetra, in 2011, and the Ph.D. degree in electrical engineering from the Indian Institute of Technology Roorkee, Roorkee, India, in 2017. He has over 4 years of teaching experience and taught subjects like Power system, Advanced Power System, Control System, Control & Instrumentation, Power System Operation and Control, Signal & System, Power System Protection, Circuit & Network Theory. He has many papers in reputed journals. His major areas of research interests include power system protection and restructuring, Protection algorithms design in smart grid and microgrid environment, Design of protection algorithms in FACTS environment.

Dr. Premalata Jena is currently Associate Professor at the Department of Electrical Engineering, Indian Institute of Technology Roorkee, Uttarakhand, India. She received her B.Tech degree in electrical engineering from the Utkal University, Bhubaneswar, India, in 2001, the M.Tech degree in Electrical Engineering from Indian Institute of Technology Kharagpur, Kharagpur, India, in 2006, and the PhD degree from the Indian Institute of Technology Kharagpur, Kharagpur, India, in 2011. She has seven years teaching experience. She is IEEE member and INAE, Young Associate. She received many awards such as Women Excellence Award and Early Carrier Research Award, SERB, DST, Gov. of India, New Delhi, INAE Young Engineer Award, POSOCO Power System Award, in 2013. She has published many papers in different reputed journals and conferences. Her fields of interest are Smart Grid, Smart grid technology and protection, Microgrid, Microgrid Protection, Signal processing application to power system relaying, Power system Protection, Protection Issues with FACTS Devices, Protection Scheme Development for a line with FACTS devices, Disturbance localization, Signal processing application for disturbance localization.

Switched Capacitor-Based Inverter with Maximum Power Point Tracking of a Photovoltaic System

Simran Priya and S. K. Jha

1 Introduction

In India, electricity is generated from several methods such as thermal, nuclear, and fossil fuels. These methods not only pollute the environment but also have low efficiency. Several researches have been conducted on Photovoltaic solar panels for the extraction of energy. A photovoltaic cell has the advantage of simplicity and convenience. A photovoltaic system converts sunlight to electricity. It is a kind of photoelectric cell whose external characteristics such as current, voltage, and resistance may vary when they are exposed to light. PV cells may be arranged to form panels. Panels can be arranged to form arrays. Since a Photovoltaic array has non-linear characteristics, hence, it is quite time consuming and expensive to get the characteristics of the panels under varied operating conditions. To overwhelm the above difficulties, simple models of a solar panel are integrated with software including MATLAB/Simulink. The energy from the sun is of noise and pollution free, and the efficiency is around 30–45%.

PV arrays have a non-linear V-I characteristics curve. A significant part of the photovoltaic system is the tracking of maximum power. Various MPPT techniques have been studied in the previous papers [1, 2]. These MPPT methods are implemented with conventional DC-DC converters. Some of the researchers used a Switched Capacitor (SC) [3]. It is a power conditioner in which the power conversion is achieved by the charging and discharging of the capacitor between the input source and the load. Switched Capacitors have certain advantages such as ruggedness, cost effectiveness, compactness, and efficiency. The integration of the PV module

S. Priya (✉) · S. K. Jha
ICE Department, Netaji Subhas University of Technology (NSIT), Dwarka 110078, Delhi, India
e-mail: priyasimranpriya@gmail.com

S. K. Jha
e-mail: sk.jha@nsut.ac.in

1

with the Switched Capacitor (SC) is a step towards economization. The absence of transformers and inductors which lead to high power densities is one of the important characteristics of SC DC-DC converters. Several control techniques for these converters are based on non-linear models. These models will perform well in a wide spectrum of operation [4]. Switched Capacitors (SC) are not only limited to DC-DC applications but can also be used for DC-AC and AC-DC [5] conversions. However, comparing to DC-DC conversion their use for the above applications has been relatively less seen.

Reference [6] discusses the topology of an SC DC-AC inverter which has two switched capacitor (SC) subcircuits. The DC input is 12 V and the sinusoidal output is 50 Hz, 110 V rms. This paper implements the SC DC-AC-based inverter. Several researches have been conducted on SC-based DC inverters. Modular SC cell-based inverter is studied in [7]. In these two different kinds of cells are discussed [7]. One is the half-cell circuit which performs DC-AC as well as DC-DC operations, the other is the fuel cell which is used in DC-AC conversion. In [8, 9], the SC transformer is used to invert the DC input into sine output. This arrangement gives the bucked sine output in [8] whereas it gives the boosted output in [9].

Some of the researchers have shown the feasibility of using SC DC-DC converters integrated with PV modules. An enhanced SC converter for the PV system has the capability of performing local MPPT. This converter combines both switched inductor and switched capacitor technologies and has high conversion efficiency over wide ranges [10]. A gallium nitride switched capacitor-based two-stage integrated inverter is analysed in [11]. This integrated inverter is a two-stage solution in which the first is a DC-DC converter which fourfold the input voltage while another is a boost inverter with five level which is responsible for MPPT.

The two-stage process of getting the AC voltage from the PV source involves MPPT, boosting the voltage since PV voltage is less than the voltage needed at the output of the inverter, and the DC-AC inversion. These stages have combined efficiency which is equal to $\eta_1 \times \eta_2$ as given in Fig. 1. This can be achieved with single-stage topology that will track the maximum power, boost the PV voltage, and will invert the DC voltage [12]. Implementing this with single stage is not only cost-effective due to lower components used but also more efficient. This topology will have the overall efficiency of η as shown in Fig. 2.

Fig. 1 Two-stage solution and overall efficiency = $\eta_1 \times \eta_2$

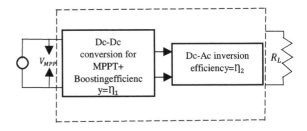

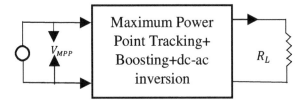

Fig. 2 Single-stage solution and overall efficiency = η

2 Proposed Topology

The schematic diagram of the Photovoltaic (PV) system integrated with a single-stage SC-based inverter is given in Fig. 3a. The model implemented in this paper performs a standalone operation.

Fig. 3 a Overall layout of PV module integrated with SC-based inverter **b** nth Switched Capacitor block of the inverter

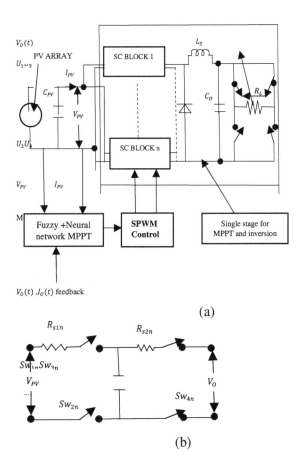

2.1 SPWM, DC-AC Inversion

The SC block is given in Fig. 3b. When switches (Sw_{1n} and Sw_{2n}) close, capacitor C_{sn} charges and when switches (Sw_{3n} and Sw_{4n}) close capacitor C_{sn} discharges into the load. Due to simplicity, Sinusoidal Pulse Width Modulation (SPWM) is utilized to generate drive signals for the Switched Capacitor inverter so as to control the SC blocks. The sinusoidal reference signal of low frequency with amplitude V_r is compared with the triangular carrier wave of high frequency with amplitude V_c. The ON and OFF duration of switches (Sw_{1n} and Sw_{2n}) and (Sw_{3n} and Sw_{4n}), respectively, are almost equal. Hence, $\alpha_1 T_s \approx \alpha_2 T_s$; here α_1 and α_2 are generated by SPWM. The modulation index is defined as

$$M = V_r / V_c \tag{1}$$

The capacitor C_s switches between the input source V_{PV} and the output V_O; the pulse width increases or decreases in a sinusoidal pattern as per the value of M. The inverter section is that where power conversion and high-frequency switching are done. Here, the unfolding section comprises switches from U_1 to U_4. Since the unfolding circuit operates at a low frequency, it unfolds the alternate sinusoidal waveforms to generate the bidirectional sinusoidal output across the load R_L. An inductor L_S is used in the output side to acquire a lesser value of THD.

2.2 MPP Tracking

The characteristics curve of the PV panel may shift under varying temperatures, partial shading conditions, varying illumination intensities, etc. To implement the MPPT when the PV array operates at the unique operating point (MPP), several conventional methods have been used in the past such as P&O [1]. It is one of the oldest and easiest techniques to track the Maximum power. As M increases, SPWM pulse width increases, hence more power can be drawn and vice versa.

2.3 Operation of SC-Based Inverter

The basic block diagram of the Switched Capacitor inverter is discussed previously. To eliminate ripples in DC voltage (V_{PV}), capacitor C_{PV} is connected in parallel across the PV module. When input switches (Sw_{1n} and Sw_{2n}) are ON and the output switches (Sw_{3n} and Sw_{4n}) of SC blocks are OFF, then capacitor C_s of all the SC blocks are charged in parallel which is connected in parallel across the photovoltaic panel. The inductor L_s freewheels through the diode D_o and charges the capacitor C_o. There is also a dead time in which all the input and output switches are OFF.

Now, when input switches (Sw_{1n} and Sw_{2n}) are OFF and the output switches (Sw_{3n} and Sw_{4n}) of SC blocks are OFF, then all the capacitors C_s connected are discharged into the circuit.

3 Proposed MPPT Technique

P&O is one of the easiest and oldest methods of MPPT. The value of $V_{o,\text{rms}}$ depends on the generation from the PV module and the load on the inverter. Pradeep and Vivek used conventional technique P&O for tracking MPP and obtained THD < 5% [3]. This paper implemented the Fuzzy Logic Controller (FLC) in combination with Error Reduction Neural Network as given in Fig. 4 to perform the MPPT of Photovoltaic panel so as to obtain a lesser value of THD at the inverter.

3.1 Fuzzy Logic Controller

FLC is popular over the last decade and has the benefit of handling non-linearity [1]. FLC usually consists of three levels which are Fuzzification, forming rule base, and Defuzzification. Based on the triangular membership function, numerical input (crisp) values get converted into fuzzy variables in fuzzification. The fuzzy subsets used are Positive Big (PB), Positive Small (PS), Zero Error (ZE), Negative small (NS), and Negative Big (NB). Table 1 shows the fuzzy rule base formation.

Fig. 4 Proposed MPPT technique

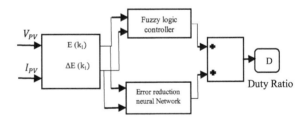

Table 1 Formation of rule base

E	ΔE				
	NB	NS	ZE	PS	PB
NB	ZE	ZS	NB	NB	NB
NS	ZE	ZE	NS	NS	NS
ZE	NS	ZE	ZE	ZE	PS
PS	PS	PS	PS	ZE	ZE
PB	PB	PB	PB	PB	ZE

In this paper, two inputs are given to FLC which are error ($E(k_i)$) and change in error ($\Delta E(k_i)$) which is defined by

$$E\left(k_i = \frac{P(k_i) - P(k_i - 1)}{V(k_i) - V(k_i - 1)}\right) \tag{2}$$

$$\Delta E(k_i) = E(k_i) - E(k_i - 1) \tag{3}$$

FLC output is usually the change in duty ratio (ΔD).

3.2 Neural Network (NN)

The basic layout of the neural network is given in Fig. 5. This paper studied the Fuzzy Logic Controller and Error Reduction Neural Network together in conjunction. There are usually three layers of NN: input, hidden, and the output layers. This system has input parameters, error and change in error which are $E(k_i)$ and $\Delta E(k_i)$, respectively. The number of nodes in each of the three layers are user dependent and it may vary. The weights W_{jk} between the nodes have to be decided throughout the training process of the neural network. The output is generally a duty cycle and is given to the power converter.

Fig. 5 Neural Network

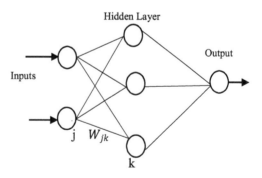

Table 2 Specifications and component selection of the integrated model

Input specifications: 60 V/70 W PV module								
Output specifications: V_o(rms) = 110 V ± 10%, 50 Hz sine wave, THD < 2%								
C_{PV}	n	Input MOSFET	Output MOSFET	C_s	L_S	R_C	C_O	THD
1800 μF	4	$R_{S1} = 0.4$ Ω	$R_{S2} = 1.6$ Ω	33 μF	850 μH	0.2 Ω	0.75 μF	0.46%

4   Results

A constant voltage is given to the input of the Switched Capacitor inverter. The target of this paper is to design an SC-based inverter that operates on a PV module. A capacitor C_{PV} is connected across the PV panel where

$$C_{PV} = \frac{P_{PV}}{4\pi V_{PV} \Delta V_{PV} f_o} \tag{4}$$

Here, V_{PV} is the input PV voltage and ΔV_{PV} is the maximum ripple allowed, P_{PV} is the maximum power of the Photovoltaic system. C_{PV} chosen is 1800 μF (Table 2). n SC blocks are required to be connected to boost the input PV voltage. Here, n is taken as 4 to meet the desired specifications. The required V_o(rms) obtained here is 110 V ± 10% and I_o(rms) = 0.8 ± 10%. Total Harmonic Distortion (THD) gives the measurement of harmonic distortions in the signal. The THD is determined with the help of the MATLAB/Simulink model. This model is used to analyse the Fast Fourier Transform (FFT) of V_O(t) which computes the THD. The THD achieved in this study is 0.46. V_O(t) and I_O(t) are obtained at the inverter as shown in Figs. 6 and 7.

The FFT analysis using the FLC and Neural Network MPPT technique is shown in Fig. 8. The result obtained is differentiated with the conventional P&O technique for the same specifications in which the THD obtained at the output is 2.35% as shown in Fig. 9.

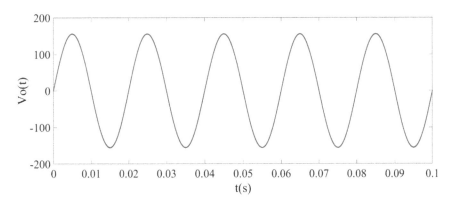

Fig. 6 Waveform of $V_O(t)$ at the inverter with $P_O(\text{max}) = 65.4$ W

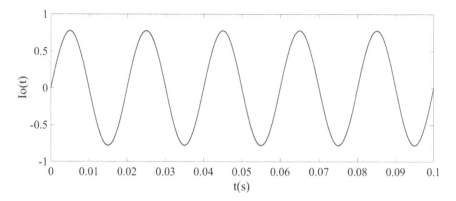

Fig. 7 Waveform of $I_O(t)$ at the inverter with $P_O(\max) = 65.4$ W

Fig. 8 FFT analysis of $V_O(t)$ using Fuzzy Logic Controller and Error Reduction Neural Network

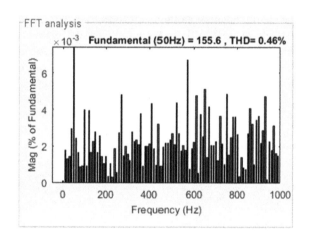

Fig. 9 FFT analysis of $V_O(t)$ using P&O

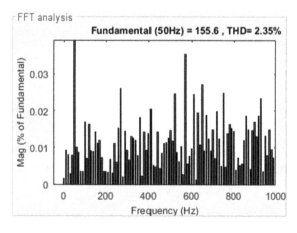

5 Conclusion

This paper has studied the performance of an SC-based inverter. Due to certain advantages of the SC inverter such as ruggedness and compactness, it is integrated with the PV module. A Simulink model is studied which has the capability of boosting the DC voltage, MPPT and results in high inversion efficiency. The SPWM control scheme was used to generate a sinusoidal output of 50 Hz, 110 V rms. Total Harmonic Distortion (THD) < 2% is attained. Also, a comparison between the proposed MPPT and the P&O is observed. THD using the proposed FLC and Error Reduction Neural Network technique is better in contrast to the conventional P&O method. The model is implemented only for R load. For RL load, the applied MPPT technique does not work as it increases the harmonic distortions at the output of the inverter.

References

1. T. Esram, P.L. Chapman, Comparison of photovoltaic array maximum power point tracking techniques. IEEE Trans. Energy Convers. **22**(2), 439–449 (2007). https://doi.org/10.1109/TEC.2006.874230
2. H. Patel, V. Agarwal, Maximum power point tracking scheme for PV systems operating under partially shaded conditions. IEEE Trans. Industr. Electron. **55**(4), 1689–1698 (2008). https://doi.org/10.1109/TIE.2008.917118
3. P.K. Peter, V. Agarwal, Photovoltaic module-integrated stand-alone single-stage switched capacitor inverter with maximum power point tracking. IEEE Trans. Power Electron. **32**(5), 3571–3584 (2017). https://doi.org/10.1109/TPEL.2016.2587118
4. H. Sira-Ramirez, R. Silva-Ortigoza, Control design techniques in power electronics devices. Springer, London. (2006). https://doi.org/10.1007/1-84628-459-7
5. P. Lin, L. Chua, Topological generation and analysis of voltage multiplier circuits. IEEE Trans. Circ. Syst. **24**(10), 517–530 (1977). https://doi.org/10.1109/TCS.1977.1084273
6. A. On-Cheong Mak, Ioinovici, , Switched-capacitor inverter with high power density and enhanced regulation capability. IEEE Trans. Circ. Syst. I: Fundamental Theory Appl. **45**(4), 336–347 (1998). https://doi.org/10.1109/81.669056
7. K. Zou, M.J. Scott, J. Wang, Switched-capacitor-cell-based voltage multipliers and DC–AC inverters. IEEE Trans. Ind. Appl. **48**(5), 1598–1609 (2012). https://doi.org/10.1109/TIA.2012.2209620
8. K. Ishimatsu, I. Oota, F. Ueno, A DC-AC converter using a voltage equational type switched capacitor transformer, in *13th Annual Applied Power Electronics Conference and Exposition*, vol. 2 (1998), pp. 603–606. https://doi.org/10.1109/APEC.1998.653961.
9. F. Ueno, I. Oota, I. Harada, K. Ishimatsu, A dc-ac converter using a tapped capacitor string for lighting electroluminescence. SPEC, 5–8 (1994)
10. A. Blumenfeld, A. Cervera, M.M. Peretz, Enhanced differential power processor for PV systems: resonant switched-capacitor gyrator converter with local MPPT. IEEE J. Emerg. Select. Top. Power Electron. **2**(4), 883–892 (2014). https://doi.org/10.1109/JESTPE.2014.2331277
11. M.J. Scott, K. Zou, E. Inoa, R. Duarte, Y. Huang, J. Wang, A Gallium Nitride switched-capacitor power inverter for photovoltaic applications. In *Twenty-Seventh Annual IEEE Applied Power Electronics Conference and Exposition (APEC)* (Orlando, FL, 2012), pp. 46–52. https://doi.org/10.1109/APEC.2012.6165797

12. S. Jain, V. Agarwal, A single-stage grid connected inverter topology for solar PV systems with maximum power point tracking. IEEE Trans. Power Electron. **22**(5), 1928–1940 (2007). https://doi.org/10.1109/TPEL.2007.904202

Power Quality Problems in Grid-Connected Electric Vehicle (EV) Fast Charging Stations

Gagandeep Sharma⃝ and Vijay K. Sood⃝

1 Introduction

Electric Vehicles (EVs) will help in the future reduction of Green House Gases (GHG) emissions. According to IEA Global EV Outlook 2019, there was a 63% increase in EVs in the year 2018 compared to the previous year [1, 2]. The wider public adoption of EVs is somewhat muted due to the driving range anxiety [3]. Other barriers are lack of suitable charging infrastructure, which results in longer charging times. The three different kinds of presently available charging solutions for EV users are [4]:

Level 1: It is also known as a home-based or overnight charging system. It requires a 120 V grounded outlet and can add only about a 40-mile range with an overnight charge. It is suitable for the low-range daily usage by customers.

Level 2: These chargers are typically installed in public places or upgraded homes and is the most popular charging technique, and requires a charging unit with a 240 V outlet. It adds about a 180-mile range during an 8 h charging period. The charging rate depends upon the vehicle's acceptance rate and current rating.

Level 3: These chargers are also known as DC fast chargers. These chargers are for public places (i.e., apartment building and/or shopping plaza parking). It can add 80–90 miles of driving range in approximately 30 min. It can act as a fueling station for EVs. Most of the research in the field of EV charging is focused on these kinds of chargers. There are different standards available to govern DC fast chargers. The most famous among these are CHAdeMO standard (Japanese), CCS system (European and American), and Tesla connectors. The Society of Automotive Engineers USA (SAE), Japanese Auto Standards Development Organization, and

G. Sharma (✉) · V. K. Sood
Ontario Tech. University, Oshawa, Canada
e-mail: gaganeep.sharma@ontariotechu.net

V. K. Sood
e-mail: Vijay.Sood@ontariotechu.ca

© The Author(s), under exclusive license to Springer Nature Singapore Pte Ltd. 2021
J. Kumar and P. Jena (eds.), *Recent Advances in Power Electronics and Drives*, Lecture Notes in Electrical Engineering 707,
https://doi.org/10.1007/978-981-15-8586-9_2

Tesla electric automobiles are working to develop reliable and efficient connectors [5].

Therefore, due to DC FCS being a leading solution to popularize EV mobility [6], this study is focused on the DC FCS and its impact on the power quality of the electric grid connected to it.

The rest of the paper is arranged in the following way. In Sect. 2, the design of DC FCS is discussed. dq-SRF and UTC control strategies for the main converter are presented in Sect. 3. Power quality standards and terminology are covered in Sect. 4. In Sect. 5, simulation results and comparison tables are given. Finally, a conclusion is presented in Sect. 6.

2   Design of Common DC Bus Architecture for FCS

A grid-connected common DC Bus architecture for FCS is proposed in this study. It is assumed that 10 eV bays for charging are installed and that each EV bay has a 62 kWh (based upon a Nissan Leaf S Plus 2019) battery rating. Since it is desired that 80% of battery charging is required in about 25 min, the criteria used to calculate the rating of the station is $P = E/t$, where $P =$ charging power in watts, $E =$ Battery Energy in Wh, $t =$ charging time in hours [7], a charging station with a rating of 118 *10 kW, i.e., 1.18 MW is required. Considering some contingency reserves for different kinds of vehicles to be charged, the FCS power rating is, therefore, estimated to be about 2 MVA. Other essential parameters such as rated capacity, DC bus voltage, and size of DC bus capacitance are calculated with the help of formulas given in Table 1 [8–10].

In this era of smart grids, the inclusion of this kind of system raises the possibility to manage islanded, V2G, V2H, and V2V operations. Distributed energy resources may also become part of this kind of system.

In this grid-connected system, nearly Unity Power Factor (UPF) operation should be maintained at the Point of Common Coupling (PCC). Control system algorithms can be employed to reach this aim. The system should meet power quality standards, and IEEE 519-1992 (revised in 2014) standards are taken as a reference for the measured harmonic values. In this paper, only the charging operation and its impacts on power quality are considered.

Common DC bus architecture with 10 EVs is proposed in this paper. It has a common DC bus connected to the host utility grid through an AC-DC converter and a distribution transformer (DT), as shown in Fig. 1 [11, 12]. This architecture consists of three different stages: (1) Grid to DT, (2) AC-DC conversion with 2-Level VSC and Common DC Bus (CDCB), and (3) DC-DC conversion.

Table 1 Formulas to calculate important values for Fast Charging Station (FCS)

Value	Formula
Rated capacity of charging station (S_R)	$S_R = \frac{K_{load} N_{slot} P_{ev}}{\cos \Phi}$ where $K_{load} = 1.1$ (overload factor which takes into account the overload due to transients) N_{slot} = Number of available charging slots (taken as 10) P_{ev} = Charging power rate of an EV (Max.) $\cos \Phi$ = Power factor of the system (0.95)
DC bus voltage (V_{dc})	$V_{dc} \leq \frac{V_{min}^{bat}}{m_{min}}$, where V_{min}^{bat} = minimum battery voltage m_{min} = minimum modulation index
Capacitance value (C_{dc}) [9]	$C_{dc} = \frac{S_R * 2nt * \Delta p * \cos \Phi}{v_{dc}^2 * \Delta v}$ where t = AC voltage wave period n = multiple of t Δp = DC power range of change Δv = allowable DC bus voltage range of change in percentage

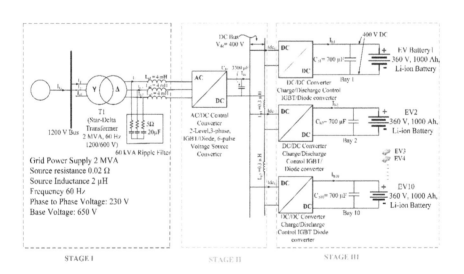

Fig. 1 Common DC bus architecture

2.1 Stage I (Grid to DT)

In stage I (Fig. 1), the host utility grid is connected to a Distribution Transformer (DT) and filters. The assumed electric grid rating is 2 MVA, RS = 0.02 Ω, XS = 2 μH, 60 Hz on the basis of 10 bays FCS. The utility grid is connected to the common

DC bus through a 2 MVA, 1200/600 V, 60 Hz DT. An AC filter of rating 60 kVA, comprising a 5 Ω resistor and 20 μF capacitor in parallel, and with a 4 mH series smoothing reactor is used for filtering purposes. The ripple filter is important to filter out distortions from the supply waveform as the load is nonlinear in nature. This filter consists of capacitors C_{r1} to C_{r3}, small damping resistors R_{d1} to R_{d3}, and smoothing series inductors L_{g1}–L_{g3}.

The values of these components can be calculated by using Eq. (1) [13]:

$$L_f = \frac{0.013 V_{LMAX}}{2\pi f I_{cr}} \tag{1}$$

where L_f = Filter Inductance, V_{LMAX} = Max. Line Voltage, f = Frequency in Hz, I_{cr} = Critical Current. Value of filter capacitance can be calculated by using Eq. (2):

$$C = \frac{1}{4\Pi^2 f^2 L_f} \tag{2}$$

These passive filters are vital to suppress the high-frequency harmonics produced by the converters. LC filter with a small damping resistor is a basic design and can be improved with LCL, LLCL-, LCL-LC- type filters. The change in configuration of the filter makes it a complex and expensive solution [14, 15].

2.2 Stage II (AC to DC Conversion with 2-Level VSC and CDCB)

A two-level, 3-phase VSC is used as the main converter. This DC supply is fed to the common DC bus voltage of 400 V_{DC} with the support of a 3300 μF DC link capacitor. This system component rating is designed for a maximum of 10 EVs. In this architecture, the Common DC bus is unipolar in nature. There are many advantages of the DC bus as it increases the stability and reliability of the system. It is easy to add renewable energy resources to FCS with a common DC bus [16]. The efficient control strategy for a 2-level VSC is significant for reliable and fast charging. Many control methods are proposed in the literature, i.e., (a) Instantaneous Reactive Power Theory (IRPT) (b) Synchronous Reference Frame Theory (SRFT), (c) Current compensation using DC bus regulation, (d) Computation based on per phase basis, (e) Scheme based on neural network techniques, and (f) Adaline-Based Control Algorithm [17–20]. IRPT and SRFT are the most commonly used control strategies for maintaining the Unity Power Factor (UPF) at the source side. Reference source currents are generated by the control method and are used to generate switching pulses for the VSC. In this work, two control strategies are used and compared. Both of these control strategies are based on the SRFT, but there is a difference in the implementation of these control methods. These control strategies are explained in the next section.

2.3 Stage III (DC to DC Conversion with EV Batteries)

Each EV bay consists of a DC-DC converter with a DC filter capacitor of 700 µF. The EVs can then be connected to this stage. A 0.3 µH inductance is assumed to be part of the leakage/parasitic inductance between bays. The battery is a 360 V, 1000 Ah Li-ion battery.

A DC-DC converter is required to control the charging/discharging of the battery. This converter may be either on-board or off-board. These converters permit bidirectional power flow from the EV battery and help in absorbing the regenerative braking energy during driving. In the charging mode, this bidirectional converter acts as a boost converter, and during the discharging/braking mode, it works as a buck converter [21–24]. There are many available topologies for these kinds of converters. In this work, the charging stage can be operated with the help of the Constant Current–Constant Voltage (CC–CV) method and can be divided into two modes of operation. In mode I, the current remains constant, and the voltage rises up to a preset value. In mode II, a constant voltage is delivered by the charging circuit, and the current starts decreasing slowly. The buck-boost converter operation can be understood from Fig. 2. In Fig. 2a, the boost converter operation is shown, which works when there is charging of the EV battery termed as the Grid to Vehicle (G2V) mode or regenerative braking during the driving cycle of the vehicle. In it, the upper IGBT and lower parallel diode work. This operation charges the battery during either the fast charging operation or when there is braking of the vehicle. It saves energy during the driving cycle, and the system becomes highly efficient. Similarly, in Fig. 2b, the buck converter operation is shown, which works during battery discharging, i.e., Vehicle to Grid (V2G) operation and acceleration mode of the driving cycle [25]. In it, the lower IGBT and upper diode conduct consecutively. This bidirectional operation makes the system efficient and economical. In this work,

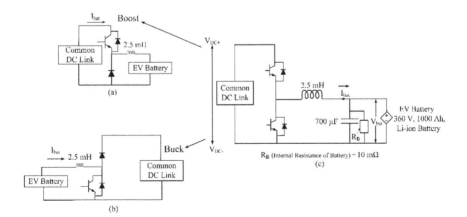

Fig. 2 Buck-boost converter operation

only the charging operation is studied, so the converter works in the boost operation mode.

3 Control Strategy for VSC in EVFCS

In this study, a 2-level VSC is used because of its simple structure and well-adopted technology. In this work, the dq-SRF (Synchronous Reference Frame) theory and Unit Template Control (UTC) methods are used and compared. The dq-SRF is the commonly used synchronizing strategy for converters based on the Phase-Locked Loop (PLL) technique, while UTC is a comparatively new algorithm that uses simple unit templates.

3.1 The dq-SRF (Synchronous Reference Frame) Control Strategy

In this method, a 3-phase frame (d-q-0) that is rotating at synchronous speed is considered as the reference frame. This frame consists of a two-phase axis, namely d- and q-axes. As the speed of this frame is same as the system frequency or stationary frame, the d-q frame is assumed as the DC quantity under steady-state conditions. The implementation of the dq-SRF method can be understood from Eqs. (3) and (4) for real and reactive power [26].

$$P(t) = \frac{3}{2}\left[v_d(t)i_d(t) + v_q(t)i_q(t)\right] \tag{3}$$

$$Q(t) = \frac{3}{2}\left[-v_d(t)i_d(t) + v_q(t)i_q(t)\right] \tag{4}$$

In Eqs. (3) and (4), if $v_q = 0$, the real and reactive power components $P(t)$ and $Q(t)$ are proportional to i_d and i_q, respectively. Due to this simplicity, the dq-SRF strategy is extensively employed for controlling the VSC. The block diagram of the dq-SRF strategy is shown in Fig. 3.

In Fig. 3, three grid currents (i_a, i_b, i_c), DC bus voltage (V_{DC}) of the VSC, and PCC voltages (V_{sa}, V_{sb}, V_{sc}) are sensed as feedback signals. The grid currents in the three phases are converted into the dq0 frame using Park's transformation as per Eq. (5):

$$\begin{bmatrix} i_d \\ i_q \\ i_o \end{bmatrix} = \frac{2}{3}\begin{bmatrix} \cos\theta & -\sin\theta & \frac{1}{2} \\ \cos\left(\theta - \frac{2\pi}{3}\right) & -\sin\left(\theta - \frac{2\pi}{3}\right) & \frac{1}{2} \\ \cos\left(\theta + \frac{2\pi}{3}\right) & \sin\left(\theta + \frac{2\pi}{3}\right) & \frac{1}{2} \end{bmatrix}\begin{bmatrix} i_a \\ i_b \\ i_c \end{bmatrix} \tag{5}$$

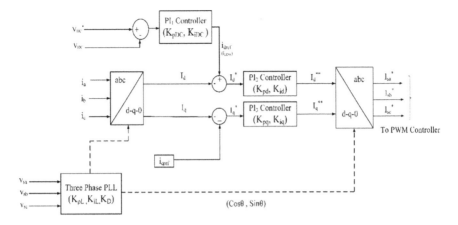

Fig. 3 Block diagram of dq-SRF control strategy

The PCC voltages are synchronized with these signals by using a 3-phase PLL. After that, the DC components of current are extracted, and the system can be operated in the UPF operation. The I_{qref} is set to zero to obtain the UPF operation. The UPF operation is desired for control operation, and the reference currents are obtained by the reverse Park's transformations as per Eq. (6):

$$
\begin{bmatrix} i_{sa}^* \\ i_{sb}^* \\ i_{sc}^* \end{bmatrix} = \begin{bmatrix} \cos\theta & \sin\theta & 1 \\ \cos\left(\theta - \frac{2\pi}{3}\right) & \sin\left(\theta - \frac{2\pi}{3}\right) & 1 \\ \cos\left(\theta + \frac{2\pi}{3}\right) & \sin\left(\theta + \frac{2\pi}{3}\right) & 1 \end{bmatrix} \begin{bmatrix} i_d^* \\ i_q^* \\ i_0^* \end{bmatrix} \tag{6}
$$

These reference signals are further supplied to the PWM controller to generate gate pulses. In the dq-SRF control strategy, mining of synchronizing components (Sin θ, Cos θ) is done by using Park and Inverse Park transformations.

3.2 UTC Strategy

UTC algorithm is also based on the SRFT theory, but it uses a unit template instead of PLL to extract the synchronizing components. The overall execution time becomes small. In the UTC, an indirect current control method is used to obtain reference signals for PWM switching of the VSC. The 3-phase voltages at PCC along with DC bus voltage of VSC are used for implementing this control algorithm. These sensed voltages from PCC and the DC bus are used to derive 3-phase reference supply currents, as shown in Fig. 4.

In Fig. 4, V_{sa}, V_{sb}, and V_{sc} are the 3-phase voltages that are sensed at the PCC, and a BPF is used to filter out any harmonics. Similarly, the DC bus voltage of VSC is also used as a feedback signal for the control strategy. BPF plays a vital role in

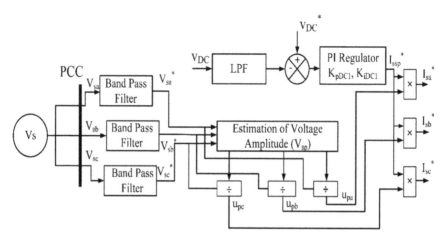

Fig. 4 Block diagram of UTC strategy [27]

real-time implementation. The self-supporting DC bus is realized using a PI voltage regulator, and V_{DC} is the voltage of the bus, and V_{DC}^* is reference DC value. This PI voltage regulator gives I_{ssp}^* as the amplitude of in-phase components of reference supply currents. This amplitude is calculated by using average DC bus voltage (V_{DC}) from the DC bus, and its reference value is set as V_{DC}^*. A comparison of these two values results in a voltage error that is fed to the voltage regulator. In the voltage regulator, proportional (K_p) and Integral (K_i) gain constants are set to achieve the desired DC bus voltage. The error signal is set as per Eq. (7):

$$V_{DCerror} = V_{DC}^* - V_{DC} \tag{7}$$

The 3-phase unit current vectors (u_{pa}, u_{pb}, u_{pc}) are obtained to be in-phase with the filtered (through BPF) PCC voltages (V_{sa}^*, V_{sb}^*, V_{sc}^*). These unit vector values are further multiplied with I_{ssp}^* to obtain the reference supply currents as follows:

$$I_{sa}^* = I_{ssp}^* u_{pa}, \ I_{sb}^* = I_{ssp}^* u_{pb}, \ I_{sc}^* = I_{ssp}^* u_{pc} \tag{8}$$

These reference values are fed to the PWM signal generator to compare with grid currents and provide the gate signals. In (8), u_{sa}, u_{sb}, and u_{sc} are in-phase unit currents obtained from the operation of estimated voltage values as

$$U_{pa} = V_{sa}^*/V_{sp}, \ U_{pb} = V_{sb}^*/V_{sp}, \ U_{pc} = V_{sc}^*/V_{sp} \tag{9}$$

where V_{sp} is the amplitude of supply voltage and is calculated by using Eq. (10)

$$V_{sp} = \left\{ \frac{2}{3} \left(V_{sa^*}^2 + V_{sb^*}^2 + V_{sc^*}^2 \right) \right\}^{\frac{1}{2}} \tag{10}$$

This value is further used to calculate reference direct and quadrature values to generate a gate signal for VSC. These reference values are fed into the PWM signal generator to compare with grid currents and generate gate signals [28, 29].

Higher rating and number of conversion stages make the system complex. This complexity increases power quality problems in the FCS, in which harmonics are predominant. The reason for these harmonics is the nonlinear nature of converters and this nonlinearity increases with the number of conversion stages. So, there are strict requirements for Total Harmonic Distortions (THD) as per power quality standards and grid code [30, 31]. The problem of power quality is discussed next.

4 Power Quality Problems Associated with EV FCS

EVs are new, nonlinear dynamic loads that may cause power quality-related problems in the electric grid depending on the strength of the grid and distribution network at the PCC. Harmonics produced by the loads may increase I^2R losses in the winding of the distribution transformer, which may affect its aging and life span.

In the FCS, EVs behave like highly nonlinear loads due to the use of power electronic converters. The rating of FCS is relatively high as compared to other electrical loads; therefore, PQ problems are significant. PQ problems are further subdivided into long- and short-term voltage variations and interruptions, transient electromagnetic disturbances, and steady-state harmonics and inter-harmonics [32].

There are standards maintained by many leading organizations and societies to govern the power quality issues in the power system. SAE, IEEE, IEC, American National Standard Institute (ANSI), British Standards (BS), Information Technology Industry Council (ITIC), and Computer Business Equipment Manufacturing Association (CBEMA) are prominent organizations that are maintaining PQ standards [33]. IEEE 519-2014 is a leading standard to follow the set the limits for current and voltage harmonics. These limits are shown in Tables 2 and 3.

The calculation of THD$_v$ and THD$_i$ is vital to implement the measures to safeguard the power system from PQ problems.

5 Results and Discussion

Both the control strategies are employed to the common DC bus architecture of FCS with different transformer configurations in the MATLAB/Simulink® environment. The analysis is done for (a) Steady state and (b) Dynamic state. In Figs. 5, 6, 7, 8, 9, and 10, results of star-delta (Y-Δ) configuration of a DT operating in steady state are compared.

In Fig. 5(a), State of Charge (SoC) using both the control strategies is compared. It is shown by zooming in Fig. 5(b) that battery charges faster in case of UTC rather than the dq-SRF control strategy.

Table 2 IEEE Standard 519-2014: Current distortion limits for general distribution systems (129–69000 V) [34, 35]

Maximum harmonic current distortion (in percent of I_L)						
Individual harmonic order (odd harmonics)						
I_{sc}/I_L	$h < 11$	$11 \le h < 17$	$17 \le h < 23$	$23 \le h < 35$	$35 \le h$	TDD (%)
$<20^*$	4.0	2.0	1.5	0.6	0.3	5.0
20 to <50	7.0	3.5	2.5	1.0	0.5	8.0
50 to <100	10.0	4.5	4.0	1.5	0.7	12.0
100 to <1000	12.0	5.5	5.0	2.0	1.0	15.0
>1000	15.0	7.0	6.0	2.5	1.4	20.0

Even harmonics are limited to 25% of the odd harmonics above
Current distortions that result in a DC offset, for example, half-wave converters, are not allowed
I_{SC} = maximum short-circuit current at PCC and I_L = maximum demand load current (fundamental frequency component) at PCC

Table 3 IEEE Standard 519-2014: Voltage distortion limits [36, 37]

Bus voltage at PCC	Individual voltage distortion (%)	Total voltage distortion (%)
1 kV and below	5.0	8.0
1 kV–69 kV	3.0	5.0
69.001–161 kV	1.5	2.5
161.000 kV and above	1.0	1.5

(a)   **(b)**

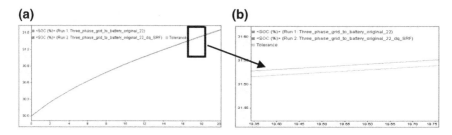

Fig. 5 Comparison of SoC of dq-SRF control strategy and UTC strategy (blue line)

(a)   **(b)**

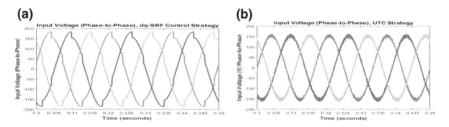

Fig. 6 Comparison of input voltage with (**a**) dq-SRF control strategy (**b**) UTC strategy

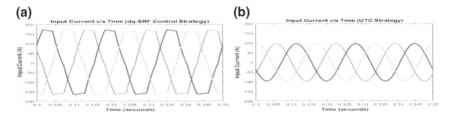

Fig. 7 Comparison of input current (**a**) dq-SRF control strategy with (**b**) UTC strategy

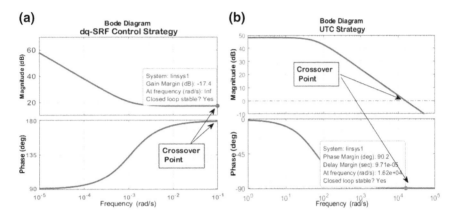

Fig. 8 Comparison of stability (**a**) dq-SRF control strategy with (**b**) UTC strategy

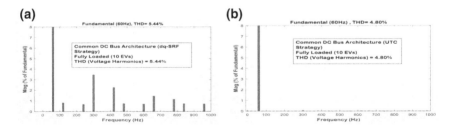

Fig. 9 Comparison of THD (voltage harmonics) for control strategies (**a**) dq-SRF control strategy, THD = 5.44% and (**b**) UTC, THD = 4.80% with fully loaded (10 EVs) common DC bus in Star-Delta Transformer Configuration

In Fig. 6, input voltage waveforms for both the control strategies are compared. The voltage waveform of UTC is relatively sinusoidal as compared to the dq-SRF waveform. UTC waveform has distortions, but notches in the case of dq-SRF strategy can create an unbalance in the system.

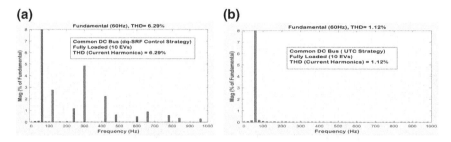

Fig. 10 Comparison of THD (current harmonics) for control strategies (**a**) dq-SRF control strategy, THD = 6.29% and (**b**) UTC strategy, THD = 1.12% with fully loaded (10 EVs) common DC bus in Star-Delta Transformer Configuration

In Fig. 7, input currents in case of both the strategies are compared. A waveform using UTC strategy is more stable and balanced as compared to the dq-SRF control strategy.

In Fig. 8, closed-loop stability analysis using Bode plots of both the control strategies is presented. It can be seen that both the control strategies are stable, but UTC strategy gives better response as observed from phase and gain margins. The phase margin is 90.2° at gain crossover frequency in UTC strategy while gain margin is 17.4 dB during phase crossover frequency. Comparative analysis shows that UTC strategy is more stable and gives faster response in the closed-loop stability analysis.

In Figs. 9 and 10, voltage and current distortions (i.e., THD_v and THD_i) are compared in the case of star-delta (Y-Δ) DT configuration. It can be observed from the FFT analysis that the UTC strategy gives better results in terms of power quality. In the case of THD_i (Fig. 10), harmonics are higher in the case of dq-SRF, which is unacceptable as per IEEE 519-2014 standards [38]. An interesting result observed in the dq-SRF case is that even-order harmonics are predominant, which results in an unbalance in the resultant waveforms.

In Table 4, THD_v and THD_i for dq-SRF and UTC control strategies are compared by considering different DT configurations. It has been observed that THD_i is unacceptable in the case of dq-SRF as per IEEE 519, and there is more unbalance in the waveforms due to second-order harmonics, as second-order harmonics cancel out and THD_v magnitude is lesser.

In Figs. 11, 12, 13, and 14, the results of the star-delta configuration of DT during the dynamic state caused by the three-phase (short-circuit) fault are shown. The fault state is created for a small duration (3 cycles), and stability of the system for both the control strategies is compared.

In Fig. 11, Input voltage waveform for both the control strategies during the dynamic state are compared. The voltage waveform of UTC gets stable faster than the dq-SRF control strategy.

In Fig. 12, Input Current in case of both the strategies during the dynamic state is compared. It can be seen that the waveform of UTC gets stable in 0.01 s as compared to 0.035 s in case of the dq-SRF control method.

Table 4 Comparison table for control strategies using different DT configurations

Harmonics during battery charging (G2V), fully loaded bus (10 EVs)

Control strategies used			dq-SRF control strategy		UTC strategy	
Type	Transformer connection configuration	Harmonics type (THD)	Percentage harmonics (THD)	Off-set component	Percentage harmonics (THD)	Off-set component
Unipolar common DC bus charging	*Star-Delta*	Voltage	5.44	Nil	4.80	0.02
		Current	6.29	Nil	1.12	0.02
	Delta-Delta	Voltage	5.44	0.01	4.89	0.01
		Current	6.20	0.07	1.13	0.03
	Star-Star	Voltage	5.44	0.07	4.89	0.08
		Current	6.20	0.07	1.13	0.03
	Delta-Star	Voltage	5.44	0.01	4.88	0.01
		Current	6.21	0.09	1.12	0.03

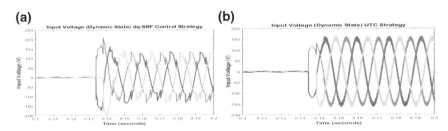

Fig. 11 Comparison of Input Voltage (dynamic-state) with (**a**) dq-SRF control strategy (**b**) UTC strategy

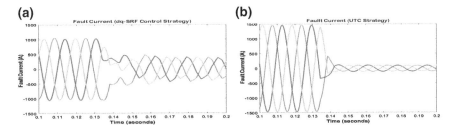

Fig. 12 Comparison of input current (dynamic-state) with (**a**) dq-SRF control strategy (**b**) UTC strategy

In Figs. 13 and 14, voltage and current (THD$_v$ and THD$_i$) harmonics during the dynamic state are compared in the case of star-delta (Y-Δ) DT configuration. It can be seen from Fig. 13, THD$_v$ is higher in the case of the dq-SRF control strategy as compared to UTC, while THD$_i$ is higher in the case of UTC as compared to the

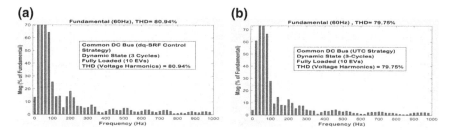

Fig. 13 Comparison of THD (Voltage Harmonics) during dynamic-state for control strategies (**a**) dq-SRF control strategy, THD = 80.94% and (**b**) UTC, THD = 79.75% with Fully loaded (10 EVs) common DC Bus in Star-Delta Transformer Configuration

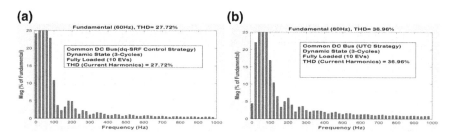

Fig. 14 Comparison of THD (Current Harmonics) during dynamic-state for control strategies (**a**) dq-SRF control strategy, THD = 27.72% and (**b**) UTC, THD = 36.96% with Fully loaded (10 EVs) common DC Bus in Star-Delta Transformer Configuration

dq-SRF control strategy. But the current waveform is more unbalanced in the case of dq-SRF, so the UTC strategy is more stable than the dq-SRF strategy and the response is faster.

6 Conclusion

In this paper, an electric vehicle fast charging station based on the common DC bus architecture is studied. In this architecture, a 2-level AC-DC VSC followed by a DC-DC converter is utilized. Two different control strategies for the VSC are employed and compared under steady-state and dynamic conditions. The two different control strategies are the dq-SRF and UTC control strategies. The two strategies are compared in terms of the following variables: Input voltage, Input current, stability, THD_v, and THD_i. It is found that the UTC strategy is more robust, stable, and has lesser impact on THD_v and THD_i. The settling time of UTC is also faster than the dq-SRF control strategy.

Different transformer configurations are also compared to study the impacts on PCC. The star-delta configuration of the transformer is found to be better than other configurations in terms of power quality.

Appendices

(A) **dq-SRF Control Strategy**: $K_{pq} = 4$, $K_{iq} = 0.002$, $K_{pd} = 1.85$, $K_{id} = 0.002$, $K_{pDC} = 1.9$, $K_{iDC} = 0.002$, PLL gains ($K_{pL} = 180$, $K_{iL} = 3200$, $K_D = 1$), $V_{DCref} = 400$ V.

(B) **UTC Strategy**: $K_{pDC1} = 2$, $K_{iDC1} = 1$, Band Pass Filters: Lower Passband Edge Frequency $= 2*pi*50$ rad/s, Upper Passband Edge Frequency $= 2*pi*70$ rad/s, LPF Passband Edge Frequency $= 2*pi*10$ rad/s, $V_{DC}^* = 400$ V.

References

1. US department of Energy Efficiency and Renewable Energy Report, https://www.energy.gov/eere/electricvehicles/reducing-pollution-electric-vehicles. Last accessed 19 March 2020
2. International Energy Agency, Global Outlook 2019 report, https://www.iea.org/reports/global-ev-outlook. Last accessed 19 March 2020
3. J.Y. Guo, J. Lu, The battery charging station location problem: impact of users' range anxiety and distance convenience. Transp. Res. Part E: Logistics Transp. Rev. **114**, 1–18 (2018)
4. Union of Concerned Scientists, Electric vehicle charging, types, time, cost and saving report, https://www.ucsusa.org/resources/electric-vehicle-charging-types-time-cost-and-savings. Last accessed 22 March 2020
5. Electric Vehicle Communication Standards, Testing and Validation—Phase I: SAE J2847/1. U.S. Department of Energy Report September 2011. https://www.pnnl.gov/main/publications/external/technical_reports/PNNL-20913.pdf. Last accessed 22 March 2020
6. CHAdeMO, What is CHAdeMO, https://www.chademo.com/about-us/what-is-chademo/. Last accessed 22 March 2020
7. M.P. Kazmierkowski, K. Zymmer, Power electronic architecture of supply systems for electric vehicle charging. Baztech Prace Instytutu Elektrotechniki J. **278**, 7–19, (2018)
8. M.R. Khalid, M.S. Alam, A. Sarwar, M.S. Jalim Asghar, A comprehensive review on electric vehicle charging infrastructures and their impacts on power-quality of the utility grid. Elsevier eTransportation **1**, 100006, (2019)
9. P. Thomas, F.M. Chacko, Electric vehicle integration to distribution grid ensuring quality power exchange, in *2014 International Conference on Power Signals Control and Computations (EPSCICON)*, Thrissur, pp. 1–6, (2014)
10. A. Arancibia, K. Strunz, Modeling of an electric vehicle charging station for fast DC charging, in *IEEE International Electric Vehicle Conference (IEVC)* (2012), pp. 1–6
11. S. Rivera, B. Wu, S. Kouro et al., Electric vehicle charging station using a neutral point clamped converter with bipolar DC bus. IEEE Trans. Ind. Electron. **62**(4), 1999–2009 (2015)
12. T. Mishima, E. Hiraki, T. Tanaka, M. Nakaoka, A new soft-switched bidirectional DC-DC converter topology for automotive high voltage DC bus architectures, in *IEEE Vehicle Power and Propulsion Conference* (Windsor 2006), pp. 1–6
13. P. Rastogi, M. Borage et al., Estimation of size of filter inductor and capacitor in 6-pulse and 12-pulse diode bridge rectifier, in *Proceedings of 2015 National Power Electronics Conference (NPEC-2015)* (IIT, Bombay, 2015)
14. S. Haghbin, T. Thiringer, M. Alatalo, R. Karlsson, An LCL filter with an active compensation for a fast charger station, in *2017 IEEE International Conference on Environment and Electrical Engineering and 2017 IEEE Industrial and Commercial Power Systems Europe (EEEIC/I&CPS Europe)* (Milan, 2017), pp. 1–5

15. S. Jiang, Y. Liu, Z. Mei, J. Peng, C. Lai, A magnetic integrated LCL–EMI filter for a single-phase SiC-MOSFET grid-connected inverter. IEEE J. Emerg. Select. Topics Power Electron. **8**(1), 601–617 (2020)
16. D. Kumar, F. Zare, A. Ghosh, DC microgrid technology: system architectures, AC grid interfaces, grounding schemes, power quality, communication networks, applications, and standardizations aspects. IEEE Access **5**, 12230–12256 (2017)
17. B. Singh, J. Solanki, A comparison of control algorithms for DSTATCOM. IEEE Trans. Indus. Electron. **56**(7), 2738–2745 (2009)
18. C.J. O'Rourke, M.M. Qasim, M.R. Overlin, J.L. Kirtley, A geometric interpretation of reference frames and transformations: dq0, Clarke and Park. IEEE Trans. Energy Convers. **34**(4), 2070–2083 (2019)
19. B. Singh, J. Solanki, An implementation of an adaptive control algorithm for a three-phase shunt active filter. IEEE Trans. Indus. Electron. **56**(8), 2811–2820 (2009)
20. M. Sabarimuthu, N. Senthilnathan, T. Manigandan, An adaptive current control strategy for a three phase Shunt Active Filter, in *2012 IEEE Students' Conference on Electrical, Electronics and Computer Science* (Bhopal, 2012), pp. 1–5
21. J. Patel, H. Chandwani, V. Patel, H. Lakhani, Bi-directional DC-DC converter for battery charging—discharging applications using buck-boost switch, in *2012 IEEE Students' Conference on Electrical, Electronics and Computer Science* (Bhopal, 2012), pp. 1–4
22. I. Lee, J. Kim, A high-power DC-DC converter for electric vehicle battery charger, in *2017 IEEE 3rd International Future Energy Electronics Conference and ECCE Asia (IFEEC 2017—ECCE Asia)* (Kaohsiung, 2017), pp. 1861–1866
23. S. Chakraborty, H.-N. Vu, M.M. Hasan, D.-D. Tran, M.E. Baghdadi, O. Hegazy, DC-DC converter topologies for electric vehicles, plug-in hybrid electric vehicles and fast charging stations: State of the art and future trends. Energies **12**(8), 1569 (2019)
24. S. Chakraborty, S. Goel, I. Aizpuru, M. Mazuela, R. Klink, O. Hegazy, High-fidelity liquid-cooling thermal modeling of a WBG-based bidirectional DC-DC converter for electric drivetrains, in *2019 21st European Conference on Power Electronics and Applications (EPE '19 ECCE Europe)* (Genova, Italy, 2019), pp. 1–8
25. T.N. Gücin, M. Biberoğlu, B. Fincan, Constant frequency operation of parallel resonant converter for constant-current constant-voltage battery charger applications. J. Modern Power Syst. Clean Energy **7**(1), 186–199 (2019)
26. A. Yazdani, R. Iravani, *Voltage Sourced Converters in Power System: Modeling, Control and Applications* (Wiley-IEEE Press, USA, 2010)
27. B. Singh, A. Chandra, K. AL Haddad, *Power Quality: Problems and Mitigation Techniques* (Wiley, West Sussex, UK, 2015)
28. R. Kumar, B. Singh, Grid interactive solar PV-based water pumping using BLDC motor drive. IEEE Trans. Ind. Appl. **55**(5), 5153–5165 (2019)
29. J. Bangarraju, V. Rajagopal, A. Jayalaxmi, Unit template synchronous reference frame theory based Control algorithm for DSTATCOM. J. Inst. Eng. India Ser. B **95**(2), 135–141 (2014)
30. S.A.R. Iunnissi, Architecture and control of an electric vehicle charging station using a bipolar DC bus. Ph.D. Thesis, Ryerson University, Toronto, ON (2016)
31. R.C. Dugan, M.F. McGranaghan, S. Santoso, H.W. Beaty, *Electrical Power Systems Quality*, third edn (McGraw Hill, 2012)
32. IEEE Recommended Practice for Powering and Grounding Electronic Equipment: Redline, in IEEE Std. 1100-2005 (Revision of IEEE Std 1100-1999) Redline ed., pp. 1–703, (2006)
33. E. Fuchs, M.A. Masoum, *Power Quality in Power Systems and Electrical Machines*, 2nd edn. (Elsevier, USA, 2015)
34. M.F. McGranaghan, S. Santoso, Challenges and trends in analyses of electric power quality measurement data, in *EURASIP J. Adv. Signal Process* (Special Issue on Emerging Signal Processing Techniques for Power Quality Applications) (2007)
35. M.H.J. Bollen et al., Bridging the gap between signal and power. IEEE Signal Process. Mag. **26**(4), 12–31 (2009)

36. A. Chudy, P.A. Mazurek, Electromobility: the importance of power quality and environmental sustainability. J. Ecol. Eng. **20**(10), 15–23 (2019)
37. IEEE draft guide for applying harmonic limits on power systems, in *IEEE P519.1/D12*, July 2012, pp. 1–124, (2015)
38. IEEE recommended practice and requirements for harmonic control in electric power systems, in IEEE Std. 519-2014 (Revision of IEEE Std 519-1992), pp. 1–29, (2014)

Study and Implementation of 1-Phase DVR for Power Quality Enhancement

CH. S. Balasubrahmanyam and Om Hari Gupta

1 Introduction

With the evolving power electronics (PE) technology, the implementation of control strategies has become so easy [1]. Simultaneously, a lot of harmonics are also injected because of the PE-based loads [2]. Apart from harmonics, other power quality disturbances are also present in the distribution system because of the continuous variations in the nature and demand of the loads [3]. There are several critical and sensitive loads which require good quality supply for their proper functioning [4]. One way of maintaining the proper voltage at the load end is to connect a DVR in series with the line [5]. The idea behind connecting the DVR is to inject the error voltage in between the supply and load so as to maintain load voltage magnitude. Also, using DVR, the waveform distortion problem can be addressed [6]. From the literature, some of the control strategies for the implementation of DVR are: using MPC to control the quasi z-source inverter-based DVR [7], unit vector template generation (UVTG)-based DVR [8], and synchronous reference frame theory (SRFT)-based DVR using photovoltaic system [9]. In the SRFT method of DVR control, if the system is unbalanced, negative sequence components may disturb the controller and that leads to malfunction, so, DSC pre-filter used PNSE is incorporated to extract the instantaneous symmetrical components (ISC) from grid voltages. It also uses multi-loop feedforward controller to enhance the DVR operation [10]. A new improved control strategy for voltage quality uses self-recovery of energy and to minimize the active power consumption during self-recovery of energy stage and another is to allow smooth switching in dynamic process [11]. A DC-DVR is illustrated to compensate

CH. S. Balasubrahmanyam (✉) · O. H. Gupta
Department of Electrical Engineering, NIT Jamshedpur, 831014 Jamshedpur, India
e-mail: chbalu16@gmail.com

O. H. Gupta
e-mail: omhari.ee@nitjsr.ac.in

© The Author(s), under exclusive license to Springer Nature Singapore Pte Ltd. 2021
J. Kumar and P. Jena (eds.), *Recent Advances in Power Electronics and Drives*, Lecture Notes in Electrical Engineering 707,
https://doi.org/10.1007/978-981-15-8586-9_3

the voltage sag at sensitive DC loads in power grid and an isolated DC-DC converter is made to work as a DC-DVR [12]. A DVR is to recompense -ve sequence voltage and harmonics in voltage based on a new double closed-loop control strategy and in this the outer voltage loop is based on PMR controller and the inside current loop is done using proportional controller and the tuning parameters are based on root locus so that the stability can be improved [13]. DVR is implemented to reduce voltage-related problems based on in-phase compensation strategy and illustrated by simulation in EMTDC of PSCAD [14]. A dual-function DVR (DF-DVR) is implemented with recloser in power system, this not only improves power quality but also limits the current during fault conditions to optimal value [15]. In this paper, based on error, the reference is generated and the VSC is controlled using hysteresis voltage controller.

2 Working and Control of DVR

The block diagram of proposed DVR is represented by Fig. 1. The appropriate location of DVR is to connect before the load bus which demands stiff voltage profile for high performance of sensitive loads. For harmonics compensation, an active source is not necessarily needed but for sag and swell compensation, an active source is compulsory. So DC source is employed. As the considered power system is 1-phase AC, DC from the source should be converted into AC to perform compensation, for this VSC is required. Filter circuit is to remove the high-frequency disturbances in the injected voltage. A series voltage injection transformer is employed to inject voltage into the system at desired level for compensation. Firing pulses are generated using the control logic as mentioned in Fig. 2 and are given to the VSC for satisfactory operation. Therefore, voltage at the load bus is the algebraic sum of voltage at source (or grid) and voltage injected by DVR into the system as mentioned in the Eq. (1).

Fig. 1 Representation of typical DVR using DC voltage source

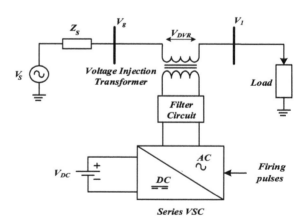

Fig. 2 Flow chart for the
generation of firing pulses

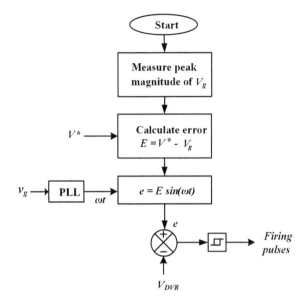

Figure 2 represents the flow diagram of controlling scheme of DVR. The grid bus and load bus voltages (i.e. V_g and V_l) are measured and the peak magnitude of load bus voltage (V_g) is obtained. A peak magnitude of specific reference voltage (V^*) is predetermined for the desired load bus to which sensitive loads are connected. The actual peak magnitude of grid bus voltage is continuously compared with the peak magnitude of specific reference load bus voltage. From this, the error (E) is found and this error in peak magnitude form is multiplied with a template which is generated using grid voltage phase (ωt) obtained from PLL block. This product (E $sin(\omega t)$) is the reference voltage which is to be tracked by the VSC to compensate the voltage . Hysteresis voltage control is used to generate pulses for the series VSC corresponding to reference generated. The compensated load voltage V_1 is given below in (1).

$$V_g + V_{\text{DVR}} = V_1 \tag{1}$$

Since the existence of non-linear type loads and connected to grid, they draw harmonic components of current from the system. Due to this, the bus voltages get distorted. This distorted bus voltage shows its significant impact on the sensitive loads. So, these distortions must be eliminated from the bus voltages. The harmonics in the bus voltages are cancelled out by injecting the same harmonic voltages out of phase in to the system using series VSC of DVR. Therefore, the load bus voltage will be free from distortions that results in good performance of sensitive loads.

3   Performance of DVR and Simulation Results

For performance investigation of 1-phase DVR, MATLAB-based simulation is carried out. A typical DVR configuration as given in Fig. 1 which describes the connection of DVR through the voltage injection transformer connected in series to the system. For sag and swell compensation, intentionally a decrement of 30% of nominal voltage and an increment of 30% of nominal voltage are created at grid side (V_S) as mentioned in Fig. 3a which are known as voltage sag (0.1–0.35 s) and voltage swell (0.6–0.85 s), respectively. Now, DVR is expected to inject the needed voltage in series with the system voltage to compensate the sag or swell in the system. In Fig. 3b, the generated reference voltage which is to be tracked by the VSC of DVR and the actual DVR voltage are depicted and they both are nearly matched which is desirable. Therefore, the nominal voltage is maintained at the load bus. The presented DVR is taking less than half a cycle to restore it to the nominal voltage whenever there is sag or swell in the system voltage.

As it is mentioned that one of the functions of DVR is to compensate harmonics in system voltage, here, Fig. 4 illustrates results for harmonics compensation in system voltage. Intentionally 10% of 3rd and 8% of 5th harmonics are introduced to the system voltage from 0.2 to 0.4 s and the corresponding waveform is shown in Fig. 4a to show how the system voltage is distorted when it consists of harmonics. Figure 4b describes the tracking of the generated reference voltage by the VSC of

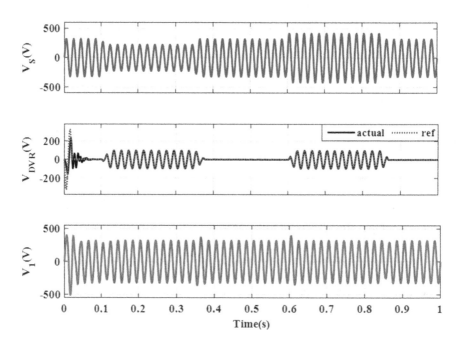

Fig. 3 Performance of DVR against sag and swell problems

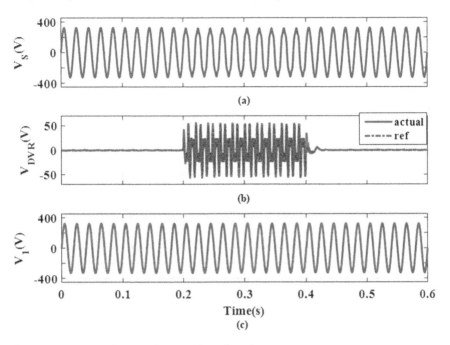

Fig. 4 Performance of DVR against waveform distortions

Table 1 Complete system data

Parameter/Device	Rating
Voltage	230 V
Frequency	50 Hz
Voltage Injection Transformer	230/230 V, 50 Hz 2.5 kVA
Load	230 V, 2.64 kW

DVR for harmonics compensation. So, the 3rd and 5th harmonics are excluded from the system voltage and only fundamental component of system voltage is appeared at the desired bus which is clear from the Fig. 4c. The DVR is very fast in response to harmonics compensation of system voltage. The complete system data is shown in Table 1.

4 Conclusion

A discussion about the requirement of power quality enhancer has been included in this paper and then the working and control of DVR have been added. The DVR presented in this study uses the error voltage obtained from the magnitude difference of reference and actual grid bus voltages. A series voltage is injected correspondingly. The control scheme is quite simple and easy to implement. The simulation results have been obtained for both the sag and swell problems as well as for waveform distortions. The results proved the effectiveness of the presented 1-phase DVR technique.

References

1. O.H. Gupta, M. Tripathy, ERF-based fault detection scheme for STATCOM-compensated line. Int. Trans. Electr. Energy Syst. (2016). https://doi.org/10.1002/etep.2314
2. K.A.G. Dos Santos, P.M. Duarte, P.F. Ribeiro, P.M. Silveira, The impact of non-sinusoidal voltages on the harmonic generation of power electronics converters, in *Proceedings of International Conference on Harmonics and Quality of Power, ICHQP* (IEEE Computer Society, 2016), pp. 762–767. https://doi.org/10.1109/ICHQP.2016.7783404
3. N. Ashraf, T. Izhar, G. Abbas, A.R. Yasin, A.R. Kashif, Power quality analysis of a three phase ac-to-dc controlled converter, in *2017 International Conference on Electrical Engineering, ICEE 2017* (Institute of Electrical and Electronics Engineers Inc., 2017, pp. 1–6. https://doi.org/10.1109/ICEE.2017.7893435
4. Bueno-Contreras, H., Ramos, G.A.: Optimal Control of an UPQC to assure Power Quality in Electric Distribution Grids. In: 2019 IEEE Workshop on Power Electronics and Power Quality Applications, PEPQA 2019 - Proceedings. pp. 1–6. Institute of Electrical and Electronics Engineers Inc. (2019). https://doi.org/10.1109/PEPQA.2019.8851538
5. A.H. Abed, J. Rahebi, H. Sajir, A. Farzamnia, Protection of sensitive loads from voltages fluctuations in Iraqi grids by DVR, in *Proceedings - 2017 IEEE 2nd International Conference on Automatic Control and Intelligent Systems, I2CACIS 2017* (Institute of Electrical and Electronics Engineers Inc., 2017), pp. 144–149. https://doi.org/10.1109/I2CACIS.2017.8239048
6. S.K. Singh, S.K. Srivastava, Enhancement in power quality using dynamic voltage restorer (DVR) in distribution network, in *Proceedings of 2017 International Conference on Innovations in Information, Embedded and Communication Systems, ICIIECS 2017* (Institute of Electrical and Electronics Engineers Inc., 2018), pp. 1–5. https://doi.org/10.1109/ICIIECS.2017.8275918
7. M. Trabelsi, P. Kakosimos, H. Komurcugil, Mitigation of grid voltage disturbances using quasi-Z-source based dynamic voltage restorer, in *Proceedings - 2018 IEEE 12th International Conference on Compatibility, Power Electronics and Power Engineering, CPE-POWERENG 2018* (Institute of Electrical and Electronics Engineers Inc., 2018), pp. 1–6. https://doi.org/10.1109/CPE.2018.8372574
8. V.S. Karale, S. Sjadhao, M. Tasare, G.M. Dhole, UVTG Based dynamic voltage restorer for mitigation of voltage sag, in *Proceedings - 2nd International Conference on Computing, Communication, Control and Automation, ICCUBEA 2016* (Institute of Electrical and Electronics Engineers Inc., 2017), pp. 1–6. https://doi.org/10.1109/ICCUBEA.2016.7860138
9. S.S. Rao, P.S.R. Krishna, S. Babu, Mitigation of voltage sag, swell and THD using dynamic voltage restorer with photovoltaic system, in *2017 International Conference on Algorithms, Methodology, Models and Applications in Emerging Technologies, ICAMMAET 2017* (Institute of Electrical and Electronics Engineers Inc., 2017), pp. 1–7. https://doi.org/10.1109/ICAMMAET.2017.8186668
10. A. Karthikeyan, D.G. Abhilash Krishna, S. Kumar, C. Nagamani, Design and analysis of multi-loop feed forward control schemes for DVR under distorted grid conditions, in *2017 14th IEEE India Council International Conference, INDICON 2017* (Institute of Electrical and Electronics Engineers Inc., 2018), pp. 1–6. https://doi.org/10.1109/INDICON.2017.8487548

11. E. Self-recovery, C. Tu, Q. Guo, F. Jiang, C. Chen, X. Li, F. Xiao, J. Gao, Dynamic voltage restorer with an improved strategy to voltage sag compensation and **4**, 219–229 (2019)
12. M. Farhadi-kangarlu, DC Dynamic Voltage Restorer (DC-DVR): a new concept for voltage regulation in DC systems, 122–127 (2018)
13. S. Zhang, Z. Zhao, J. Zhao, L. Jin, H. Wang, H. Sun, K. Liu, B. Yang, Control strategy for dynamic voltage restorer under distorted and unbalanced voltage conditions. 2019 IEEE Int. Conf. Ind. Technol. 411–416 (2019)
14. A.K. Gupta, P. Gupta, A novel control scheme for single and three phase dynamic voltage restorer using PSCAD/EMTDC. 2018 Int. Conf. Smart Electr. Drives Power Syst. 44–49 (2018)
15. F. Jiang, L. Chen, C. Tu, S. Cheng, R. Zhu, M. Liserre, Coordination of dual-functional dynamic voltage restorer and recloser in power distribution system **3**, 5–10

Study and Simulation of a PMSG-Based Wind Turbine

Sriparna Das and Om Hari Gupta

1 Introduction

Nowadays, the motto of every country is to increase the production of energy by using renewable sources by running the industry for generating energy from turbines whose blades are rotated by wind, from solar cells whose energy is stored from sun's rays, and from the movement of the tide, waves, etc. Both, the enormous power availability in the form of wind and increment in load demand have led to a significant increase in the production of energy from the wind turbines. At the end of 2010, the production became 180,850 MW but in the year 2019, we can see the power production increased to 622,704 MW.

The increased production in power generation from wind is shown in Table 1. From this table, we can see that due to the increase in demand every year, the power generation gets increased. Wind turbines, whose power generation capacity is high, run on the principle of adjustable speed generators (ASGs). The small power-producing wind turbine uses the constant speed "Danish concept" shown in Fig. 1 which tells about the direct conversion of wind energy to electrical energy using a squirrel cage IM attached with the AC power grid. With the help of a fixed gear ratio, the rotating shaft of the generator is connected with a turbine. While some of the induction generators use pole-adjustment winding to run at various synchronous speeds, at a certain operating condition, the Danish turbine has to run at a certain speed only. In the case of a fixed-speed turbine, the performance depends on the mechanical sub-circuits. When the wind strikes the blade every time, there will be a wide deviation in the output power. For sustaining this kind of operation, the stiff

S. Das (✉) · O. H. Gupta
Department of Electrical Engineering, NIT Jamshedpur, Jharkhand 831014, India
e-mail: dassriparna97@gmail.com

O. H. Gupta
e-mail: omhari.ee@nitjsr.ac.in

© The Author(s), under exclusive license to Springer Nature Singapore Pte Ltd. 2021
J. Kumar and P. Jena (eds.), *Recent Advances in Power Electronics and Drives*, Lecture Notes in Electrical Engineering 707,
https://doi.org/10.1007/978-981-15-8586-9_4

Table 1 Year-wise production of power from wind turbines [1]

Year	Onshore wind capacity (MW)	Offshore wind capacity (MW)
2010	177,794	3,056
2011	216,244	3,776
2012	261,575	5,334
2013	292,749	7,171
2014	340,808	8,492
2015	404,558	11,717
2016	452,485	14,342
2017	495,565	18,837
2018	540,191	23,629
2019	594,396	28,308

Fig. 1 Overview of grid connected wind turbine

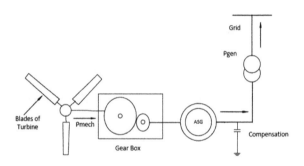

functioning of the power grid is required for making a stable operation by absorbing the mechanical stresses. Nowadays, ASGs are much preferred than FSGs, known as fixed speed generators, due to less cost and they also provide the controlling of the pitch very easily. The speed at which the generator revolves depends upon the frequency and this speed control makes the time constant value higher which lessens the complexity of the controlling of the pitch. The value of the pitch angle remains unchanged at a lower value of the wind speed. When the wind speed is increasing in nature to limit the output power pitch angle controlling technique is generally preferred. ASGs help in reducing the mechanical stress and thus the pulsating nature of the torque. As the toque is free from pulsating due to elasticity that evolved in the system, power quality gets improved. The efficiency of the system improves and the noise gets reduced. Moreover, islanding can also be allowed easily in ASGs than FSGs. So ASGs are preferred over FSGs.

In Fig. 1, we can see that the basic block diagram showing a wind turbine connected to the grid. The blades rotate and provide mechanical power to the generator unit via a gearbox. The energy transfer takes place from mechanical to electrical form with the help of a generator unit followed by a transformer for increasing the voltage as per the required level and is connected to the grid [2]. In Fig. 2, we can easily

Fig. 2 Varying electrical
power corresponding to
change in turbine speed [3]

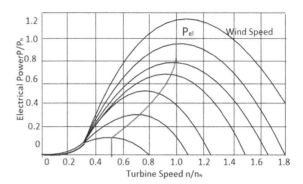

see how, with the change in turbine speed, the output power is varied. The red line
depicted in the graph shows the maximum power point tracking of a wind energy
conversion system.

In this paper the authors demonstrated at first the types of wind turbine generators
and why the authors selected the PMSG method in their simulation. Then there is
present the boost converter design and a discussion on one of the techniques of MPPT
which the authors have included and at last the inverter design, and finally the paper
is concluded with simulation results and observation.

Direct-in-line Adjustable Speed Generator: In this case, a generator running
at synchronous speed is employed to obtain AC with changing frequency. So, using
a converter of high range is essential to convert the changing frequency to some
constant value. This is employed up to 1.5 MW and there are also some disadvantages.
The use of a power converter which is rated at 1 p.u is costly. To make the output
free from ripple at the end of the inverter, the filter is used, making the design
tougher, and also the efficiency of the converter is considered to be important. If
the efficiency of the converter decreases, the power generated also becomes less—
affecting the reduction in the whole system efficiency. Figure 3a shows the model of
a direct-in-line ASG.

Doubly-Fed Induction Adjustable Speed Generator System: In this case, the
stator windings are directly attached with the grid while the rotor windings are
attached with grid via back-to-back converter blocks. By decoupling, the capacity

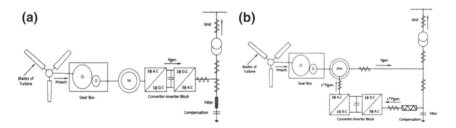

Fig. 3 Block diagram model of **a** direct-in-line ASG **b** DFIG-based ASG system

of the converter can be varied and made 30% of the wind turbine. So, the cost of the converter, as well as harmonic filter, is reduced. The disadvantages of using a direct-in-line adjustable speed generator are overcome by the doubly-fed induction generator. It consists of a converter block attached with the rotor winding, working on all four quadrants. The rating of the inverter becomes 25% of the power produced so it lessens the inverter costs. As the power becomes 25%, the filters also become 0.25 p.u of the whole system so the filter design becomes easier and the costs also reduce. Because of using IGBT whose switching and heat losses are less than other devices, the efficiency of the model improves. As DFIG is working on a four-quadrant operation, the power factor controlling scheme can be achieved at very optimum or low costs. Figure 3b depicts the model of the DFIG-based ASG system.

The requirement of gearbox, in case of DFIG, for high-speed control, makes the system bulky and costly. The replacement of DFIG by PMSG will serve the purpose effecively. The use of PMSG is a common and a modern method nowadays and as it is associated with low speed, a gear-box is not required for control. It has high efficiency as movable magnets are used [4]. If permanent magnets are used, there is no requirement of extra electrical energy. So, in the exciter, the copper loss is absent. As there is the absence of a commutator and brush arrangement, mechanical losses cannot take place, making the system compact by nature. Due to the usage of a high-power magnet, the rotor windings are not required. This helps the generator to be small in size and light in weight. As the rotor winding is absent, no current can flow in the rotor circuit, so heat produced is much less in the rotor circuit. The heat that is produced in the stator can easily be cooled down as the stator is static and it is present on the periphery of the generator. The insulation of the winding used and the magnet last for a long time. The absence of mechanical arrangements decreases the failure chance of the instruments and also lessesn the noise associating with them. By getting these advantages, PMSG can easily be replaced by DFIG but can be run with less wind speed.

2 Methodology

The model is created in MATLAB 2015a. The pitch angle, generator speed, and wind speed are taken as inputs to the wind turbine. The turbine shaft rotates and the output is torque and mechanical speed. This is fed as an input of a permanent magnet synchronous generator which gives us a three-phase supply. The three-phase supply is converted to single-phase DC by using rectifier units. Here, we measure the voltage and current. Then as per the required voltage output, we are designing the boost converter to increase the voltage. We know only the frequency of the supply and the input voltage; from there as per the requirement of the output voltage we are designing the inductor, capacitor, and resistor values which help to step up the voltage. After boosting, we are attaching a capacitor that acts as a filter in reducing the ripple. Then the single-phase supply is required to be converted into three-phase AC supply by the use of the inverter unit, and the inverter controller unit is designed.

We are changing or converting it from DC to AC; as the loads that are attached are AC in nature, we have made this conversion [5].

3   Designing of a Boost Converter

To design the boost converter, the parameters that are available are rectified output voltage from PMSG which serves as an input to the boost converter, frequency, and load current. So, we can use the following formulas to find out the values of the inductor, capacitor, and resistance. They are as follows:

$$V_0 = \frac{V_{\text{in}}}{(1 - D)} \tag{1}$$

$$\text{Load Resistance} = \frac{V_0}{I_0} \tag{2}$$

$$\text{Duty Cycle} = 1 - \frac{V_{\text{in}}}{V_0} \tag{3}$$

$$\text{Capacitance } (C) = \frac{I_0 D}{f_s V_s} \tag{4}$$

$$\text{Inductor } (L) = \frac{V_s D}{f_s I_0} \tag{5}$$

By using these five sets of formulas, we can easily boost up the voltage that is required at the consumer end [6].

In Fig. 6, it can be seen that a block named MATLAB Function is used, in which the code is written for providing duty to the inverter. The logic behind the code comes from the flow chart of the Perturb and Observe method, that is P&O, for maintaining the output voltage constant. This process is taken as the P&O method only requires the final output state and the previous to the final output state and is not dependent on any parameters like surroundings, humidity, etc. The process is quite fast in obtaining the condition, which has been discussed subsequently.

4   Perturb and Observe Method

This technique is most widely used among all other MPPT methods where we can easily track or find out the maximum power point for keeping the boost converter output constant and the output also maximizes. In this method, the current power reading P_k and the previous reading P_{k-1} are found out, and they both are compared.

First, a small perturbation is introduced in the system and with that change in perturbation, the duty cycle is changed to track the maximum power point. If $P_k > P_{k-1}$,
then the duty cycle is increased and shifts in the right direction to get that point while
if $P_k < P_{k-1}$ then the duty cycle is decreased and shifts in the left direction to reach
that point [8]. The algorithm is discussed in Fig. 7.

5 Inverter Controller Design

In this design, the CCVS PWM technique is applied. CCVS stands for current control
voltage source where the increment and decrement of current depend on the change
of DC voltage. The controller compares the DC output voltage with the threshold
voltage, and changes the output current accordingly. Here, DC is fed to the universal
and the pulses are given to this IGBT from a PWM 2-level pulse generator. The three-
phase voltage is fed to PLL where we can get the phase as well as the frequency.
Then the voltage is transformed from abc domain to dq0 domain, and a discrete
PID controller is used where we have set the operating range and we have tuned
the controller to get the value of the constant. Then again, it is converted to the abc
domain which we give to the PWM generator for generating pulses for the IGBT [9].
In Fig. 8, we can see the simulated model of an Inverter Controller and the output is
shown in Fig. 9.

6 Simulation Results

Each block has been simulated in MATLAB 2015a and the output is taken. The wind
speed is varied from 6 m/s to 8 m/s, and we are seeing the MPPT with the change in
wind speed. Here, we are keeping the value of density of air $\rho = 1.08$ m^3/kg, the value
of gear ratio as 5, the radius as 1.525 m, and finally the speed as 20 revolutions per
second. In the case of PMSG used the stator, the following parameters are considered,
phase resistance as 1.5 Ω, the armature inductance as 10 mH, flux linkage as 0.119
δWb, the inertia J as 2 J(kg m^2) and the number of poles as 4. The three-phase round
rotor is used. Regarding the converter, two resistors of 1 O, Co = 3500 mF, C1 =
500 mF, and inductance of 200 mH are used. Switching frequency is considered to
be 5000 Hz. The complete model of WECS is shown in Fig. 5; the designing of
the converter is shown in Fig. 6 whereas the boosted current waveform presented
in Fig. 4a, b gives the information about voltage. The MPPT method is explained
and the algorithm I shown in Fig. 7. The design of the inverter controller along with
its waveforms is presented in Figs. 8 and 9, respectively. When the wind speed gets
increased, it is observed that the power fed to the grid, gets increased and accordingly
the magnitude of voltage and current of the inverter output gets increased [10]. We
can see that the tip speed ratio increases with the increase in time up to 2 s then it
becomes constant when it reaches 2 s (Fig. 10).

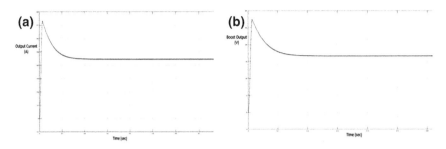

Fig. 4 a Output current and **b** output voltage waveform of boost converter

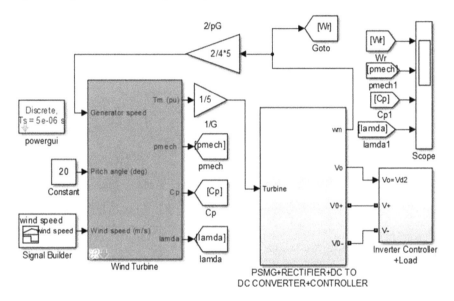

Fig. 5 Simulink model of WECS system

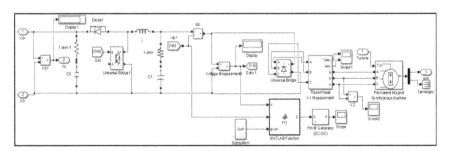

Fig. 6 Simulink model of PMSG associated with rectifier unit, boost converter, and controller unit [7]

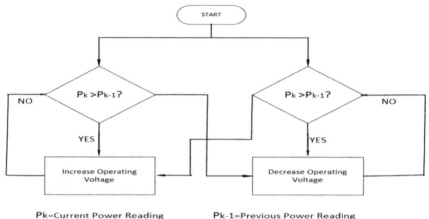

Fig. 7 Perturb and observe algorithm

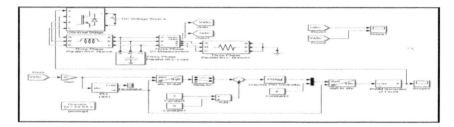

Fig. 8 Inverter controller design in Simulink

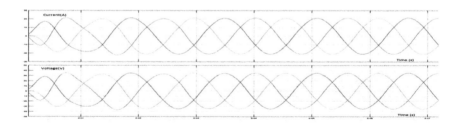

Fig. 9 Output waveform of current and voltage of inverter

7 Conclusion

A detailed study of the WECS system is done by describing one of the MPPT techniques [11] and also by implementing it in MATLAB. Each block has been separately simulated and the concerning results are also disclosed. The exact values and parameters that are used in this study have been given in the simulation section.

Fig. 10 With changing wind speed, the maximum power point tracking

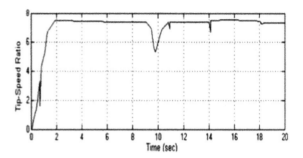

The design of the boost converter [12] along with the design of the inverter controller [13] is depicted in this paper and also how it is connected to the grid [14] is shown. As the world is moving toward using renewables, this can also be implemented into a real project from a simulated module. The effects of the generation of power by varying wind speed [15] have been clearly described in this paper.

References

1. Report on renewable energy, *International Renewable Energy Agency (IRENA)*. [Online]. Available: http://www.irena.org/wind
2. Y. Zhang, L. Zhang, Y. Liu, Implementation of maximum power point tracking based on variable speed forecasting for wind energy systems. Processes **7**(3), 158 (2019)
3. S. Müller, M. Deicke, R.W. De Doncker, Doubly fed induction generator systems for wind turbines. IEEE Ind. Appl. Mag. **8**(3), 26–33 (2002)
4. M. Elzalabani, F. H.Fahmy, A. E.-S. A. Nafeh, G. Allam, Modelling and simulation of tidal current turbine with permanent magnet synchronous generator. TELKOMNIKA Indones. J. Electr. Eng. **13**, 1 (2015)
5. M. Kesraoui, N. Korichi, A. Belkadi, Maximum power point tracker of wind energy conversion system. Renew. Energy **36**(10), 2655–2662 (2011)
6. S.S. Mohammed, D. Devaraj, Simulation of incremental conductance MPPT based two phase interleaved boost converter using MATLAB/Simulink, in *Proceedings of 2015 IEEE International Conference on Electrical, Computer and Communication Technologies, ICECCT 2015* (2015)
7. A. H. Cuong, H. Tran, F. Nollet, N. Essounbouli, Maximum power point tracking techniques for wind energy systems using three levels boost converter, in *2018 7th International Conference on Clean and Green Energy-ICCGE 2018*, (2018), pp. 1–9
8. R.F. Coelho, F.M. Concer, D.C. Martins, A MPPT approach based on temperature measurements applied in PV systems, in *2010 9th IEEE/IAS International Conference on Industry Applications, INDUSCON 2010* (2010)
9. N. Dong, H. Yang, J. Han, R. Zhao, Modeling and parameter design of voltage-controlled inverters based on discrete control. Energies **11**(8), 2154 (2018)
10. A.Y. Anuj Kumar Palariya, A. Choudhary, Modelling, control and simulation of MPPT for wind energy conversion using Matlab/Simulink. Eng. J. Appl. Scopes 1(2), 9–13 (2016)
11. K. Kobayashi, I. Takano, Y. Sawada, A study on a two stage maximum power point tracking control of a photovoltaic system under partially shaded insolation conditions, in *2003 IEEE Power Engineering Society General Meeting (IEEE Cat. No.03CH37491)*, pp. 2612–2617

12. P. Sahu, D. Verma, S. Nema, Physical design and modelling of boost converter for maximum power point tracking in solar PV systems, in *2016 International Conference on Electrical Power and Energy Systems (ICEPES)* (2016), pp. 10–15
13. T. Bhattacharjee, M. Jamil, A. Jana, Designing a controller circuit for three phase inverter in PV application, in *2018 International Conference on Electrical, Electronics, Communication, Computer and Optimization Techniques (ICEECCOT)* (2018), pp. 882–886
14. J. Li, M. Zhang, Z. Li, T. Zhang, Q. Zhang, C. Chi, Study of grid planning method considering multiple energy access, in *2018 International Conference on Smart Grid and Electrical Automation (ICSGEA)* (2018), pp. 59–62
15. M. Yesil, A. Kose, E. Irmak, Effects of change in wind energy generation on the power grid angle, in *2019 1st Global Power, Energy and Communication Conference (GPECOM)* (2019), pp. 364–367

A Multilevel Boost Converter-Fed High-Frequency Resonant Inverter for Induction Heating by Using ADC Control

A. Kumaraswamy, Ananyo Bhattacharya, and Pradip Kumar Sadhu

1 Introduction

The time-varying magnetic field developed due to the current flowing inside the working coil results in induced Foucault currents in the nearer workpiece at the same frequency stated by Faraday [1]. Lenz law illustrates the fact that coil current produces a magnetic field which is in the opposite direction of the changes in the magnetic flux. With this addition, the electromagnetic principle became one of the important laws in electromagnetism. According to Joules law, heat is produced by passing an electric current through the conductor; heat produced is ohmic heat [2]. The heat developed from electrical energy is known as electric heating. Typically the applications of electric heating include industrial processes, medical treatment, and cooking. The heat generated through energy conversion processes is associated with energy losses. Electric heating is generally classified into two categories, namely high-frequency heating and power frequency heating. A method of producing heat at a specific location on a metallic object is known as induction heating [3]. Heating technology has been popular for over a hundred years, due to its inherent advantages in the industries, medical and residential applications.

With the development of solid-state devices, the frequency of the inverters can be increased to a very high value with fewer components and good efficiency, and hence the cost involved in induction heating for domestic and industrial

A. Kumaraswamy (✉) · A. Bhattacharya
Department of Electrical Engineering, NIT Jamshedpur, Jamshedpur 831014, Jharkhand, India
e-mail: 2018rsee018@nitjsr.ac.in

A. Bhattacharya
e-mail: ananyo.ee@nitjsr.ac.in

P. K. Sadhu
Indian Institute of Technology (Indian School of Mines), Dhanbad, India
e-mail: pradip@iitism.ac.in

© The Author(s), under exclusive license to Springer Nature Singapore Pte Ltd. 2021
J. Kumar and P. Jena (eds.), *Recent Advances in Power Electronics and Drives*, Lecture Notes in Electrical Engineering 707,
https://doi.org/10.1007/978-981-15-8586-9_5

Fig. 1 Conventional induction heating block diagram

applications has reduced drastically. The Various inverter topologies are used in IH appliances because of the advantages like economical, green environment, less unit consumption, better heat conversion, accurate process of heating, more reliable and less electromagnetic noise. These merits can be achieved with solid-state switching devices, soft-switching inverters, high-frequency circuit components, and analog and digital control devices [4].

Under the abovementioned conditions, soft switching high-frequency inverter topologies are essential for IH applications. All these inverters should have a simple configuration, low cost, wide range of operations, and high efficiency. The high-frequency soft-switching inverter with voltage control uses Insulated Gate Bipolar Transistor, Metal Oxide Semiconductor Field Effective Transistor (MOSFET), and its alternate models are used for practical requirements.

The conventional process of induction heating in which low-level AC frequency is converted into high-level AC frequency is as shown in Fig. 1. It contains a Converter, a DC-Link Filter, and a Resonate High-Frequency Inverter. Electro Magnetic Compatibility (EMC) Filter compensates the input current harmonics. The converter (Full Wave Rectifier) is used to convert unchanging AC to unchanging DC. A DC-link filter is used to boost the voltage as well as reduce DC harmonics. High-frequency resonant inverter converts boosted DC to AC with high frequency and then is connected to load [5].

The proposed paper is sectioned as follows: the family of Multi-resonant converters is expressed in Sect. 2. The Full-bridge-based operation is selected; its detailed implementation and its extension to the rest of the scheme are illustrated in Sect. 3. In Sect. 4, the implementation and experimental results of the proposed topology are explained, and paper conclusion is summarized in Sect. 5.

The present scheme of a Multilevel Boost Converter (MBC)-based high-frequency resonate inverter is depicted in Fig. 2. In this scheme, source voltage (V_s) is coupled to a full-bridge diode rectifier [6, 7] thereby connecting to a two-level boost converter [8, 9]. DC-link voltage (V_d) is modified by changing the duty cycle of the switched boost converter which consists of inductor (L_s), three capacitors (C_1, C_2, and C_3), and three diodes (D_1, D_2, and D_3). A full-bridge inverter having four MOSFETs with an anti-parallel diode converts boosted DC to AC with high frequency then is coupled to load.

The R_r and L_r depict the equivalent resonating resistance and inductance of the workpiece and coil referred to primary. The resonating capacitor (C_r) is used to make the circuit underdamped. For normal operation of inverters, the resonate frequency is chosen below the switching frequency, therefore switching losses are high. To reduce

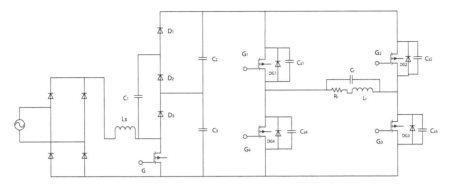

Fig. 2 Proposed topology

these turn ON losses zero voltage switching with Asymmetrical Duty Cycle (ADC) is implemented.

The analysis of the proposed scheme is explained with these assumptions:

- Solid-state devices have ideal characteristics.
- The supply voltage is fixed during switching.
- The resonate frequency is always lesser than the switching frequency.

The resonant coil frequency is given by

$$F_r = \frac{1}{\sqrt{2\pi L_r C_r}} \tag{1}$$

The Quality factor of the coil is given by

$$Q = \frac{1}{R}\sqrt{\frac{L_r}{C_r}} \tag{2}$$

The Multilevel Boost Converter inductance [10] equation is

$$L_S = \frac{V_s D}{\Delta I_{in} f_s} \tag{3}$$

The Multilevel Boost Converter capacitance [10] equation is

$$C_1 = C_2 = C_3 = \frac{P_{rated}}{2\omega V_{dc}\Delta V_{dc}} \tag{4}$$

2 MBC Operation Modes

The operation of MBC is classified into two modes are depicted in Fig. 3.

In Mode-I, MOSFET G is turned OFF and the capacitor C_3 is charged through boost inductor L_s and Diode D_3. Hence, the capacitor C_3 reaches a voltage equal to the sum of source voltage and inductor voltage (V_L) through charging. Hence, diode D_3 is reverse biased; it will not conduct. Capacitors C_2 and C_3 get charged through diode D_1 to a voltage equal to source voltage, inductor voltage, and capacitor C_1 voltage. In Mode-II, MOSFET G is switched ON; complete current passes through boost inductor L_s charging it from AC source voltage; capacitor C_3 discharges through diode D_2, capacitor C_1, and MOSFET G. Therefore, the capacitor C_1 voltage reaches a voltage in capacitor C_3.

Hence, at the end of two modes, all the capacitors in the multilevel boost converter get charged equal to the boost converter voltage. For example, if the MOSFET G is operated with 50% duty cycle, then the voltage of the boost inductor is increased to twice the input source voltage. Hence, each capacitor voltage reaches twice the power supply voltage. Since two capacitors (C_2 and C_3) are feeding the inverter circuit, the DC-link potential across the inverter will be four times the AC source voltage. This provides the advantage of employing low voltage source as input sources for the proposed topology. Various proposed conversion parameters are listed in Table 1.

The voltage (V_c) across the capacitor is

Table 1 Proposed converter parameter

S. No.	Item	Symbol	Value (unit)
1	Input RMS voltage	V_s	230 V
2	Boosting inductance	L_s	1.33 mH
3	Boosting capacitors	C_1, C_2, C_3	100 uF
4	Parasitic capacitors	$C_{s1}, C_{s2}, C_{s3,1}\ C_{s4}$	2 uF
5	Load resistance	R_r	2.43 ohm
6	Load inductance	L_r	53.927 uH
7	Load capacitance	C_r	0.2 uF
8	Duty cycle	D	50%
9	Switching frequency	F_s	50 kHz
10	Resonance frequency	F_r	47 kHz

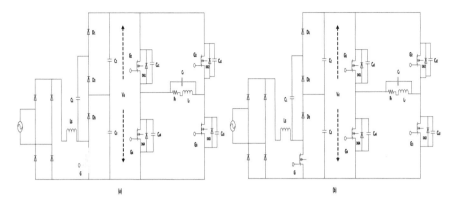

Fig. 3 Operation modes proposed topology. **a** Mode I (MOSFET G OFF), **b** Mode II (MOSFET G ON)

$$V_C = \frac{V_S}{1 - D} \tag{5}$$

The DC-link voltage (V_D) equation of the two-level boost converter is

$$V_D = \frac{2V_S}{1 - D} \tag{6}$$

DC-link current (I_D) equation is

$$I_D = \frac{4V_C}{(1 - D)R} \tag{7}$$

3 Resonant Inverter for Induction Heating

3.1 Principle of Operation

Operation of resonant inverter explained is by the following modes in Fig. 4. Mode I: At this mode, MOSFET G_2 and G_4 are initially in OFF condition, and also MOSFET G_1 and G_3 are in OFF condition. The negative current flows through parasitic capacitors C_{s1}, C_{s2}, C_{s3}, and C_{s4}. Mode II: In this mode, MOSFETs G_2 and G_4 are still switched OFF; negative current passing through C_{s1} and C_{s3} is shifted to parallel diodes D_{s1} and D_{s3}. In Mode III, MOSFETs G_1, G_3 are switched ON, anti-parallel diodes D_{s1} and D_{s3} are switched OFF, and ZVS condition is achieved. Output voltage is equal to V_d. Mode IV: In this mode, MOSFET G_1 is still ON, G_3 is not conducting because of the β shifting angle as shown in Fig. 5. It is used to control the output

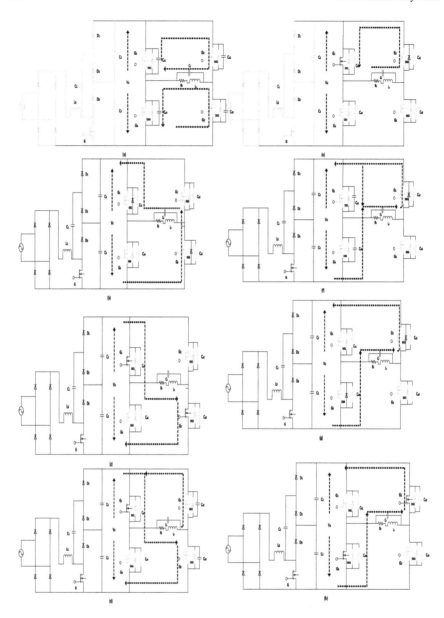

Fig. 4 High-frequency resonant inverter operation intervals. **a** Mode I; **b** Mode II; **c** Mode III; **d** Mode IV; **e** Mode V; **f** Mode VI; **g** Mode VII, and **h** Mode VIII

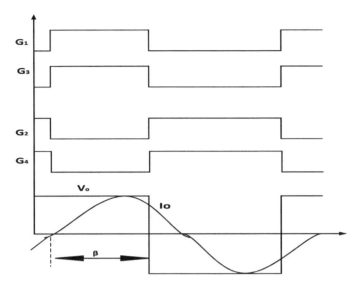

Fig. 5 ADC control

power. In this mode, capacitor C_{s3} starts getting charged and capacitor C_{s2} starts discharging. The current flow through the resonant tank is in the same direction, and voltage across the resonant tank is decreased to zero. Mode V: In this mode, MOSFET G_1 is still conducting and parallel diode D_{s2} is forward biased; at this time, voltage across the tank is zero, and the output current is decreasing. Mode VI: In this mode, all MOSFETs are turned OFF; at this time, voltage across C_{s1} is increased V_d and voltage across capacitor C_{s4} is decreased to zero. The potential difference across the resonant tank is $(-V_d)$. Mode VII: In this mode, all MOSFETs are not conducting during dead time. The parallel diodes D_{s2} and D_{s4} are positive biased, the load current is decreased to zero, and load across voltage is $-V_d$. Mode VIII: In this mode, the gating signal is given to MOSFETs G_2 and G_4 under the ZVS operation; these are conducting and the body diodes D_{s2}, D_{s4} are turned OFF, which will obtain negative values of voltage and current.

3.2 Asymmetrical Duty Cycle Control

In this control, MOSFETs G_1 and G_3 are operating at duty cycle D, MOSFETs G_2 and G_4 are operating at duty cycle $(1 - D)$. Changing the duty cycle can achieve the required output power. The main advantages of this control are it minimizes switching loss and input current [11]. By using this technique, asymmetrical output waveforms are obtained. Figure 5 illustrates the gating signals of MOSFETs G_1, G_3, G_2, and G_4, and output voltage and current waveforms.

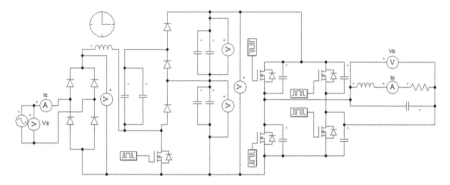

Fig. 6 Simulation circuit

4 Results and Simulated Circuit

The simulation of the proposed Multilevel boost full-bridge resonate inverter with ADC control is performed in PSIM and different parameters as shown in Table 1. In Fig. 6, the switching frequency of the MOSFET is incorporated in the Multilevel Boost Converter and chosen to be 100 kHz and the resonant inverter to be 50 kHz Fig. 7 depicts the RMS value of output current (Io), output voltage (Vo) of the inverter waveforms when the duty cycle of MBC is 0.5 with the RMS value of the load current (Io) is 50.2 A and output voltage (Vo) is 907 V. Table 2 and Fig. 8 depict the comparison of different results obtained in the proposed scheme of multilevel boost converter-fed resonant inverter for IH with different duty cycles of MBC. So, from the simulation results obtained, it is observed that by varying the duty cycle of the MBC, the output voltage can be varied smoothly. So the controllability of this scheme is quite good when compared with previous methods.

5 Conclusion

A profound study of a multilevel boost converter-fed induction heating using ADC control with guaranteed ZVS operation avoids the use of an EMC filter at the source, variable DC-link voltage, and high output RMS voltage. DC-link voltage is controlled by varying the duty cycle of MBC. By the control of the DC-link voltage, the heat in the workpiece can be changed. This gives an additional benefit over the switching frequency control of the inverter, for controlling the heat in the workpiece. Simulated outputs validate the performance and controllability of this scheme. As per the simulation results, output current obtained is 33–69 A. So, it can be preferred in all industrial applications for Induction Heating. The proposed configuration has merits like independent control, high voltage gain, high input power factor, high reliability, and less complexity.

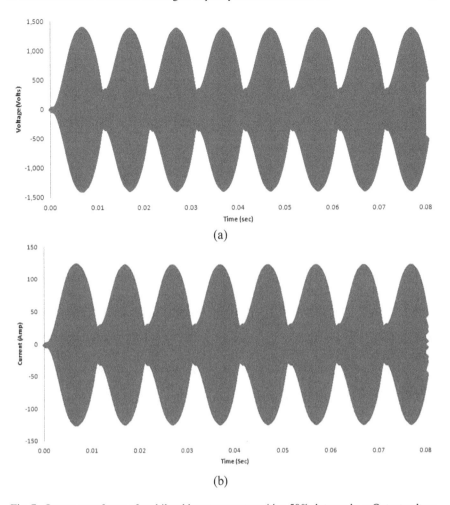

Fig. 7 Output waveforms of multilevel boost converter with a 50% duty cycle. **a** Output voltage. **b** Output current

Table 2 Output values

S. No.	Duty cycle of MBC	Output RMS voltage	Output RMS current	Input current THD	Input power factor
1	0.3	659	33	20.31	0.975
2	0.4	763	41.1	18.71	0.983
3	0.5	907	50.2	13.65	0.9872
4	0.6	1057	57.04	19.21	0.928
5	0.7	1340	69.49	37.07	0.903

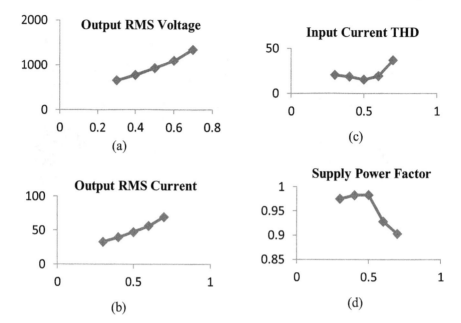

Fig. 8 Output values versus duty cycle (δ). **a** δ versus Vo, **b** δ versus Io, **c** δ versus input current THD, and **d** δ versus supply power factor

References

1. O. Lucía, P. Maussion, E.J. Dede, J.M. Burdío, Induction heating technology and its applications: past developments, current technology, and future challenges. IEEE Trans. Industr. Electron. **61**(5), 2509–2520 (2014). https://doi.org/10.1109/TIE.2013.2281162
2. A. Kumar, M. Sadhu, N. Das, P.K. Sadhu, D.R. Ankur, A survey on high-frequency inverter and their power control techniques for induction heating applications. J. Power Technol. **97**(X), 201–213 (2017)
3. P.K. Sadhu, N. Jana, R. Chakrabarti, D.K. Mittra, A unique induction heated cooking appliances range using hybrid resonant converter. Int. J. Circuits Syst. Comput. World Scienti. **14**(3), 619–630 (2005)
4. N. Park, D. Lee, D. Hyun, A power-control scheme with constant switching frequency in class-D inverter for induction-heating jar application. IEEE Trans. Industr. Electron. **54**(3), 1252–1260 (2007). https://doi.org/10.1109/TIE.2007.892741
5. A. Bhattacharya, K. Sit, P.K. Sadhu, N. Pal, A novel circuit topology of modified switched boost hybrid resonant inverter fitted induction heating equipment. Arch. Electri. Eng. **65**(4), 815–826 (2016)
6. H. Sarnago, O. Lucia, A. Mediano, J.M. Burdio, Modulation scheme for improved operation of an RB-IGBT-based resonant inverter applied to domestic induction heating. IEEE Trans. Industr. Electron. **60**(5), 2066–2073 (2013). https://doi.org/10.1109/TIE.2012.2207652
7. A. Bhattacharya, P.K. Sadhu, A. Bhattacharya, N. Pal, Voltage controlled hybrid resonant inverter an essential tool for induction heated equipment. Rev. Roum. Sci. Techn.—Électrotechn. et Énerg. **61**(3), 273–277, Bucarest (2016)
8. M. Forouzesh, Y.P. Siwakoti, S.A. Gorji, F. Blaabjerg, B. Lehman, Step-up DC–DC converters: a comprehensive review of voltage-boosting techniques, topologies, and applications. IEEE Trans. Power Electron. **32**(12) (2017)

9. J. Chen, C. Wang, J. Li, C. Jiang, C. Duan, An input-parallel–output-series multilevel boost converter with a uniform voltage-balance control strategy. IEEE J. Emerg. Select. Topics Power Electron. **7**(4), 2147–2157 (2019). https://doi.org/10.1109/JESTPE.2019.2894514

10. A. Iqbal, M.S. Bhaskar, M. Meraj, S. Padmanaban, S. Rahman, Closed-loop control and boundary for CCM and DCM of nonisolated inverting $N\times$ multilevel boost converter for high-voltage step-up applications. IEEE Trans. Industr. Electron. **67**(4), 2863–2874 (2020). https://doi.org/10.1109/TIE.2019.2912797

11. M.B. Borage, K.V. Nagesh, M.S. Bhatia, S. Tiwari, Characteristics and design of an asymmetrical duty-cycle-controlled LCL-T resonant converter. IEEE Trans. Power Electron. **24**(10), 2268–2275 (2009). https://doi.org/10.1109/TPEL.2009.2022627

Control of a Three-Phase Diode Clamped Multilevel Inverter Using Phase Disposition Modulation Scheme

Manoj Kumar Kar, Parthasarathi Giri, Neelesh Kumar Gupta, and A. K. Singh

1 Introduction

Various techniques of Pulse-Width Modulation (PWM) are applied to a conventional inverter (two-level) for obtaining an improved output. In order to obtain an output with minimum harmonic content, a high switching frequency is also applied. But, when it comes to high power with medium voltage applications, the basic inverters of two levels have high frequency operation limits. Switching losses and restrictions of device ratings account for this limitation [1]. To overcome this problem, a new group of converters called 'Multilevel Inverter' (MLI) has been developed as the means of operating with higher power applications [2–4]. Nowadays, MLIs find their use in compensation of reactive power, enhancement of stability [5], high voltage motor drives [6], active filtering [7], and in variable speed drives of induction motor for medium voltages [8]. In nutshell, MLIs are becoming an intrinsic part of motor drives in industries for medium voltage, sustainable energy systems [9], FACTS, and traction system. Different PWM methods are compared with parallel multilevel converters reducing the ripple current [10]. The output harmonic current is reduced by reducing the number of independent dc voltage sources [11, 12]. PD modulation scheme has been enacted on 5-level Cascaded H-Bridge MLI. The present and

M. K. Kar · P. Giri · N. K. Gupta · A. K. Singh (✉)
Department of Electrical Engineering, National Institute of Technology Jamshedpur, Jharkhand 831014, India
e-mail: aksnitjsr@gmail.com

M. K. Kar
e-mail: manojkar132@gmail.com

P. Giri
e-mail: parthsarathigiri@gmail.com

N. K. Gupta
e-mail: gptneelesh91@gmail.com

© The Author(s), under exclusive license to Springer Nature Singapore Pte Ltd. 2021
J. Kumar and P. Jena (eds.), *Recent Advances in Power Electronics and Drives*, Lecture Notes in Electrical Engineering 707,
https://doi.org/10.1007/978-981-15-8586-9_6

future aspects of using multilevel SHE-PWM is discussed in [13]. A novel wavelet transform-based technique is developed to minimize the power loss [14]. A novel approach of grid connected system with MLI has been discussed [15]. In this paper, PD modulation scheme has been enacted on five-level Diode Clamped MLI. The THD of both the results are then going to be compared.

2 Diode Clamped MLI

A five-level DCMLI has been depicted in Fig. 1. Here, the DC voltage source is splitted into five levels by using capacitors connected in series. If the DC source voltage is E, then the terminal voltage appearing across each capacitor is E/4 (Fig. 2 and Table 1).

In general, for any 'm' level DCMLI, the components required are

- Number of Switches $= 2p - 2$.
- Number of Capacitors $= p - 1$.
- Number of Diodes $= (p - 1)(p - 2)$.

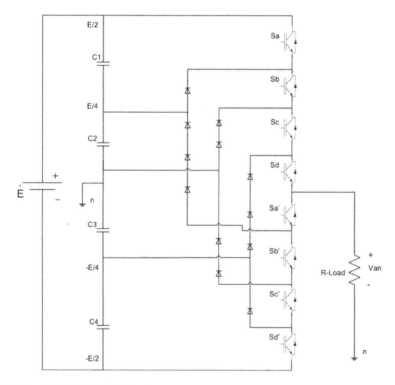

Fig. 1 Circuit for DCMLI (five level)

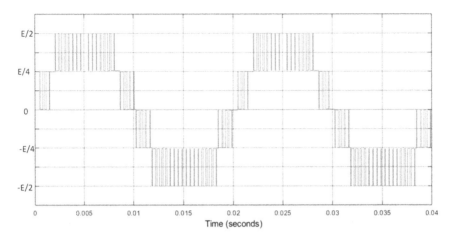

Fig. 2 Output of DCMLI (five level)

Table 1 Voltage levels for five-level DCML

Output voltage 'Van'	Sa	Sb	Sc	Sd	Sa'	Sb'	Sc'	Sd'
E/2	1	1	1	1	0	0	0	0
E/4	0	1	1	1	1	0	0	0
0	0	0	1	1	1	1	0	0
−E/4	0	0	0	1	1	1	1	0
−E/2	0	0	0	0	1	1	1	1

- Clamping diodes connected across p − 1 number of thyristors.

3 PWM Schemes for Three Phase Inverter

In PWM Schemes, a reference signal is compared with an appropriate carrier signal of high frequency to generate gate pulse for the switches. For three-phase inverters, there are two methods through which the leg-voltage waveforms are generated. For each of the method, we require three 120° phase-shifted sinusoidal modulating waves.

- In the first case, a single set of carrier signals is compared with the three sinusoidal modulating signals. This method is called single-phase modulation.
- In the second case, three different sets of carrier signals, 120° displaced to each other, are compared with the corresponding sinusoidal signal. This method is called three-phase modulation.

In this paper, we are dealing with single-phase modulation.

4 Ulti-Carrier Pwm (Mcpwm) Schemes

In order to obtain low ripple content output waveform from MLIs, it is desired to implement an appropriate modulation scheme.

Among the various available schemes, an important family of modulation scheme, i.e., MCPWM is preferred because of its simple design and easier implementation.

In all MCPWM techniques, the signal used for modulation is a sine wave, whereas triangular waves are used as the carrier signals. Sometimes, saw-tooth signals can also be used instead of triangular signals. Generally, these carrier signals contain same amount of amplitude and frequency. 'q-1' carrier signals are needed for 'q' level output.

The various MCPWM schemes can be grouped as 'Level-Shifted' and 'Phase-Shifted'. The earlier schemes are of three types: PD, POD, and APOD.

If all the carriers, i.e., triangular signals are in the same phase, the scheme is called PD modulation (Fig. 3). For POD modulation, all the carriers over the horizontal axis remain in phase among them while in opposition (180° phase-shifted) with those underneath the horizontal axis (Fig. 4). In APOD technique, all the adjacent carriers

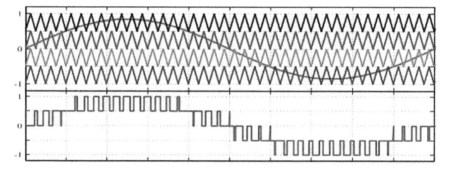

Fig. 3 PD modulation scheme

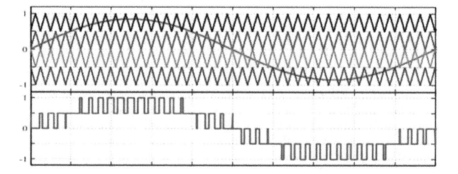

Fig. 4 POD modulation scheme

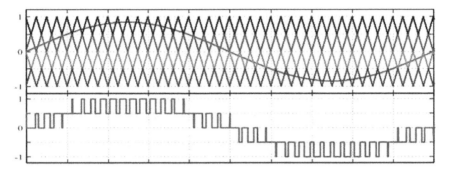

Fig. 5 APOD modulation scheme

are 180° opposite to each other (Fig. 5). For phase-shifted PWM, an appropriate phase-shift is made between different carrier signals. In Level-Shifted PWM, the triangular carriers are shifted in a vertical way unlike the phase-shift used in Phase-Shifted PWM.

5 Advantages of PD Modulation Scheme

- The line voltage is optimized in PD modulation because the harmonics of carrier frequency are made as common-mode voltage of the carrier frequency and hence gets canceled out in line voltage [11]. This is not the case with other dispositions and hence PD modulation is chosen over other disposition techniques.
- Traditional carrier-based schemes, discontinuous modulations, and Space-Vector Modulation (SVM); all of these are described within the same frame by PD modulation technique.
- This approach is topology independent, i.e., it is applicable to all type of MLIs.

6 Simulation Results and Discussions

The MATLAB results have been shown. The frequency of the carrier signals used is 2.5 kHz. The carrier signals are compared with three sinusoidal waves producing gate signal for the switches of DCMLI. The DC voltage fed to the inverter is 400 V. The five-leveled limb voltages have been shown in Fig. 6. These limb voltages are used to obtain the line and phase voltages.

The corresponding waveforms for phase and line voltages are given in Figs. 7 and 8, respectively. The output voltage has THD of 17.25%. A low-pass filter of second order has improved the THD to 0.28% (Fig. 9).

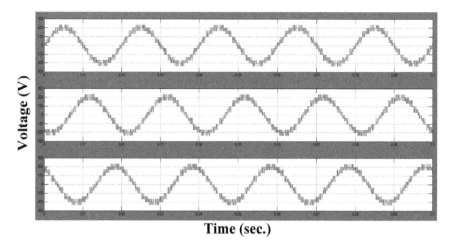

Fig. 6 Output voltage (V_{P-N}) of the inverter

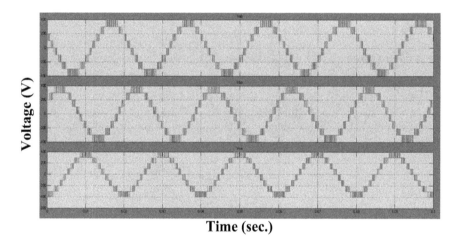

Fig. 7 Output voltage (V_{L-L}) of the inverter

7 Conclusion

The PD scheme strategy, which is based on MCPWM technique, has been proposed and successfully implemented to five-level DCMLI using MATLAB environment. The basic operating principle and switching pattern of DCMLI is explained in detail. A five-level DCMLI is simulated using a second-order LPF in order to reduce the harmonic contents. It is clear from the obtained result, using PD technique has given better output results in three-phase DCMLI (THD = 17.25%) compared to the case when PD technique is used in Cascaded H-Bridge (THD = 26.69%). The filter used here has improved the THD to 0.28%.

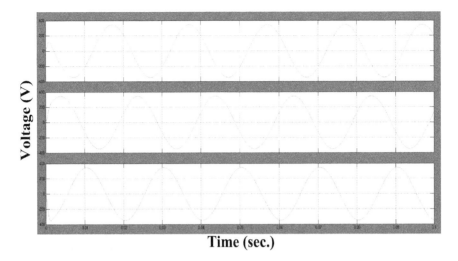

Fig. 8 Filtered output voltage (V_{L-L}) of the inverter

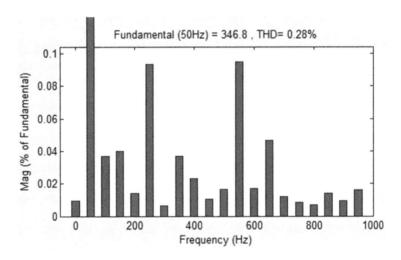

Fig. 9 Frequency spectra of output line voltage waveform after filtering

References

1. M.H. Rashid, Power Electronics: Circuits, Devices and Applications (Pearson Education, Inc., 2009)
2. J.S. Lai, F.Z. Peng, Multilevel converters–a new breed of power converters. IEEE Trans. Ind. Appl. **32**, 509–517 (1996)
3. L. Tolbert, F.Z. Peng, T. Habetler, Multilevel converters for large electric drives. IEEE Trans. Ind. Appl. **35**, 36–44 (1999)

4. R. Teodorescu, F. Beaabjerg, J.K. Pedersen, E. Cengelci, S. Sulistijo, B. Woo, P. Enjeti, Multi-level converters—a survey, in Proceedings of the European power Electronics Conference (EPE 99), (Lausanne, switzerland, 1999)
5. T. Meynard, H. Foch, Multilevel choppers for high voltage applications. Eur. Power Electron. J. 2(1), 45–50 (1992)
6. M. Malinowski, K. Gopakumar, J. Rodriguez, M.A. Perez, A survey on cascaded multilevel inverters. IEEE Trans. Power Electron. 57(7), 2197–2206 (2010)
7. M. Young, The Technical Writer's Handbook (University science, Mill Valley, CA, 1989)
8. J. Rodriguez, S. Bernet, B. Wu, J.O. Pontt, S. Kouro, Multilevel voltage source converter topologies for industrial medium-voltage drives. IEEE Trans. Ind. Electron. 54(6), 2930–2945 (2007)
9. J.I. Leon, S. Kouro, S. Vazquez, R. Portillo, L.G. Franquelo, J.M. Carrasco, J. Rodriguez, Multidimensional modulation technique for cascaded multilevel converters. IEEE Trans. Ind. Electron. 58(2), 412–420 (2011)
10. B. Cougo, T. Meynard, F. Forest, E. Labouré, Optimal PWM method for flux reduction in inter cell transformers coupling double three-phase systems, in Proceedings of the Conference on Electron Puissance Future, 2010, pp. 1–6
11. S.R. Kathalingam, P. Karantharaj, Comparison of multiple carrier disposition PWM techniques applied for multilevel shunt active filter. J. Electr. Eng. 63(4), 261–265 (2012)
12. V. Sivanesan, C. Rajan, Design and analysis of pd-pwm multilevel inverter with reduced switch count. Int. J. Latest Trends in Eng. Technol. 10(2), 89–95 (2018)
13. M.S.A. Dahidah, G. Konstantinou, V.G. Agelidis, A review of multilevel selective harmonic elimination PWM: formulations solving algorithms, implementation and applications. Power Electron. IEEE Trans. 30(8), 4091–4106 (2015)
14. D.P. Garapati, V. Jegathesan, M, Veerasamy, Minimization of power loss in newfangled cascaded H-bridge multilevel inverter using in-phase disposition PWM and wavelet transform based fault diagnosis. Ain Shams Eng. J. 9(4), 1381–1396 (2018)
15. N. prabaharan, P.E. Campana, A R Ann Jerin, K. Palanisamy, A new approach for grid integration of solar photovoltaic system with maximum power point tracking using multi-output converter. Energy Procedia 159, 521–526 (2019)

New Asymmetric 25-Level Inverter Topology with Reduced Switch Count

V. Thiyagarajan

1 Introduction

Multilevel inverters are highly reliable and good quality voltage source power converters to interconnect the dc system with ac system. It can be formed by different arrangements of much lower rated power electronic switches and dc sources to synthesize multistep staircase voltage waveform [1]. With higher number of output levels, the output waveform looks similar to the sinusoidal waveform with minimum harmonic content and thus reduces the filter requirements [2]. The three main classifications of the multilevel inverters used in commercial applications are neutral point clamped, flying capacitor, and cascaded H-bridge [3]. However, there are certain issues in conventional inverter topologies such as need of large number of dc sources, switching devices, and its associated gate driver circuits.

In recent years, many efficient topologies have been presented to overcome these drawbacks [4–6]. A new hybrid single-phase multilevel inverter topology is presented in [1]. In this topology, the output levels are generated by switching the several dc sources in series/parallel combinations. The proposed inverter contains an auxiliary circuit to increase the output levels by generating a new step between two output levels. By properly introducing the switching scheme for the polarity changing unit, the zero level is obtained at the output. A new extendable type multilevel inverter with combination of different ratings of both unidirectional and bidirectional switches is presented in [2]. This topology reduces the number of switches and extending the use for high-voltage application. However, this topology lack the attribute of combining two dc sources in parallel. Also, this topology does not own the load sharing capability. An improved multistep converter is proposed in [3] with a basic unit consisting of three sources, four unidirectional switches, and one bidirectional

V. Thiyagarajan (✉)
Sri Sivasubramaniya Nadar College of Engineering, Tamil Nadu, Kalavakkam 603 110, India
e-mail: thiyagarajanv@ssn.edu.in

© The Author(s), under exclusive license to Springer Nature Singapore Pte Ltd. 2021   67
J. Kumar and P. Jena (eds.), *Recent Advances in Power Electronics
and Drives*, Lecture Notes in Electrical Engineering 707,
https://doi.org/10.1007/978-981-15-8586-9_7

switch and is connected with H-bridge circuit containing four unidirectional switches. The main drawback of this inverter circuit is higher switching stress and total standing voltage (TSV). A 17-level asymmetric type inverter consists of four dc sources and ten switches is proposed in [4]. This basic circuit can be connected in series to create higher output levels. But, because of high variety of voltage sources and TSV value, this topology is not suitable for high-voltage applications. Another asymmetric topology with three sources and 12 switches is proposed in [5]. This inverter creates only 21-level output voltage. A new inverter topology with three sources, six unidirectional switches, and one bidirectional switch is presented in [6]. This paper also presented the optimal selection of number of basic units cascaded to achieve higher number of output levels while minimizing number of sources, switches, and TSV vale. However, all these topologies fails to create higher output levels during asymmetric operation.

This paper aims to present a new 25-level asymmetric inverter circuit with three dc sources and thirteen switches. Section 2 proposed a new 25-level inverter topology. Comparison study is done in Sect. 3. The strategy for generating the switching signals in presented in Sect. 4. Section 5 presents the simulation results, and conclusion is given in Sect. 6.

2   Proposed 25-Level Inverter

The suggested multilevel inverter topology with three dc voltage sources and 13 switches is shown in Fig. 1. This circuit is able to achieve 25-levels at the output employing only three dc sources during asymmetric operation. In asymmetric operation, the dc sources magnitude are different in order to create higher output levels with same circuit components as compared with symmetric operation. In order to avoid short circuit, the pairs of switches (S_2, S_3), (S_4, S_5), (S_8, S_{10}), (S_{12}, S_{13}), (S_1, S_2, S_4), (S_5, S_6, S_7), (S_7, S_8, S_9), (S_{10}, S_{11}, S_{12}) should not turn on simultaneously.

Fig. 1 Proposed 25-level inverter

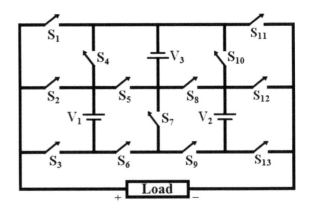

The number of output levels depends upon the magnitude of dc sources. Thus, the magnitudes of V_1, V_2, and V_3 are selected in the ratio 1:3:8 to create 25 output levels. Some of the sample output levels obtained at the inverter output terminals are shown in Fig. 2. The main feature of the proposed topology is creating negative levels without additional H-bridge. To achieve zero level at the output, the pairs of the switches (S_1, S_{11}), (S_3, S_6, S_9, S_{13}), (S_2, S_5, S_8, S_{12}) are turned on. The cost of the inverter depends on the total standing voltage (TSV) value of the switches. The blocking voltage for switch S_{13} and S_7 is $5V_{dc}$ and $6V_{dc}$, respectively. For S_3, the blocking voltage is $7V_{dc}$ and is $8V_{dc}$ for the switches S_2, S_5, S_8, S_{12}. Similarly, for switches S_1, S_4, S_6 and S_9, S_{10}, S_{11}, the blocking voltage is $9V_{dc}$ and $11V_{dc}$, respectively. Therefore, the total standing voltage (TSV) across the switches of the proposed 25-level inverter topology is $110V_{dc}$.

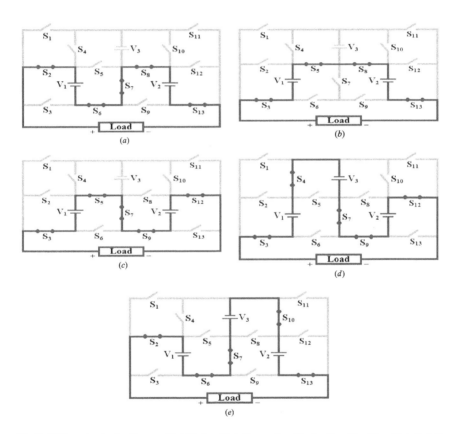

Fig. 2 Different output voltages **a** $V_1 + V_2$, **b** $-V_1 + V_2$, **c** $-V_1 - V_2$, **d** $-V_1 - V_2 - V_3$, **e** $V_1 + V_2 + V_3$

Table 1 Comparative analysis

Topology	Negative level	No. of sources	No. of switches	ON-state switches	Output level
[1]	H-bridge	4	12	5	13
[2]	Inherent	3	8	4	13
[3]	Inherent	3	9	4	15
[4]	Inherent	4	10	4	17
[5]	H-bridge	3	12	4	21
[6]	Inherent	6	15	5	25
Proposed	Inherent	3	13	5	25

3 Comparison Study

Table 1 presents the comparative analysis between the proposed 25-level inverter topology with other asymmetrical topologies presented in the literature.

This comparative study is investigated based on the total number of sources, switches, and ON-state switches. As shown in Table 1, the proposed topology inherently creates negative output levels, however, the topologies presented in [1, 5] need an additional polarity changing unit to produce negative output levels. The ratio of the total circuit components (i.e., total number of dc sources and total number of switches) to the output level of an inverter is given by k.

$$k = \frac{\text{Total number of circuit components}}{\text{Output level}} \tag{1}$$

The cost of the inverter depends on the value of 'k'. If 'k' is high, then cost of the inverter is more. If 'k' is low, then cost of the inverter is less. For the topology presented in [1], this ratio is equal to 1.23. However, for the topologies presented in [2–4, 6], the ratio of total number of circuit components to the output level generated is between 0.8 and 0.84. The value of 'k' is equal to 0.7 for the inverter topology presented in [5]. However, for the proposed inverter, the value of 'k' is equal to 0.64. This shows that the proposed inverter topology creates 25-level output with minimum circuit components. Also, the size, cost, and complexity of the inverter are reduced as compared with the other presented topologies.

4 Switching Pulse Generation

The main purpose of modulation technique is to generate a stepped output waveform with adjustable voltage amplitude, frequency, and phase fundamental component of the sinusoidal steady state. It is used to control the load voltage and current and also helps to minimize the total harmonic distortion (THD) and switching losses.

The conventional sinusoidal pulse width modulation (PWM) technique is used with multiple carrier signals to control each power electronics switches. In this technique, multiple carrier signals are compared with sinusoidal reference signal and the required switching pulses are generated whenever the carrier signal is greater than the reference signal. For an 'm' level inverter, the required number of carrier signal is $\frac{m-1}{2}$. The switching losses of an inverter mainly depends on the type of carrier signals and its range of frequency. The amplitude of the carrier signals is determined by

$$V_{Cj} = V_{\max}\left(\frac{j - 0.5}{\frac{m-1}{2}}\right) where, \ j = 1, 2, 3 \dots, m \qquad (2)$$

where $V_{\max}$ is the maximum amplitude of the reference sinusoidal signal.

5 Simulation Results

The performance of the proposed 25- level asymmetrical inverter is simulated using Simulink/Matlab. The magnitude of the voltage sources is in the ratio $V_1 : V_2 : V_3 = 1 : 3 : 8$ and $V_{dc} = 10$ V. The 25-level output voltage waveform and its THD is shown in Fig. 3. As expected, the output voltage is of staircase waveform with minimum THD. The output voltage and current waveforms of the proposed 25-level inverter for different load parameters are shown in Fig. 4. For pure resistive load, R = 50 Ω, the load voltage THD and load current THD are equal to 3.27%. Also, it is observed that the load current THD is varying between 0.12 and 3.27% as the power factor of the load varies from 0 to 1.

6 Conclusion

In this paper, a new 25-level inverter topology with reduced number of dc sources and switches is recommended. The proposed inverter uses three dc sources and 13 switches to generate 25-level voltage at the output. The proposed inverter structure is compared with several other asymmetrical topologies which indicate the reduction of both the number of sources and switches. Additionally, the proposed inverter offers higher efficiency and lower switching losses. Finally, the inverter performance is analyzed by generating 25-level output voltage using Matlab/Simulink. The proposed inverter is suitable for low- and medium-voltage applications including grid connected renewable energy system and hybrid microgrid.

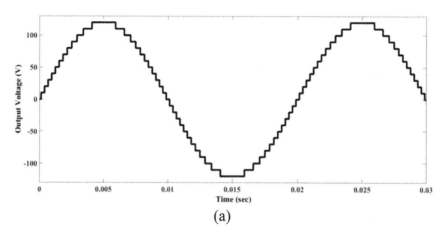

(a)

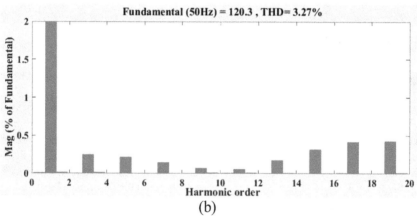

(b)

Fig. 3 25-level **a** output voltage waveform **b** THD

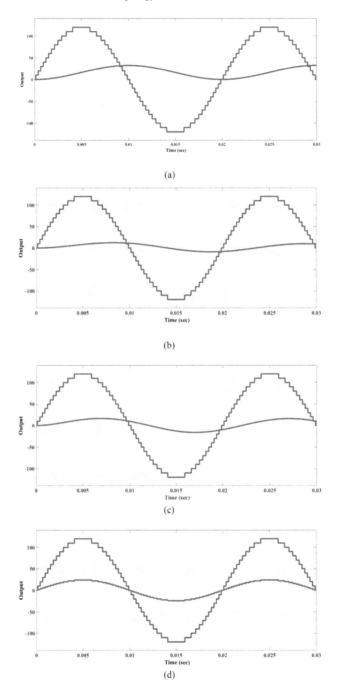

Fig. 4 Output waveforms **a** pf = 0 **b** pf = 0.4 **c** pf = 0.7 **d** pf = 1

References

1. H.M. Bassi, Z. Salam, A new hybrid multilevel inverter topology with reduced switch count and dc voltage sources. Energies **12**(977), 1–15 (2019)
2. V. Thiyagarajan, P. Somasundaram, K. Ramash Kumar, Simulation and analysis of novel extendable Multilevel inverter topology. J. Circuits Syst. Comput. **28**(06), 1950089-1-25 (2020)
3. M.D. Siddique, S. Mekhilef, A new multilevel inverter topology with reduce switch count. IEEE Access **7**, 58584–58594 (2019)
4. V. Thiyagarajan, P. Somasundaram, Modeling and analysis of novel Multilevel inverter topology with minimum number of switching components. CMES **113**(4), 461–473 (2017)
5. V. Thiyagarajan, P. Somasundaram, New asymmetric 21-level inverter with reduced number of switches. J. Eng. Res. (TJER) **16**(1), 18–27 (2019)
6. M.D. Siddique, S. Mekhilef, Optimal design of a new cascaded multilevel inverter topology with reduced switch count. IEEE Access **7**, 24498–24510 (2019)

Design and Analysis of Resonant Inverter-Based Induction Heating Equipment

Anshuman Baruah, Atanu Samanta, Abhigyan Dutta, Ananyo Bhattacharya, and Rahul Raman

1 Introduction

In the present scenario of environmental awareness, the induction heating (IH) systems have been able to successfully hold a broad range of applications in the industries as well as household equipment. The conventional heating process uses fossil fuel that contributes significantly to the pollutant percentage in the atmosphere. Use of electricity in industries for heating is the most suitable practice as compared with other conventional methods of using non-renewable sources of energy. Induction heating (IH) system works on the principle that when a high frequency alternating current is fed to an inductor coil it sets up an alternating magnetic field which cuts the workpiece and produces the heating effect in the workpiece [1]. The contactless, fast and pollution-free characteristics of induction heating systems are rising its popularity in the market. Lucia et al. [2] presented a review paper based on induction heating technology and its importance in today's industrial sector. Besides, it discusses the advancement of different circuitry topologies for the development of induction heating technology as a technology of the future. An overview of its applications is also well described. The innovation of new technologies and advancements in the power electronics domain encourages manufacturers to implement this technology in the domestic sector. Han et al. [3] present the concept of flexible induction

A. Baruah (✉) · A. Samanta · A. Dutta
Jorhat Engineering College, Jorhat, Assam, India
e-mail: anshumandadly@gmail.com

A. Samanta
e-mail: iamatanu.samanta@gmail.com

A. Bhattacharya
National Institute of Technology Jamshedpur, Jamshedpur, India

R. Raman
Indian Institute of Technology (ISM) Dhanbad, Dhanbad, India

© The Author(s), under exclusive license to Springer Nature Singapore Pte Ltd. 2021
J. Kumar and P. Jena (eds.), *Recent Advances in Power Electronics and Drives*, Lecture Notes in Electrical Engineering 707,
https://doi.org/10.1007/978-981-15-8586-9_8

heating based on the magnetic resonant coupling (MRC) mechanism. The proposed induction heating technique uses one resonant coil in the heater and one in the workpiece that can significantly strengthen the coupling effect and hence enhances the heating effect. Use of such design to yield strong magnetic field may result in injection of high amount of harmonics to flow back into the supply system. The domestic induction heating system requires efficient power converters with precise power control. A resonant inverter with reverse biased IGBT is used in [4] for domestic induction heating system. Additionally, they have proposed a modulation technique to have linear output power control and high efficiency.

Since induction heating systems require high current to set up the eddy current in the metal workpiece, a high frequency power conversion circuit is very essential. Ahmed [5] presented a soft-switching power conversion circuit for high-power induction heating applications that can be operated from single-phase as well as three-phase systems. Millan et al. [6] presented a modified half-bridge series resonant inverter topology with two selective harmonic operation techniques that can be used for ferromagnetic as well as non-ferromagnetic objects.

A theoretical model of multiple-coil inductors which is supplied by resonant half-bridge inverter is discussed in [7]. The proposed method uses an adaptable heating zone with two or three concentric inductive coils each one fed from an independent output of a power multi-inverter stage, commonly a dual half-bridge series resonant inverter. It is mainly used to decrease the efficiency losses which occur when a small workpiece is heated with a larger heating zone. Moreover, with the proposed model a homogeneous power distribution in the load can be obtained. Authors of [8–10] have discussed various topologies based on asymmetrical voltage cancellation to improve the efficiency in full bridge but have not obtained a satisfactory result in reducing flicker emissions completely. To minimize the switching losses, most of the above-mentioned methods operate at a frequency higher than the resonating frequency to obtain zero voltage switching (ZVS) transition [4]. As a result, maximum heating cannot be obtained. The need for optimization in the efficiency of the induction heating application is also well discussed in this paper. Nowadays induction heating technology is the backbone of many industries. The authors of the paper [11] presented an analytical methodology to get an optimum efficiency of induction heating systems using both planer and solenoidal cases.

Use of high frequency inverter generally introduces two types of harmonic currents viz. switching frequency harmonics and high frequency harmonics. Hence, there is always a need to eliminate these harmonics before it feeds back into the grid and deteriorates the power quality. The increase in the use of induction heating systems has raised concerns about harmonic effect resulting in deterioration of the power quality. The authors of [12] propose methods based on the active filter and the passive filter to overcome the problem of harmonics on the power grid. However, THD with the proposed method shows around 22.1% which can still have an impact on the grid voltage and poor function of the system utilities.

This paper is mainly focused on the designing of a high frequency resonant inverter for the induction heating equipment with utilization of shunt active power filter to eliminate the harmonics generated and thereby improving the overall quality of the power. Furthermore, the proposed hysteresis controller is designed in a way ensuring effective operation of shunt active power filter (SAPF). The paper is organized as follows: Sect. 2 Proposed Circuit Configurations, Sect. 3 Working of SAPF, Sect. 4 Results and Discussions, Sect. 5 Conclusion.

2 Proposed Circuit Configuration

The proposed block diagram of the IHE with hysteresis controller-based filter circuit is shown in Fig. 1.

It simply represents the source connected to the load block through the rectifier circuit. The filter and controller block are connected to the main bus bar. High frequency resonant inverter comprising RLC will be behaving like a non-linear load generating high frequency harmonics.

The simulation diagram required for analysis of induction heating equipment is designed in PSIM software. Additionally, the SAPF circuit along with the hysteresis controller topology is also presented. An analysis is being made one with the use of filter circuit and one without it to signify the importance of harmonic elimination.

Figure 2 represents induction heating equipment without the filter circuit with equivalent circuit parameters. It consists of a three-phase sinusoidal ac source with

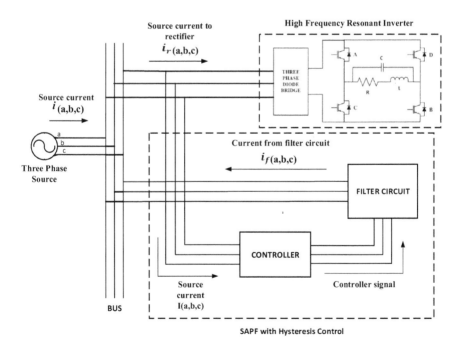

Fig. 1 The proposed block diagram of IHE

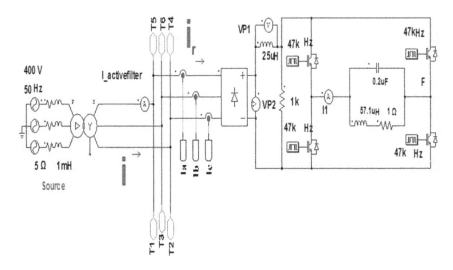

Fig. 2 Sub-circuit simulation diagram of the proposed induction heating equipment

voltage magnitude of 400 V and frequency of 50 Hz. The chosen resistance and inductance in series with the source has a value of 5 Ω, 0.1 mH, respectively. The output from Δ-Y transformer is fed to the three-phase diode rectifier that acts as input to the high frequency full-bridge inverter. A parallel combination of capacitor (C) and coil of inductance (L) and resistance (R) is the load to the high frequency inverter with values 0.2 μF, 57.1 μH and 1 Ω, respectively.

A shunt active power filter (SAPF) is attached in parallel to the high frequency inverter with induction heating load is shown in Fig. 3.

The SAPF is designed with capacitance of 0.2 μF and resistance 0.1 Ω. The hysteresis controller shown in Fig. 4 uses system current and voltage as reference values. A triangular-wave voltage with frequency of 10,020 Hz and duty cycle of 0.5 is fed to the comparator to generate the gate pulses for IGBT switches of the SAPF.

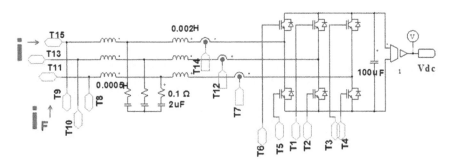

Fig. 3 Sub-circuit simulation diagram of the proposed SAPF

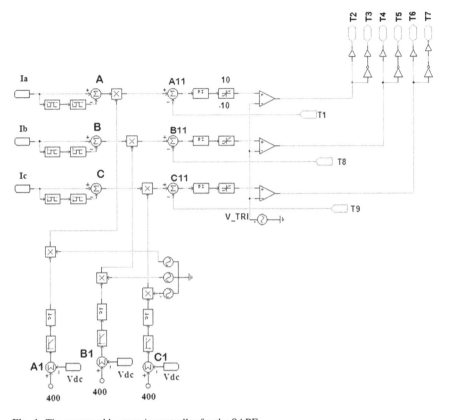

Fig. 4 The proposed hysteresis controller for the SAPF

3 Working of SAPF

To compensate the harmonic current generated due to IHE, the SAPF must operate. The gate pulse required for the switching action of IGBT can be deduced as follows:

Let, the reference currents be I_a, I_b, I_c, respectively.

$$I_a = (I_{ha})\sqrt{2}\sin\omega t + i_{ha}$$

$$I_b = (I_{hb})\sqrt{2}\sin(\omega t - 120°) + i_{hb}$$

$$I_c = (I_{hc})\sqrt{2}\sin(\omega t + 120°) + i_{hc}$$

These currents consist of fundamental component and harmonic component of high frequency, i.e. i_{ha}, i_{hb}, i_{hc}.

Cascading of bandpass filter reduces the bandwidth and Q-factor is sharpened. The 2nd order bandpass filter allows only mid-range frequencies to pass through. For phase A, from Fig. 4,

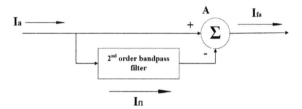

At summer A,

$$I_{FA} = I_a - I_{f1} = \left[\left\{ (I_{ha}) \sqrt{2} \sin \omega t + i_{ha} \right\} - I_{f1} \right] \tag{1}$$

Similarly for other phases,

$$I_{FB} = \left[\left\{ (I_{hb}) \sqrt{2} \sin(\omega t - 120°) + i_{hb} \right\} - I_{f2} \right] \tag{2}$$

$$I_{FC} = \left[\left\{ (I_{hc}) \sqrt{2} \sin(\omega t + 120°) + i_{hc} \right\} - I_{f3} \right] \tag{3}$$

where I_{FA}, I_{FB}, I_{FC} are currents mid-range frequency currents from summer A. Taking, reference sinusoidal voltages from source viz. V_a, V_b, V_c, i.e.

$$V_a(t) = V_r \sin \omega t + v_{ha}(t)$$

$$V_b(t) = V_r \sin(\omega t - 120°) + v_{hb}(t)$$

$$V_c(t) = V_r \sin(\omega t + 120°) + v_{hc}(t)$$

where V_r is the source rms voltage and v_{ha}, v_{hb}, v_{hc} is the component with moderate frequency in time-domain.

Now, at summing point A1, from Fig. 4,

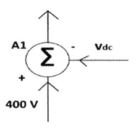

For phase A,

$$V_a(t) = [\{V_r \sin \omega t + v_{ha}(t)\} - v_{dc}] \tag{4}$$

Similarly, for other phases

$$V_b(t) = [\{V_r \sin(\omega t - 120°) + v_{ha}(t)\} - v_{dc}] \tag{5}$$

$$V_c(t) = [\{V_r \sin(\omega t + 120°) + v_{ha}(t)\} - v_{dc}] \tag{6}$$

The input fed to the PI controller is passed before through a bandpass filter that stops frequency above a specified range in kHz.

Equations (4), (5) and (6) converting to s-domain,

$$V_a * (s) = \left[V_r \frac{s.\sin(0) + \omega.\cos(0)}{s^2 + \omega^2} + v_{ha}'(s) \right] \tag{7}$$

where $v_{ha}'(s)$ is component with moderate frequency in s-domain.

Similarly,

$$V_b * (s) = \left[V_r \frac{s.\sin(-120°) + \omega.\cos(-120°)}{s^2 + \omega^2} + v_{hb}'(s) \right] \tag{8}$$

$$V_c * (s) = \left[V_r \frac{s.\sin(120°) + \omega.\cos(120°)}{s^2 + \omega^2} + v_{hc}'(s) \right] \tag{9}$$

Equations (7), (8), (9) can be written as

$$V_a * (s) = V_r \frac{\omega}{s^2 + \omega^2} + v'_{ha}(s) \tag{10}$$

$$V_b * (s) = V_r \frac{s\left(-\frac{\sqrt{3}}{2}\right) + \omega\left(-\frac{1}{2}\right)}{s^2 + \omega^2} + v'_{hb}(s) \tag{11}$$

$$V_c * (s) = V_r \frac{s\left(\frac{\sqrt{3}}{2}\right) + \omega\left(\frac{1}{2}\right)}{s^2 + \omega^2} + v'_{hc}(s) \tag{12}$$

Let, proportional constant of PI controller is k_p and integrational constant is k_e.

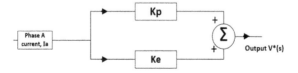

$$\text{Output, } V*(s) = V(t).\left\{k_p + k_e \int V(t)dt\right\}$$

Using Laplace transform,

$$= V_s\left\{k_p + \frac{k_e}{s}\right\} \tag{13}$$

Equations (10), (11), (12) can be written as, respectively,

$$V_a * (s) = \left\{V_r\frac{\omega}{s^2 + \omega^2} + v'_{ha}(s)\right\}.\left\{k_p + \frac{k_e}{s}\right\} \tag{14}$$

$$V_b * (s) = \left\{V_r\frac{s\left(-\frac{\sqrt{3}}{2}\right) + \omega\left(-\frac{1}{2}\right)}{s^2 + \omega^2} + v'_{hb}(s)\right\}.\left\{k_p + \frac{k_e}{s}\right\} \tag{15}$$

$$V_c * (s) = \left\{V_r\frac{s\left(\frac{\sqrt{3}}{2}\right) + \omega\left(\frac{1}{2}\right)}{s^2 + \omega^2} + v'_{hc}(s)\right\}.\left\{k_p + \frac{k_e}{s}\right\} \tag{16}$$

At multiplier 2,

$$V_a * (s).1 = \left\{V_r\frac{\omega}{s^2 + \omega^2} + v'_{ha}(s)\right\}.\left\{k_p + \frac{k_e}{s}\right\}$$

$$V_b * (s).1 = \left\{V_r\frac{s\left(-\frac{\sqrt{3}}{2}\right) + \omega\left(-\frac{1}{2}\right)}{s^2 + \omega^2} + v'_{hb}(s)\right\}.\left\{k_p + \frac{k_e}{s}\right\}$$

$$V_c * (s).1 = \left\{V_r\frac{s\left(\frac{\sqrt{3}}{2}\right) + \omega\left(\frac{1}{2}\right)}{s^2 + \omega^2} + v'_{hc}(s)\right\}.\left\{k_p + \frac{k_e}{s}\right\}$$

Equations (1), (2), (3) at summer A is fed to multiplier.
For phase A,

$$= I_{FA}V^*_a(s)$$

$$= \left\{V_r\frac{\omega}{s^2 + \omega^2} + v'_{ha}(s)\right\}.\left\{k_p + \frac{k_e}{s}\right\}.I_{FA}$$

For phase B,

$$= I_{FB} V^*{}_b(s)$$

$$= \left\{ V_r \frac{s\left(-\frac{\sqrt{3}}{2}\right) + \omega\left(-\frac{1}{2}\right)}{s^2 + \omega^2} + v'{}_{hb}(s) \right\} \cdot \left\{ k_p + \frac{k_e}{s} \right\} \cdot I_{FB}$$

For phase C,

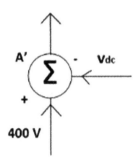

$$= I_{FC} V^*{}_c(s)$$

$$= \left\{ V_r \frac{s\left(\frac{\sqrt{3}}{2}\right) + \omega\left(\frac{1}{2}\right)}{s^2 + \omega^2} + v'{}_{hc}(s) \right\} \cdot \left\{ k_p + \frac{k_e}{s} \right\} \cdot I_{FC}$$

where I_{FA}, I_{FB}, I_{FC} are currents mid-range frequency currents from summer A. Again at summer A11, from Fig. 4,

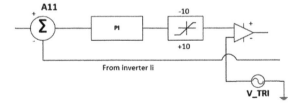

For phase A,

$$= \left\{ V_r \frac{\omega}{s^2 + \omega^2} + v'{}_{ha}(s) \right\} \cdot \left\{ k_p + \frac{k_e}{s} \right\} \cdot I_{FA} - I_i \tag{17}$$

Fig. 5 Comparator waveform

Similarly, other phases can also be shown as Eq. (17)
PI controller will work as

$$= \left\{ \left(V_r \frac{\omega}{s^2 + \omega^2} + v'_{ha}(s) \right) \cdot \left(k_p + \frac{k_e}{s} \right) \cdot I_{FA} - I_i \right\} \cdot \left\{ k_p + \frac{k_e}{s} \right\}$$

which is shown for phase A only.

This value is fed to the limiter with range of -10 to $+10$ V.

Now, the comparator is set up with a reference triangular voltage at negative terminal and output of limiter to positive terminal, which can be understood from Fig. 5.

When $V_{tri} > V_{output}$, that is at point A to B, comparator will set to maximum of $-V_{cc}$ and give 0 as pulse to IGBT.

When $V_{output} > V_{tri}$, that is at point 0 to A, comparator will set to maximum of $+V_{cc}$ and give 1 as pulse to IGBT.

Hence, the operation of SAPF can be analyzed from the above calculations.

4 Results and Discussions

The simulation of IH equipment without filter circuit is shown in Fig. 6.

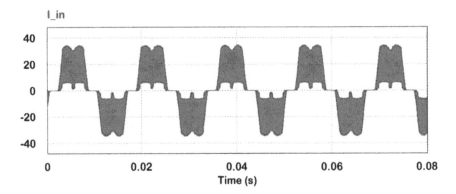

Fig. 6 Time-domain analysis of input current waveform without filter

The input current contains harmonics due to high frequency operation which results in distorted sinusoidal waveform. The amount of harmonics present in the above analysis is enough to impose serious issues like distortion of the grid voltage, excessive heating of electrical and electronics equipments. In Fig. 7, the FFT analysis of the input current clearly shows the domination of high frequency harmonics in the input sinusoidal waveform.

The simulation result of the main circuit with the shunt active power filter (SAPF) is presented in Fig. 8. Consequently in Fig. 9, FFT analysis shows that the use of SAPF with hysteresis control technique eliminates the harmonic to almost negligible percentage.

Calculations for THD is given as

$$THD = \frac{\sqrt{\sum_{n=2}^{\infty} I_{n,r.m.s}^2}}{I_{1,r.m.s}}$$

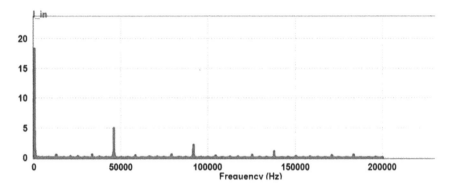

Fig. 7 FFT analysis of the input current waveform without filter

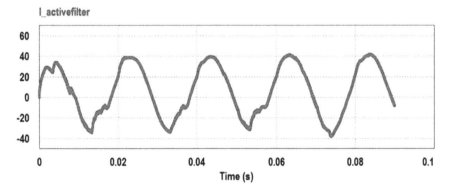

Fig. 8 Time-domain analysis of input current waveform with filter

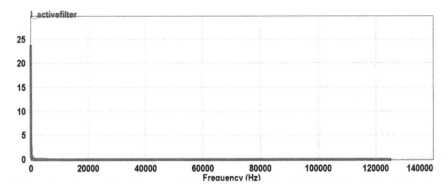

Fig. 9 FFT analysis of the input current waveform with filter

Without Filter

$$= \frac{\sqrt{4.99^2 + 1.329^2 + 1.059^2}}{18.328} \times 100\% = \mathbf{28.76\%}$$

With Filter

$$= \frac{\sqrt{2.42^2 + 1.51^2 + 0.0827^2}}{23.65} \times 100\% = \mathbf{12.47\%}$$

5   Conclusion

We have seen that harmonics play a big drawback role in any electrical circuit. One of the major effects of harmonics in power system is the increase in distortion of input current in the system. The dominant harmonic decreases the reliability of the entire system and its elimination becomes necessary to make the IHE a worthy technology. A detailed analysis of the IHE technology with and without the filter circuit is discussed to highlight the amount of distortions present in the current waveform. The hysteresis controller technique is implemented for its easy control and faster response.

In this paper, the calculation of THD with and without filter circuit is discussed which shows drastic change of distortion value from 28.76% to 12.47% in the input waveform and hence the use of shunt active power filter (SAPF) is justified. Only then the induction heating technique can replace the existing conventional heating systems in the industry and make this greener approach a successful one.

References

1. R. Raman, P. Sadhu, A. Kumar, P.K. Sadhu, Design and analysis of EMI and RFI suppressor for induction heating equipment using Vienna rectifier, in *2nd International Conference on Power, Energy and Environment: Towards Smart Technology, ICEPE 2018* (2019), pp. 2–7
2. O. Lucia, P. Maussion, E.J. Dede, J.M. Burdio, Induction heating technology and its applications: past developments, current technology, and future challenges. IEEE Trans. Ind. Electron. **61**(5), 2509–2520 (2014)
3. W. Han, K.T. Chau, Z. Zhang, Flexible induction heating using magnetic resonant coupling. IEEE Trans. Ind. Electron. **64**(3), 1982–1992 (2017)
4. H. Sarnago, O. Lucia, A. Mediano, J.M. Burdio, Modulation scheme for improved operation of an RB-IGBT-based resonant inverter applied to domestic induction heating. IEEE Trans. Ind. Electron. **60**(5), 2066–2073 (2013)
5. N.A. Ahmed, High-frequency soft-switching AC conversion circuit with dual-mode PWM/PDM control strategy for high-power IH applications. IEEE Trans. Ind. Electron. **58**(4), 1440–1448 (2011)
6. I. Millán, J.M. Burdío, J. Acero, O. Lucía, S. Llorente, Series resonant inverter with selective harmonic operation applied to all-metal domestic induction heating. IET Power Electron. **4**(5), 587–592 (2011)
7. F. Sanz-Serrano, C. Sagues, S. Llorente, Power distribution in coupled multiple-coil inductors for induction heating appliances. IEEE Trans. Ind. Appl. **52**(3), 2537–2544 (2016)
8. S. Chudjuarjeen, A. Sangswang, C. Koompai, An improved LLC resonant inverter for induction-heating applications with asymmetrical control. IEEE Trans. Ind. Electron. **58**(7), 2915–2925 (2011)
9. J.M. Burdío, L.A. Barragán, F. Monterde, D. Navarro, J. Acero, Asymmetrical voltage-cancellation control for full-bridge series resonant inverters. IEEE Trans. Power Electron. **19**(2), 461–469 (2004)
10. C. Carretero, O. Lucía, J. Acero, J.M. Burdío, Phase-shift control of dual half-bridge inverter feeding coupled loads for induction heating purposes. Electron. Lett. **47**(11), 670–671 (2011)
11. I. Lope, J. Acero, C. Carretero, Analysis and optimization of the efficiency of induction heating applications with Litz-wire planar and solenoidal coils. IEEE Trans. Ind. Electron. **31**(7), 5089–5101 (2016)
12. A. Namadmalan, S. Abedi, S.H. Hosseinian, J.S. Moghani, Power quality improvement for single phase and three phase current source induction heating systems, in *2010 1st Power Quality Conference, PQC 2010*, no. 1 (2010), pp. 2580–2584

Harmonic Elimination of a T-Type Multilevel Inverter Based on Multistate Switching Cell

Manoj Kumar Kar, Mohd Imam Hasan Mansoori, Sanjay Kumar, and S. K. Gupta

1 Introduction

The requirement of converters having high performing parameters, reduced weight, and compact in size in the application of high voltage has been gradually increased. The size of L and C can be reduced by increasing switching frequency in medium range (10–25 kHz), in this order the switching loss will increase [1]. Nowadays, the switching devices are made of wide bandgap semiconductor materials such as GaN, SiC, and boron nitride (BN). WBG semiconductors permit devices to operate at higher voltage, frequency, and temperature as compared to conventional semiconductors like silicon and gallium arsenide. However, these devices are not used widely because of high cost, which has an important role in the industry. Conventional insulated-gate bipolar transistor has high-voltage blocking capability (6 kV) with high switching frequency (f_s) (10 kHZ). Therefore, IGBT is a most valuable for high-voltage applications.

For low-voltage applications, two-level topologies are the standard solution although three-level topologies are preferred for high switching frequencies. When high output voltage, high efficiency and compact size of magnetics are required, multilevel converters become very popular solution [2–6]. Researchers have found

M. K. Kar (✉) · M. I. H. Mansoori · S. Kumar · S. K. Gupta
Department of Electrical Engineering, National Institute of Technology Jamshedpur, Jamshedpur 831014, Jharkhand, India
e-mail: manojkar132@gmail.com

M. I. H. Mansoori
e-mail: imammnna786@gmail.com

S. Kumar
e-mail: sanjay.ee@nitjsr.ac.in

S. K. Gupta
e-mail: skgupta.ee@nitjsr.ac.in

© The Author(s), under exclusive license to Springer Nature Singapore Pte Ltd. 2021
J. Kumar and P. Jena (eds.), *Recent Advances in Power Electronics and Drives*, Lecture Notes in Electrical Engineering 707,
https://doi.org/10.1007/978-981-15-8586-9_9

that three-level topologies are agreeable solution in the application of low voltage with high f_s [7]. If f_s is more than 10 kHz, then the efficiency of three-level neutral point clamped converter is better than two-level inverters [8]. An active neutral-point clamped five-level (ANPC-5L) topology is introduced in [9] for medium voltage application drives. Different VSC types topologies are discussed in [10, 11]. Three-state switching cell (3SSC) concept is introduced in [12]. Current is equally distributed through the semiconductor devices and the voltage applied on the load. Different types of topologies of multilevel inverter applying the 3SSC were introduced in [13]. By studying the aforesaid topologies, it is emphasized that the current is divided among the semiconductor switching devices, hence the conduction losses and size can be reduced. Initially, 3SSC has been extended to dc–dc converters [12], and then to ac–dc and dc–ac converters [14, 15]. The 3SSC and MSSC principles were later introduced [16]. Such simple cells may be applied to the power converter topologies. The proposed multilevel topology five-level neutral point clamped multistate switching cell (5L NPC MSSC) having high efficiency and less conduction losses as compared to classical topologies [17]. In 5L NPC MSSC topology, the main reason for reducing the size is due to output filter ripple frequency of $2f_s$.

2 Methodology

2.1 Proposed Methodology

In this topology, there are eight switches T_a–T_h with their anti-parallel diodes D_a–D_h, respectively, an interphase transformer with W_1 and W_2 windings and a passive filter having inductor L_0 and capacitor C_0. The transformation ratio of the transformer is unity. The switches are connected as shown in the Fig. 1.

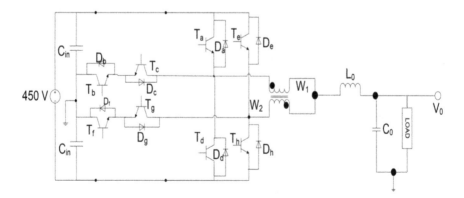

Fig. 1 Circuit diagram of T-type MLI

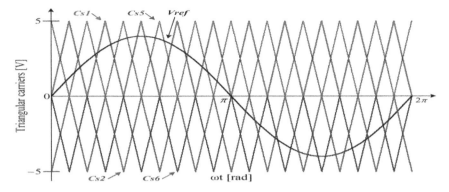

Fig. 2 SPWM technique

2.2 Modulation Technique

SPWM technique can be implemented to get interesting result. As shown in Fig. 2, carrier and reference waves are triangular and sinusoidal consecutively. The switching frequency of carrier is much more than reference. When the amplitude of reference is more than amplitude of carrier, a gating pulse will produce. There can be more than one carrier in SPWM technique as per requirement.

As shown in Fig. 2, the reference and carriers can be seen. The carriers $Cs1$ and $Cs2$ are in same phase but they are vertically shifted to each other. The carriers $Cs5$ and $Cs6$ are shifted by half time period compared to carriers $Cs1$ and $Cs2$ consecutively. The gating signals for the switches T_a, T_b, T_e, T_f can be obtained by the comparative study of sinusoidal reference and carrier $Cs1, Cs2, Cs5, Cs6$, respectively. The gating signals of devices T_c, T_d, T_g, and T_h are in complimentary fashion with switches T_a, T_b, T_e, and T_f, respectively.

By employing SPWM technique, switches T_b and T_f are always on while switches T_a and T_e are always off in +ve and −ve half cycle of modulating signal, respectively.

2.3 Modes of Operation

Based on switching devices states, there are two modes of operation.

1. Overlapping (OL) mode
2. Non-overlapping (NOL) mode.

OL mode of operation comes into the picture when switches T_a and T_e, or T_d and T_h are in on state simultaneously in +ve or −ve half cycle of the respective output voltage waveform. In NOL mode of operation, switches T_a and T_e, or T_d and T_h are not switched on simultaneously in +ve or −ve half cycle of respective output waveform. The OL and NOL modes of operation are shown clearly in the figure.

Fig. 3 OL and NOL mode
operation

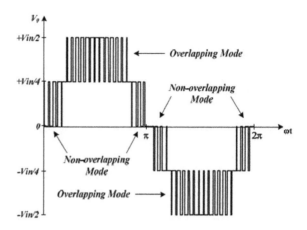

1. NOL mode of operation in positive half cycle: Switches T_b and T_f are always in
   on state.

   - 1st step: The device T_a is off and the device T_e is on. The 50% of I_L is flowing
     through input, T_e and winding W_2. And rest 50% is flowing through T_b, diode
     D_c and winding W_1. The output voltage (V_{AO}) is $Vin/4$ in this step.
   - 2nd step: The devices T_a and T_e are turned off. The 50% of I_L is flowing
     through T_b, diode D_c and winding W_1. And rest 50% is flowing through T_f,
     diode D_g and winding W_2. The V_{AO} is zero in this step.
   - 3rd step: The device T_a is on and switch T_e is off. The 50% of load current is
     flowing through input voltage, switch T_a and winding W_1. And rest 50% load
     current is flowing through switch T_f, diode D_g and winding W_2. The V_{AO} is
     $Vin/4$ in this step.
   - 4th step: Same as second step (Fig. 3).

2. OL mode of operation in positive half cycle:

   - 1st step: Same as first step of NOL mode operation.
   - 2nd step: The switches T_a and T_e are on. The 50% of I_L is flowing through
     T_a and winding W_1. And rest 50% is flowing through T_e and winding W_2.
     The V_{AO} is $Vin/2$ in this step.
   - 3rd step: Same as third step of NOL mode operation.
   - 4th step: Same as second step of NOL mode of operation.

   Similar steps are also there in negative half cycle of reference sinusoidal voltage.
Similar output voltage pulses will produce of negative amplitude.

2.4 Output Voltage

There are three possible levels in each half cycle. There are total five level of voltage $(+Vin/4, +Vin/2, 0, -Vin/2, Vin/4)$ present in one cycle of output voltage as stated in above steps accordingly.

3 Theoretical Analysis

3.1 Modulation Index

If Vpk_ref and Vpk_car are the peak amplitude of sinusoidal reference and carrier waveform, respectively. Then,

$$M = \frac{v_{pk\_ref}}{v_{pk\_car}} = \frac{2.\sqrt{2}.v_o}{v_{in}} \tag{1}$$

3.2 THD Analysis

It is the measure of harmonics content present on ac side of converter and is given by

$$THD = \sqrt{\frac{1}{g^2} - 1} \tag{2}$$

where g is the distortion factor. Distortion factor is given by

$$g = \sqrt{\frac{v_{o1}}{v_{or}}} \tag{3}$$

where $Vo1$—fundamental component of voltage (rms) Vor—output voltage (rms).
    Now, the rms output voltage without filter is given by

$$V_{Ao\_rms} = \frac{v_{in}}{2} \sqrt{\left[\frac{M}{\pi} + \frac{\sqrt{4M^2 - 1}}{\pi} + \frac{1}{\pi} \sin^{-1}\left(\frac{1}{2M}\right) - \frac{1}{2} \right]} \tag{4}$$

THD in terms of M is given by

$$THD_{VAO} = \sqrt{\frac{2}{M^2}\left[\frac{M}{\pi} + \frac{\sqrt{4M^2 - 1}}{\pi} + \frac{1}{\pi}\sin^{-1}\left(\frac{1}{2M}\right) - \frac{1}{2}\right]} - 1 \quad (5)$$

3.3 Current and Voltage Stress in Switches

There are two group of switches. Switches T_a, T_d, T_e, and T_h are in one group and switches T_b, T_c, T_f, and T_g are in another group. The current and voltage stress are calculated for switches T_e and T_f.

a. T_a, T_d, T_e, and T_h switch group: The average value of current is given by

$$I_{Te\_avg} = \frac{\sqrt{2M}}{8}I_0 \quad (6)$$

And the rms value of current is given by

$$I_{Te\_rms} = \frac{\sqrt{M}}{3\pi}I_0 \quad (7)$$

The maximum voltage (V_{max}) that appears across the switch.

$$V_{Te\_max} = V_{in} \quad (8)$$

b. T_b, T_c, T_f, and T_g switch group: The mean current is given by

$$I_{Tf\_avg} = \left[\frac{1}{2\pi} - \frac{M}{8}\right]\sqrt{2}I_0 \quad (9)$$

And the rms value of current is given by

$$I_{Tf\_rms} = \left[\frac{1}{4} - \frac{2M}{3\pi}\right] \cdot \frac{I_0}{\sqrt{2}} \quad (10)$$

V_{max} that appear across the switches

$$V_{Tf\_max} = \frac{V_{in}}{2} \quad (11)$$

3.4 Output Filter

LC filter is used at the o/p terminal of inverter. The values of parameters of output filter L_0 and C_0 are given below. If f_0 is the cut-off frequency of *LC* filter and ξ is the damping factor, then

$$f_0 = \frac{2f_s}{10} \tag{12}$$

$$C_0 \geq \frac{1}{4\pi \xi f_0 R_0} \tag{13}$$

$$L_0 = \frac{1}{(2\pi f_0)^2 C_0} \tag{14}$$

4 Designing

To implement the aforesaid method discussed for the proposed converter, an example 5L T-type MSSC is discussed below.

4.1 Specification

The specifications for the inverter are shown

Output power	$P_o = 5$ KW
Input voltage	$V_{in} = 450$ V
Output voltage	$V_o = 127$ V (rms)
Switching frequency	$f_s = 20$ kHz
Frequency (output)	$f_r = 50$ Hz
M	M = 0.80

4.2 Simplified Design Example

a. $T_a, T_d, T_e,$ and T_h switch group: The average current is given by

$$I_{Te\_avg} = \frac{\sqrt{2}M}{8}I_o = 5.56\,\text{A}$$

And the rms current is given by

$$I_{Te\_rms} = \sqrt{\frac{M}{3\pi}}I_o = 11.47\,\text{A}$$

V_{max} appear across the switches

$$V_{Te\_max} = V_{in} = 450\,\text{V}$$

b. T_b, T_c, T_f, and T_g switch group: The average current is given by

$$I_{Tf\_avg} = \left[\frac{1}{2\pi} - \frac{M}{8}\right]\sqrt{2I_0} = 3.29\,\text{A}$$

And the rms current is given by

$$I_{Tf\_rms} = \left[\frac{1}{4} - \frac{2M}{3\pi}\right]\frac{I_0}{\sqrt{2}} = 7.88\,A$$

V_{max} appear across the switches

$$V_{Tf\_max} = \frac{V_{in}}{2} = 225\,\text{V}$$

c. Output filter

$$C_0 = 615\,\mu\text{F}, \quad f_0 = 4\ \text{kHz}$$

$$L_0 = \frac{1}{(2\pi f_0)^2 C_0} = 180\mu H$$

d. THD.

From Eq. (5), THD = 38.4%.

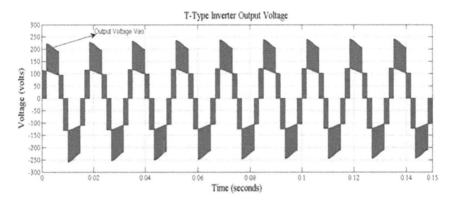

Fig. 4 Output *VAO* waveform before the filter

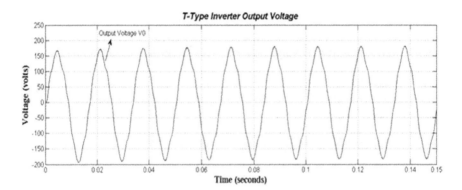

Fig. 5 Output voltage waveform after the filter

4.3 Simulation Results

The proposed converter is of 5 KW. The simulation results are shown below

The output voltage *VAO* waveform before the filter is shown in Fig. 4. The output voltage waveform after the filter is depicted in Fig. 5. The output current waveforms through the inductor and transformer are shown in Fig. 6. The *THD* versus frequency graph is shown in Fig. 7. The *THD* versus harmonics order graph is depicted in Fig. 8.

5   Conclusion

In this paper, the proposed inverter was 5L T-Type-MSSC inverter. The study and simulation have done for this inverter. In this topology, five-level voltage is obtained at the output. High ordered harmonics are eliminated by using an *LC* filter. The

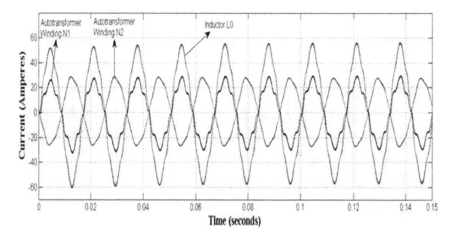

Fig. 6 Output current waveforms through the inductor and transformer

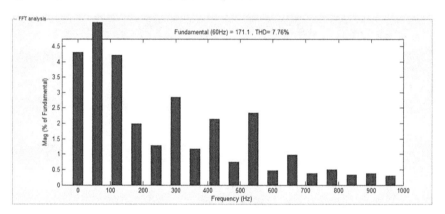

Fig. 7 *THD* versus Frequency graph

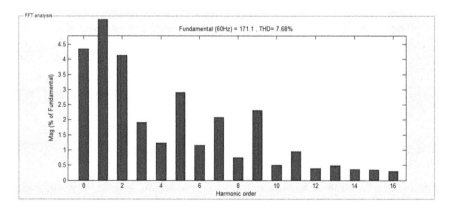

Fig. 8 *THD* versus harmonics order graph

output ripple frequency of LC filter is $2fs$, hence the size is less. The inverter has some other characteristics that the current is shared by the switches. The contact area is increased, hence switching loss decreases. Due to low switching and conduction losses, the efficiency of this inverter is high. Due to its efficient performance, it can be used in high-voltage industrial applications.

References

1. M. Schweizer, I. Lizama, T. Friedli, J.W. Kolar, Comparison of the chip area usage of 2-level and 3-level voltage source converter topologies, in *Proceedings of 36th Annual IEEE IECON*, pp. 391–396
2. B. Wu, *High-Power Converters and AC Drives* (Wiley, Hoboken, NJ, USA, 2006)
3. J. Rodríguez, J.-S. Lai, F.Z. Peng, Multilevel inverters: a survey of topologies, controls, and applications. IEEE Trans. Ind. Electron. **49**(4), 724–738 (2002)
4. J.-S. Lai, F.Z. Peng, Multilevel converters—a new breed of power converters. IEEE Trans. Ind. Appl. **32**(3), 509–517 (1996)
5. J. Rodriguez, S. Bernet, B. Wu, J.O. Pontt, S. Kouro, Multilevel voltage-source-converter topologies for industrial medium-voltage drives. IEEE Trans. Ind. Electron. **54**(6), 2930–2945 (2007)
6. J. Rodriguez et al., Multilevel converters: an enabling technology for high-power applications. Proc. IEEE **97**(11), 1786–1817 (2009)
7. R. Teichmann, S. Bernet, A comparison of three-level converters versus two-level converters for low voltage drives, traction, and utility applications. IEEE Trans. Ind. Appl. **41**(3), 855–865 (2005)
8. A. Nabae, I. Takahashi, H. Akagi, A new neutral-point-clamped PWM inverter. IEEE Trans. Ind. Appl. **IA-17**(5), 518–523 (1981)
9. F. Kieferndorf et al., ANPC-5L technology applied to medium voltage variable speed drives applications, in *Proceedings International Symposium SPEEDAM* (2010), pp. 1718–1725
10. T.B. Soeiro, J.W. Kolar, The new high-efficiency hybrid neutral point-clamped converter. IEEE Trans. Ind. Electron. **60**(5), 1919–1935 (2013)
11. T.B. Soeiro, J.W. Kolar, Novel 3-level hybrid neutral-point-clamped converter, in *Proceedings of 37th Annual IEEE IECON*, pp. 4457–4462
12. G.V.T. Bascopé, I. Barbi, Generation of a family of non-isolated DC–DC PWM converters using new three-state switching cells, in *Proceedings of IEEE PESC*, vol. 2 (2000), pp. 858–863
13. M.T. Peraça, I. Barbi, Three-level half-bridge inverter based on the three-state switching cell, in *Proceedings of. 7th IEEE/IAS INDUSCON* (Recife, Brazil, 2006), pp. 1–8
14. R.A. Camara, P.P. Praça, R.P.T. Bascopé, C.M.T. Cruz, Comparative analysis of performance for single-phase ac–dc converters using FPGA for UPS applications, in *Proceedings of 28th Annual IEEE APEC*, vol. 1 (2013), pp. 1852–1858
15. R. Hausmann, I. Barbi, Three-phase dc–ac converter using four-state switching cell. IEEE Trans. Power Electron. **26**(7), 1857–1867 (2011)
16. R.P.T. Bascopé, J.A.F. Neto, G.V.T. Bascopé, Multi-state commutation cells to increase current capacity of multi-level inverters, in *Proceedings of INTELEC* (Amsterdam, The Netherlands, 2011), pp. 1–9
17. J.A.F. Neto, R.P.T. Bascopé, C.M. Cruz, R.G.A. Cacau, G.V.T. Bascopé, Comparative evaluation of three single-phase NPC inverters, in *Proceediings of 10th IEEE/IAS INDUSCON* (Fortaleza, Brazil, 2012), pp. 1–8

Cascaded H-Bridge Based Multilevel Inverter for Power Quality Issues

Arvind Yadav and Subhash Chandra

1 Introduction

In today's interconnected power transmission systems, power quality at the point of common coupling point (PCC) should be ensured; reduction in transmission losses will maximize transmission capability, therefore a static VAR compensator has become a necessity. Distribution static compensator (DSTATCOM) has gain attention nowdays due to the use of renewable energy generation; the problem of voltage fluctuation at PCC, harmonic content in waveform, and variation in power factor due to load fluctuation should be addressed. If the renewable energy output is grid integrated, then PCC voltage monitoring and conditioning become essential. However, the output voltage obtained by solar photovoltaic modules is of lower magnitude, thus a need for an intermediate converter which can boost the low output needs to be addressed. Multilevel inverters not only can suitably boost the low output, but also offer a solution for power quality issues, a high voltage level with lower harmonic distortion can be achieved, thus can be integrated with low power sources [1–3].

The basic concept behind the multilevel inverter is to connect the power semiconductor switching devices in a series configuration to get the approximated sinusoidal output voltage waveform from several levels of dc voltage. Multilevel inverters have the advantages in terms of lower switching stress on switching devices, good harmonic performance, and better output waveform [4–6]. However, with a two-level voltage source converter, optimum system performance such as filter size, losses, and THD is not achievable, leading to poor efficiency and high cost. Multilevel inverters offer reduced filter size, lower voltage stress on switching devices, low switching losses, and lower electromagnetic interference. Basically, three main topologies of a multilevel inverter are used such as neutral point clamped multilevel inverter, cascade multilevel inverter (CMI), and flying capacitor multilevel inverter in which CMI is

A. Yadav (✉) · S. Chandra
Department of Electrical Engineering, GLA University, Mathura 281406, India
e-mail: arvindyadavpec@gmail.com

S. Chandra
e-mail: subhash.chandra@gla.ac.in

© The Author(s), under exclusive license to Springer Nature Singapore Pte Ltd. 2021
J. Kumar and P. Jena (eds.), *Recent Advances in Power Electronics and Drives*, Lecture Notes in Electrical Engineering 707,
https://doi.org/10.1007/978-981-15-8586-9_10

the most attractive topology because of its modularity in structure and its capability to reach a higher output voltage and power level [7–12]. CMI increases the reliability of the system since it uses the fewer semiconductor devices and capacitors as compared to other topologies and is capable to reduce the size and cost of the converter. The output levels in CMI are decided by N = 2k + 1, where N is the output phase voltage level and k is the number of dc sources used in a single phase; in this study an 11-level CMI is modeled for VAR compensation and harmonic elimination at PCC. This paper is organized as follows: Sect. 2 explains CMI structure, Sect. 3 describes the system configuration and control layout for CMI-based DSTATCOM, Sect. 4 elaborates the results and discussion for the described control methodology, and Sect. 5 concludes the paper.

2 Cascaded Multilevel Inverter Structure

Different topologies of a multilevel inverter are available in the literature [13–15], however, this study uses a cascaded multilevel inverter, and the structure layout is shown in Fig. 1. The basic unit is called H bridge; as per the number of levels, H bridges can be selected; it is composed of a two-leg inverter, having four switches (S1, S2, S3, S4); IGBTs and MOSFETs are suitable as per the switching frequency and voltage stress. These H bridges are connected in series so as to provide an additive voltage at the output as shown in Fig. 1. The cascaded H bridge topology has an advantage of the low voltage rating of the semiconductor switching devices. Sources connected at the cascade H bridge inverter are of the same rating as unequal source voltage will make the control circuit more complex.

The cascaded multilevel inverter uses an H bridge series connection per phase; in this

Fig. 1 Basic structure of cascaded multilevel inverter (**a**) basic unit H bridge (**b**) cascaded structure for a 11-level CMI

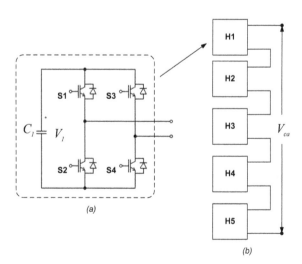

study, a capacitor is used as an input as will be suitable for reactive VAR generation. The CMI structure is unique as it avoids clamping diodes and additional capacitors for voltage balancing, thus making it suitable for most of the renewable power conversion resources such as photovoltaic and fuel cells. Compared to diode clamped and flying capacitor multilevel inverters, CMI uses the least number of components in the H bridge, and different combinations of switch triggering will lead to three different voltage levels (i.e. $+V_1$, 0, and $-V_1$). In order to obtain a high voltage, these H bridges are connected in series with each other in a phase; thus the output voltage will be the combined voltage of individual H bridge output as shown in Fig. 1.

The modulation strategy used for the multilevel inverter is based on the carrier wave switching frequency, i.e. constant switching frequency and variable switching frequency. The primitive abstraction of multi-carrier pulse width modulation (MCPWM) is to compare a modulating signal with a multi-carrier signal which may be phase shifted or level shifted. And to get the specialized feature of a multilevel inverter, i.e. lower THD and better output waveform, MCPWM is preferred [16–19]. The high switching frequency scheme is characterized by higher switching losses and the presence of high-frequency harmonics in the output voltage waveform, unlike in the constant switching frequency approach, lower and uniform switching losses and lower order harmonics are present in the output waveform. The selective harmonic elimination (SHE) technique is used to obtain the optimum triggering angle for each H bridge, as here five H bridges are used, 11-level output voltage is obtained at each phase as shown in Fig. 2. In this method (SHE), switches operate at a fundamental frequency and each switch conducts for half the fundamental period hence uniform switching losses will be there for each switching device; k − 1 harmonics can be eliminated with k H bridges.

3   System Configuration and Control

The system configuration is shown in Fig. 3, where v_s is the source voltage, v_o is the voltage at the point of common coupling (PPC), and v_c is output voltage of the cascaded multilevel inverter (CMI). Power is fed to load from the source, and CMI is connected to the PCC bus bar through a coupling reactor. In a distribution network, DSTATCOMs can improve the PCC bus power factor close to unity by compensating reactive power; voltage sag and swell due to load variation can be taken care of, and a filtered current is obtained. Here, DSTATCOM is mainly composed of a cascaded multilevel inverter on the DC side, and a coupling reactor on the AC side as shown in Fig. 3.

The harmonic components in the converter output waveform with N (=2k + 1) levels are Fourier analyzed (Eq. (1)) and the amplitude of individual harmonic can be obtained (Eq. (2)); further k − 1 harmonics can be eliminated by expanding Eq. (2) and the required triggering angle ($\alpha_1 - \alpha_5$) can be obtained.

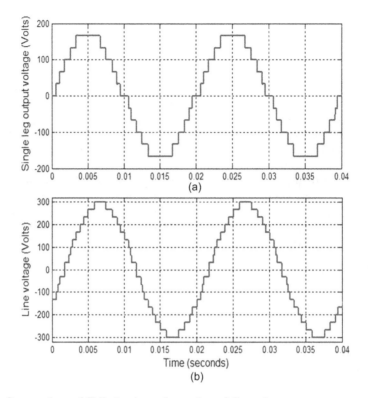

Fig. 2 Output voltage of CMI, showing **a** phase voltage, **b** line voltage

Fig. 3 System layout for
CMI-based DSTATCOM

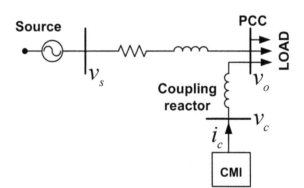

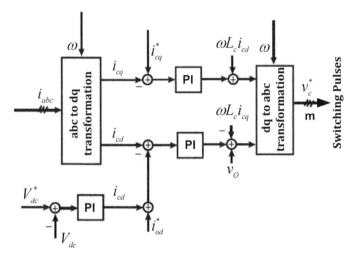

Fig. 4 Overall control structure layout for CMI-based DSTATCOM

$$V_{ca}(\omega t) = \frac{4V_{dc}}{n\pi} \left[\sum_{m=1}^{k} \cos n\alpha_m \right] \sin n\omega t \tag{1}$$

$$H(n) = \frac{4V_{dc}}{n\pi} \left[\sum_{m=1}^{k} \cos n\alpha_m \right] \quad m = 1, 2, ..., k \tag{2}$$

where V_{ca} = Phase output voltage of CMI
H(n) = Amplitude of nth harmonic
V_{dc} = Capacitor voltage of each H bridge (V_1, V_2, V_3, V_4, V_5)
k = per phase H bridge count
N = levels per phase output waveform (2k + 1)
α_m = Triggering angle for H bridge (mth cell).

The overall control description is shown in Fig. 4; the state-space equation for the configured system is given in Eq. (3); abc to dqo transformation is used to obtain dc quantities, which will be further utilized in a PI controller to produce the necessary corrections. The decoupled structure is obtained after abc to dq transformation, where the transformation matrix is given in Eq. (4), and the state-space equation containing direct axis and quadrature axis components is given in Eq. (5).

$$L_c \frac{d}{dt} \begin{bmatrix} i_{ca} \\ i_{cb} \\ i_{cc} \end{bmatrix} + r_c \begin{bmatrix} i_{ca} \\ i_{cb} \\ i_{cc} \end{bmatrix} = \begin{bmatrix} v_{oa} \\ v_{ob} \\ v_{oc} \end{bmatrix} - \begin{bmatrix} v_{ca} \\ v_{cb} \\ v_{cc} \end{bmatrix} \tag{3}$$

$$\begin{bmatrix} x_d \\ x_q \\ x_o \end{bmatrix} = \frac{2}{3} \begin{bmatrix} \cos \omega t & \cos(\omega t - \frac{2\pi}{3}) & \cos(\omega t + \frac{2\pi}{3}) \\ -\sin \omega t & -\sin(\omega t - \frac{2\pi}{3}) & -\sin(\omega t + \frac{2\pi}{3}) \\ \frac{1}{2} & \frac{1}{2} & \frac{1}{2} \end{bmatrix} \begin{bmatrix} x_a \\ x_b \\ x_c \end{bmatrix} \quad (4)$$

$$L_c \frac{d}{dt} \begin{bmatrix} i_{cd} \\ i_{cq} \end{bmatrix} + \omega L_c \begin{bmatrix} -i_{cq} \\ i_{cd} \end{bmatrix} + r_c \begin{bmatrix} i_{cd} \\ i_{cq} \end{bmatrix} = \begin{bmatrix} v_{od} - v_{cd} \\ v_{oq} - v_{cq} \end{bmatrix} \quad (5)$$

The reference current (i_d^*, i_q^*) are obtained from the active power (CMI) and reactive power (PCC bus) requirement, and the error is fed to the PI controller as shown in Fig. 4. The amount of current injected depends upon the voltage difference between the PCC bus and CMI, thus by varying the amplitude and phase angle of the generated voltage a DSTATCOM can inject or absorb the reactive power. Thus if the PCC bus voltage is greater than the CMI output voltage, then the reactive VAR is absorbed, whereas for a lower PCC bus voltage than CMI output voltage, the reactive VAR is delivered. Here, the switching angles for each H bridge are calculated offline; these angles correspond to the lowest THD for the CMI output, thus the modulation index obtained provides a unique set of triggering angles from the lookup table for an individual H bridge. Here, the optimized angles are obtained using the selective harmonic elimination technique; as k − 1 dominating lower order harmonics can be eliminated with this technique, it improves the THD of the voltage waveform.

4   Results and Discussion

In this study, CMI application for power conditioning is presented; an 11-level CMI uses five H bridges for a phase, and the harmonic spectrum of the output voltage is improved using the selective harmonic elimination modulation technique. In this study, MATLAB/Simulink/PSIM software are used for various simulations. As a power conditioning equipment, the DSTATCOM must be able to absorb and supply the required reactive power such that the PCC voltage can be maintained at 1 pu; Table 1 shows the system parameters and the 240 V of PCC bus is taken as a reference for per-unit system with switching frequency 50 Hz. Initially at t = 0, there is no load at the PCC bus; the CMI-based DSTATCOM is also in the OFF state, but under lagging or leading load, the power factor at the PCC bus will get disturbed, and thus additional reactive power is required to compensate; in such condition, CMI can be used as a power conditioner, which will help in maintaining the voltage level as well as the power factor to unity.

The PCC voltage is to be maintained at 1 pu under all loading conditions; an inductive load is switched ON at t = 0.04 s, but DSTATCOM is switched ON at t = 0.09 s; as result, a voltage sag is seen in Fig. 5 from t = 0.04 s to t = 0.09 s. As soon as DSTATCOM is ON, thereafter the PCC voltage by the action of DSTATCOM is maintained at 1 pu as shown in Fig. 5, and the corrective action by the controller is achieved in less than two cycles. The sensed signals from the load sides help in

Table 1 System configuration parameters for CMI-based DSTATCOM

Parameter	Value
Source voltage	240 V
Source impedance (R_s, L_s)	0.45 Ω, 4.8 mH
Coupling reactance	0.2 Ω, 32 mH
PCC nominal voltage	240 V
DSTATCOM kVA rating	10
Individual H bridge dc voltage	37.5 V
Individual H bridge capacitance rating	11.85 mF
Inductive load	500 + j2500
Capacitive load	400 − j3500
System frequency and switching frequency	50 Hz

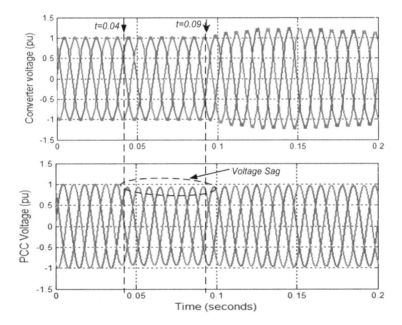

Fig. 5 Results for load variation, showing converter output voltage (v_c) (top) and PCC voltage (v_o) (bottom)

estimating active power (P) and reactive power (Q); with the help of the decoupled control structure shown in Fig. 4 obtained from the state-space equations, the direct and quadrature axis current components are estimated, and the error is fed to the PI controller to produce the necessary correction. Then the two-axis components are transformed back to three-axis components; thus the CMI required output voltage is obtained; the modulation index corresponding to CMI output will generate a set of

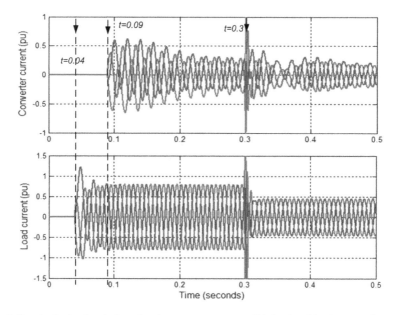

Fig. 6 Results for load variation, showing converter current (i_c) (top) and load current (bottom)

Table 2 Description of operation

Time (s)	Operation
t = 0.04	Inductive load ON
t = 0.09	CMI-based DSTATCOM ON
t = 0.3	Capacitive load ON

triggering angles for each H bridge; these triggering angles will ensure the voltage level at 1 pu.

At t = 0.3 s, a capacitive load is switched ON, as result a spike is visible at t = 0.3 s in Fig. 6. Due to capacitive load a voltage swell is seen, and the DSTATCOM is also in ON position; therefore the required modulation index and switching angles are generated by controller, and the PCC voltage is again maintained at 1 pu. PCC voltage, converter voltage, converter current and load current for the described operation in Table 2 are shown in Figs. 5 and 6, where the corrective action is clearly visible. The quadrature current component guide for VAR compensation, for the described operation of Table 2, quadrature current components of converter, and load are shown in Fig. 7. In both cases, the required action by the controller has taken place in less than two cycles, which shows the effectiveness of the described control methodology for the cascaded multilevel inverter acting as DSTATCOM. Harmonic distortion for DSTATCOM voltage and current are 4.6% and 0.5%, respectively, whereas for PCC voltage and current THD it is 0.9% and 0.2%, respectively.

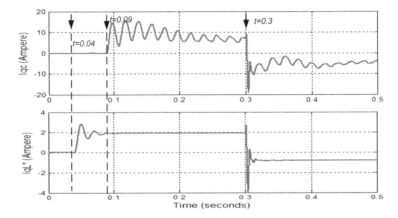

Fig. 7 Results for load variation, showing quadrature axis current (i_{qc}) fed by converter (top) and quadrature axis current (i_{qc}) for load (bottom)

5 Conclusion

This paper presents the modeling and control of a CMI-based DSTATCOM for power quality issues in a distribution network; the detailed structure of a cascaded multilevel inverter is elaborated, and the dynamic model for the connected system is presented along with the detailed control structure. The optimized triggering angle for each bridge is calculated offline, and as per the required modulation index these are fetched from the lookup table, thus lower THD in voltage waveform is achieved using the selective harmonic elimination technique. The results are shown for different loading conditions, the PCC voltage is maintained at 1 pu for inductive as well as capacitive loading, and reactive VAR is either delivered or absorbed as per the PCC bus voltage status. The obtained result validates the dynamic modeling and control methodology for the CMI-based DSTATCOM, the disturbance at the load side is effectively rejected by a controller action, THD obtained are as per IEEE standard 1547. Thus the CMI-based DSTATCOM can be used to address power quality issues in a distribution network.

References

1. M.F. Kangarlu, E. Babaei, A generalized cascaded multilevel inverter using series connection of submultilevel inverters. IEEE Trans. Power Electron. **28**(2), 625–636 (2012)
2. S. Daher, J. Schmid, F.L.M. Antunes, Multilevel inverter topologies for stand-alone PV systems. IEEE Trans. Ind. Electron. **55**(7), 2703–2712 (2008)
3. F.Z. Peng, A generalized multilevel inverter topology with self voltage balancing. IEEE Trans. Ind. Appl. **37**(2), 611–618 (2001)

4. H. Gupta, A. Yadav, S. Maurya, Dynamic performance of cascade multilevel inverter based STATCOM, in *IEEE 1st International Conference on Power Electronics, Intelligent Control and Energy Systems (ICPEICES)* (IEEE, 2016)
5. L.K. Haw, M.S.A. Dahidah, H.A.F. Almurib, SHE-PWM cascaded multilevel inverter with adjustable DC voltage levels control for STATCOM applications. IEEE Trans. Power Electron. **29**(12), 6433–6444 (2014)
6. N. Raj, et al., Development and experimental validation of fault detection and diagnosis method in SPWM modulated symmetric cascaded H-bridge multilevel inverter. Int. J. Power Electron. **11**(3), 409–426 (2020)
7. S.-A. Amamra, et al., Multilevel inverter topology for renewable energy grid integration. IEEE Trans. Ind. Electron. **64**(11), 8855–8866 (2016)
8. A. Mokhberdoran, A. Ajami, Symmetric and asymmetric design and implementation of new cascaded multilevel inverter topology. IEEE Trans. Power Electron. **29**(12), 6712–6724 (2014)
9. A. Kirubakaran, D. Vijayakumar, Development of LabVIEW-based multilevel inverter with reduced number of switches. Int. J. Power Electron. **6**(1), 88–102 (2014)
10. P. Omer, J. Kumar, B.S. Surjan, A new multilevel inverter topology with reduced switch count and device stress, in *5th IEEE Uttar Pradesh Section International Conference on Electrical, Electronics and Computer Engineering (UPCON)* (IEEE, 2018)
11. A. Yadav, J. Kumar, Harmonic reduction in cascaded multilevel inverter. Int. J. Recent. Technol. Eng. **2**(2), 147–149 (2013)
12. K. Boora, J. Kumar, A novel cascaded asymmetrical multilevel inverter with reduced number of switches. IEEE Trans. Ind. Appl. **55**(6), 7389–7399 (2019)
13. E. Najafi, A.H.M. Yatim, Design and implementation of a new multilevel inverter topology. IEEE Trans. Ind. Electron. **59**(11), 4148–4154 (2011)
14. J. Rodriguez, J.-S. Lai, F.Z. Peng, Multilevel inverters: a survey of topologies, controls, and applications. IEEE Trans. Ind. Electron. **49**(4), 724–738 (2002)
15. H. Karaca, Analysis of a multilevel inverter topology with reduced number of switches, in *Transactions on Engineering Technologies* (Springer, Dordrecht, 2014), pp. 41–54
16. B.P. McGrath, D.G. Holmes, Multicarrier PWM strategies for multilevel inverters. IEEE Trans. Ind. Electron. **49**(4), 858–867 (2002)
17. B.P. McGrath, D.G. Holmes, A comparison of multicarrier PWM strategies for cascaded and neutral point clamped multilevel inverters, in *2000 IEEE 31st Annual Power Electronics Specialists Conference, Conference Proceedings (Cat. No. 00CH37018)*, vol. 2 (IEEE, 2000)
18. H. Gupta, A. Yadav, S. Maurya, Multi carrier PWM and selective harmonic elimination technique for cascade multilevel inverter, in *2nd International Conference on Advances in Electrical, Electronics, Information, Communication and Bio-Informatics (AEEICB)* (IEEE, 2016)
19. P. Omer, J. Kumar, B.S. Surjan, Comparison of multicarrier PWM techniques for cascaded H-bridge inverter, in *IEEE Students' Conference on Electrical, Electronics and Computer Science* (IEEE, 2014)

Equivalent Two-Coil Model for a Four-Coil Wireless Power Transfer System

Anamika Das, Ananyo Bhattacharya, and Pradip Kumar Sadhu

1 Introduction

Wireless power transfer technology makes today's practical area of application more easy and secure than conventional wired power transfers, especially in the area of some crucial technologies like implantable medical devices, mining, underwater, and lot more where messy wired technology is not convenient enough. Pioneered by Nicolas Tesla near nineteenth century, wireless power transfer technologies have advanced a long way. Roughly, wireless power transmission can be of either radiation mode or non-radiation mode, depending on the methods and range for energy transfer. Radiative wireless power transfer uses radio waves for transmission and roughly used for far range. Non-radiative WPT is based on the near-range magnetic coupling of conductive coils and can be labeled for applications in short to medium range [3, 8, 13–15]. For WPT, inductively coupled and magnetic resonance-based coils are usually used because of their high performance and simplicity [9, 10]. The two configurations most used in a WPT are 2C and 4C structure. There are already a number of papers discussing their merits and demerits. From input impedance characteristics, it is seen that 2C model is apt for near or short-range and 4C model for mid-range transmissions of power [1–4, 6].

A. Das (✉) · A. Bhattacharya
Electrical Engineering Department, National Institute of Technology, Jamshedpur, Jharkhand, India
e-mail: 2019rsee002@nitjsr.ac.in

A. Bhattacharya
e-mail: ananyo.ee@nitjsr.ac.in

P. K. Sadhu
Electrical Engineering Department, Indian Institute of Technology (ISM), Dhanbad, Jharkhand, India
e-mail: pradip@iitism.ac.in

© The Author(s), under exclusive license to Springer Nature Singapore Pte Ltd. 2021   111
J. Kumar and P. Jena (eds.), *Recent Advances in Power Electronics and Drives*, Lecture Notes in Electrical Engineering 707,
https://doi.org/10.1007/978-981-15-8586-9_11

However, it is possible to redesign an equivalent 2C system for a 4C system without affecting their respective input impedance profile [4]. In this paper, the same is done and checked for equivalence in MATLAB/SIMULINK.

2 Analysis of Input Impedance of 4C System

A 4C WPT topology is represented in Fig. 1, an AC source of high frequency (10 MHz) supplies power to the source coil (SC), and the SC induces current in the transmitting coil (TC). Both the TC and receiving coil (RC) operate at the same resonant frequency. The current in load coil (LC) is caused by the RC, which supplies power to the load [9, 12].

The circuit model of 4C WPT system can be illustrated as shown in Fig. 2, considering mutual inductances between each two coils where

R_s Source internal resistance.
R_L Load resistance.
R_1 SC internal resistance.

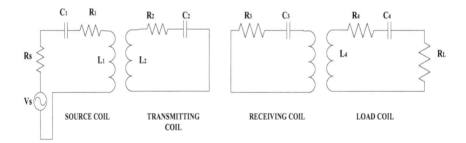

Fig. 1 Schematic diagram of a 4C WPT system

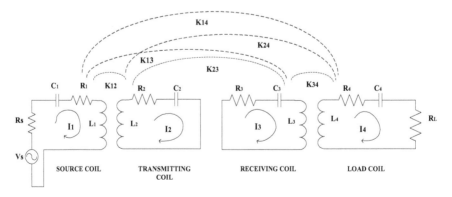

Fig. 2 Circuit model of 4C WPT

R_2 TC internal resistance.
R_3 RC internal resistance.
R_4 LC internal resistance.
L_1 SC self-inductance.
L_2 TC self-inductance.
L_3 RC self-inductance.
L_4 LC self-inductance.
V_s Voltage of high-frequency voltage source.

K_{12}, K_{23}, K_{34}, K_{13}, K_{24}, and K_{14} are the coupling coefficient between SC and TC, TC and RC, RC and LC, SC and RC, TC and LC, and SC and LC, respectively. The corresponding mutual inductances between each pair of coils are

$$M_{12} = K_{12}\sqrt{L_1 L_2}$$
$$M_{23} = K_{23}\sqrt{L_2 L_3}$$
$$M_{34} = K_{34}\sqrt{L_3 L_4}$$
$$M_{24} = K_{24}\sqrt{L_2 L_4}$$
$$M_{14} = K_{14}\sqrt{L_1 L_4}$$

I_1, I_2, I_3, and I_L are the current through SC, TC, RC, and LC, respectively. Applying KVL on the circuit in Fig. 2, the following equations can be obtained

$$(R_s + R_1 + j\omega L_1 - j/\omega C_1)I_1 - j\omega M_{12}I_2 + j\omega M_{13}I_3 - j\omega M_{14}I_L = V_s \quad (1)$$

$$(R_2 + j\omega L_2 - j/\omega C_2)I_2 - j\omega M_{12}I_1 - j\omega M_{23}I_3 + j\omega M_{24}I_L = 0 \quad (2)$$

$$(R_3 + j\omega L_3 - j/\omega C_3)I_3 + j\omega M_{13}I_1 - j\omega M_{23}I_2 - j\omega M_{34}I_L = 0 \quad (3)$$

$$(R_4 + R_L + j\omega L_4 - j/\omega C_4)I_L - j\omega M_{14}I_1 + j\omega M_{24}I_2 - j\omega M_{34}I_3 = 0 \quad (4)$$

Upon simplifying further, Eqs. (1)–(4) give expressions for I_1, I_2, I_3, and I_L, which can be further rearranged to evaluate the system efficiency.

As the flux linkages between non-adjacent coils are small enough to be negligible, for simplicity, only the mutual inductances between the adjacent coils can be considered, i.e., M_{12}, M_{23}, and M_{34}. The load is assumed to be pure resistive, that is, R_L is purely resistive. Further, the reflected impedances are calculated between adjacent coils [2, 7, 10].

Reflected impedance from LC toward RC is

$$Z_{\text{ref34}} = \frac{(\omega M_{34})^2}{Z_4} \quad (5)$$

The reflected impedance from RC toward TC is

$$Z_{\text{ref23}} = \frac{(\omega M_{23})^2}{Z_3 + Z_{\text{ref34}}} \tag{6}$$

The reflected impedance from TC toward SC is

$$Z_{\text{ref12}} = \frac{(\omega M_{12})^2}{Z_2 + Z_{\text{ref23}}} \tag{7}$$

Now, the system overall impedance as seen from input side can be expressed as

$$Z_n = Z_1 + Z_{ref12} - R_s \tag{8}$$

From the Eqs. (5), (6), (7), and (8), system input impedance (Z_{in}) can be written as

$$Z_{\text{in}} = Z_1 + \frac{\omega^2 M_{12}^2 (Z_3 Z_4 + \omega^2 M_{34}^2)}{Z_2 Z_3 Z_4 + \omega^2 M_{34}^2 Z_2 + \omega^2 M_{23}^2 Z_4} - R_S \tag{9}$$

where

$$\begin{aligned}
Z_1 &= R_s + R_1 + j\omega L_1 - j/\omega C_1 \\
Z_2 &= R_2 + j\omega L_2 - j/\omega C_2 \\
Z_3 &= R_3 + j\omega L_3 - j/\omega C_3 \\
Z_4 &= R_4 + R_L + j\omega L_4 - j/\omega C_4
\end{aligned} \tag{10}$$

At resonant frequency ω_r, the input impedance of the system is transferred into

$$Z_{\text{inresonance}} = \text{Re}(Z_{\text{in}}) \tag{11}$$

A schematic circuit diagram of 4C WPT is depicted in Fig. 3, and components with values considered for analyzing its input characteristics are listed in Table 1.

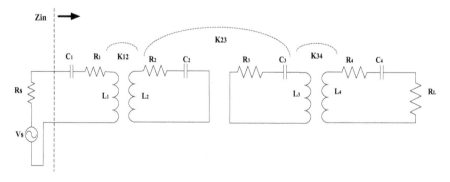

Fig. 3 Schematic circuit model for calculation of Z_{in} for 4C system

Table 1 Components of the 4C system

Components	Value	Components	Value
R_s, R_L	50 Ω	C_1, C_4	253 pF
L_1, L_4	1 μH	k_{12}, k_{34}	0.7
R_1, R_4	0.1 Ω	k_{23}	0.1
L_2, L_3	20 μH	f	10 MHz
C_2, C_3	12.67 PF	R_2, R_3	0.5 Ω

Fig. 4 Variation of real part of Z_{in} w.r.t. coupling coefficient for 4C WPT system

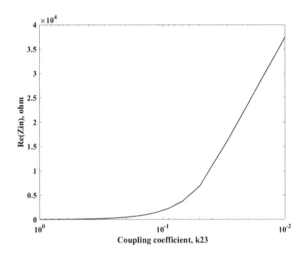

The variation of real part of input impedance Zin of 4C WPT with coupling coefficient is plotted in Fig. 4.

It is clear from Fig. 4 that real Re (Z_{in}) of 4C WPT decreases with increasing value of coupling coefficient k and can be calculated by the relation expressed in Eq. 12.

$$k = \frac{1}{\left[1 + 2^{2/3} \cdot D^2/r_1 r_2\right]^{3/2}} \tag{12}$$

wherein D is the linear separation from TC to RC. r_1 and r_2 are the radius of loop of TC and RC, respectively. Equation no 12 shows that the variation of magnetic coupling coefficient between a pair of coils is inversely proportional to the physical distance between them [5]. At the 4C system, real part of Z_{in} rises with decline in coupling coefficient and rises to its peak value when the distance is too long. This is why 4C scheme is apt for mid-range application [1].

3   Equivalent 2C Model for a 4C WPT System

Using the decoupling method, a pair of mutually coupled coils may be replaced by
a T-type combination of three inductors [4, 7, 11]. When the dotted terminals are
at the opposite sides (as shown in Fig. 5), the value of those three inductors can be
calculated as $L_1 + M$, $L_2 + M$, and $-M$.

At resonance angular frequency ω_r

$$C = -\frac{1}{\omega_r^2 L} \tag{13}$$

$$L = -\frac{1}{\omega_r^2 C} \tag{14}$$

From (13) and (14), it is found that an inductance (L) with a negative value may
be replaced by a positive value of capacitance (C) and vice versa. The values of C
and L can be determined from the equations listed as in next page [4].

$$L_1' = L_1 + M - \frac{1}{\omega_r^2 C_1} = k_{12}\sqrt{L_1 L_2} = M \tag{15}$$

$$C_M = \frac{1}{\omega_r^2 M} \tag{16}$$

$$C_2' = \frac{1}{\frac{1}{C_2} - \omega_r^2 M} = \frac{C_M C_2}{C_M - C_2} \tag{17}$$

Using the equations, the equivalent 2C LCC circuit can be calculated from both
transmitting and receiving side (Fig. 6). The same can be calculated for the pair of
SC and LC.

$$C_1' = C_4' = C_M \tag{18}$$

$$C_2' = C_3' \tag{19}$$

Fig. 5 Equivalent T circuit
for a pair of mutually
coupled coils

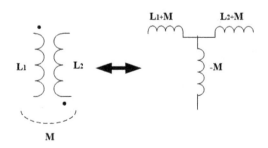

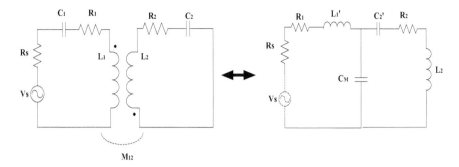

Fig. 6 Equivalence of SC and TC to LCC circuit

$$L_1' = L_4' \tag{20}$$

Components of complete equivalent 2C circuit calculated from Eqs. (15)–(20) are given in Table 2. The corresponding equivalent circuit is shown in Fig. 7.

From the 2C structure displayed in Fig. 8, the Z_{in}' (input impedance) can be also calculated using the equations below

$$Z_4' = \left(R_4 + R_L + j\omega L_4'\right) || \frac{1}{j\omega C_4'} \tag{21}$$

Table 2 Components of the 2C System

Components	value	Components	value
R_S, R_L	50 Ω	C_2', C_3'	15.02 pF
L_1', L_4'	4.13 μH	R_2, R_3	0.5 Ω
C_1', C_4'	81.01 pF	k_{12}, k_{34}	0.7
R_1, R_4	0.1 Ω	f	10 MHz
L_2, L_3	20 μH	k_{23}	0.1

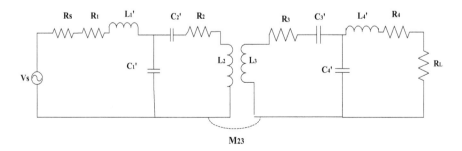

Fig. 7 Complete LCC 2C equivalent circuit for 4C WPT system

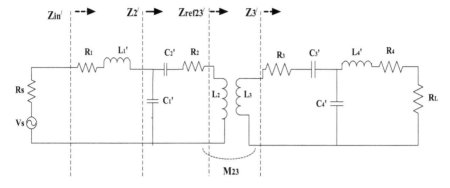

Fig. 8 Schematic circuit model for calculation of Z_{in} for 2C system

$$Z_3' = Z_4' + R_3 + \frac{1}{j\omega C_3'} \tag{22}$$

Considering the reflected impedance from receiving side as seen from transmitting side [2]

$$Z_{ref23}' = \frac{(\omega_r M_{23})^2}{Z_3'} \tag{23}$$

$$Z_2' = \left(Z_{ref23}' + R_2 + \frac{1}{j\omega_r C_2'} \right) \Big\| \frac{1}{j\omega_r C_1'} \tag{24}$$

$$Z_{in}' = R_1 + j\omega_r L_1' + Z_2' \tag{25}$$

Variation in Re (Z_{in}') (as discussed in Eq. 11) in 2C WPT with coupling coefficient is depicted in Fig. 9. Clearly, real part of the Z_{in}' of 2C model increases with increase in the coefficient of coupling (k) which means that lesser the distance between coils, larger the transmission efficiency [1]. This validates the concept that 2C model is more apt for near-range transmission.

4 Simulation and Result Analysis

A 4CWPT scheme simulated in MATLAB/SIMULINK is based on circuit shown in Fig. 2 with components listed in Table 1. The input voltage applied is 200 V sinusoidal waveform with a frequency of 10 MHz. With input supply and value of k_{23} unaltered, a simulated model of its corresponding 2C LCC model (based on Fig. 7 and components with values listed in Table 2) is also simulated and both are analyzed thereafter.

Fig. 9 Variation of real part of Z_{in}' w.r.t. coupling coefficient for 2C model

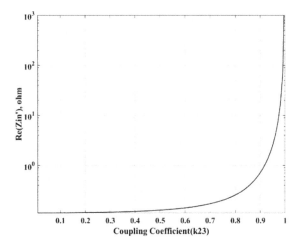

Fig. 10 Instantaneous output voltage profile of 4C resonant WPT scheme

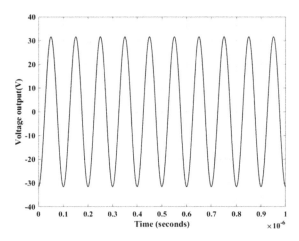

From Figs. 10, 11, 12, and 13, it is noticed that the instantaneous output profile of voltage and current for both the 4C configuration and its redesigned 2C counterpart are nearly similar. A comparison between RMS values of Output voltage (V_{or}), Current (I_{or}), Input power (P_{in}), Output power (P_{out}), and power transfer efficiency (η) of both the schemes are shown in Table 3.

Fig. 11 Instantaneous
output current profile of 4C
resonant WPT scheme

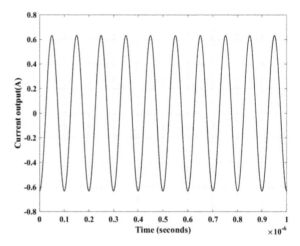

Fig. 12 Instantaneous
output voltage profile of 2C
LCC model of resonant WPT
scheme

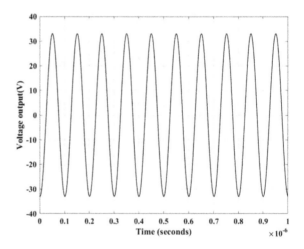

5   Conclusion

In this article, resonant wireless power transfer (WPT) system for a 4C model and
its corresponding 2C LCC has been designed and simulated in MATLAB Simulink
with same value of AC input voltage 200 V with 10 MHz frequency and coefficient
of coupling between TC and RC taking as $k_{23} = 0.1$. Also, the variation in real part
of Z_{in} w.r.t. coupling coefficient for both 2C and 4C systems is investigated. Here, it
is found that efficiency of 4C scheme is 94.78% with input power of 10.54 W and
output power of 9.994 W. And in its corresponding 2C LCC model an efficiency
of 93.22% is recorded with input power of 11.72 W and output power of 10.92 W.
Hence, the LCC model of 2C proves to be closely equivalent to its corresponding 4C

Fig. 13 Instantaneous output current profile of 2C LCC model of resonant WPT scheme

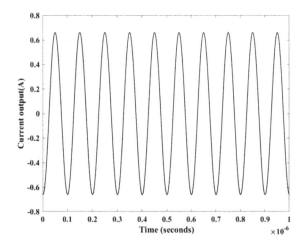

Table 3 Comparison between outputs of 4C and 2C models

WPT model	V_{or} (V)	I_{or} (A)	P_{in} (W)	P_{out} (W)	η (%)
4C model	22.35	0.4471	10.54	9.994	94.78
Equivalent 2C LCC model	23.37	0.4674	11.72	10.92	93.22

model in view of output and power transfer efficiency retaining its input impedance profile.

References

1. R. Huang, B. Zhang, Frequency, impedance characteristics and HF converter of two coil and four coil wireless power transfer. IEEE J. Emerg. Sel. Topics Power Electron. **3**(1), 177–183 (2015)
2. X. Wei, Z. Wang, H. Dai, A critical review of wireless power transfer via strongly coupled magnetic resonances. Energies **7**(7), 4316–4341 (2014)
3. S.Y.R. Hui, W. Zhong, C.K. Lee, A Critical review of recent progress in mid-range wireless power transfer. IEEE Trans. Power Electron. **29**(9), 4500–4511 (2014)
4. Z. Liu, H. Zhao, C. Shuai, S. Li, Analysis and equivalent of four-coil and two coil systems in wireless power transfer, in *2015 IEEE PELS Workshop on Emerging Technologies: Wireless Power (2015 WoW)*
5. M. Ruhul Amin, R.B. Roy, Design and simulation of wireless stationary charging system for hybrid electric vehicle using inductive power pad in parking garage, in *The 8th International Conference on Software, Knowledge, Information Management and Applications (SKIMA 2014)*
6. G.A. Covic, J.T. Boys, Modern trends in inductive power transfer for transportation applications. IEEE J. Emerg. Sel. Topics Power Electron. **1**(1), 28–41 (2013)
7. Y. Zhang, in *Key Technologies of Magnetically-Coupled Resonant Wireless Power Transfer* (Springer Science and Business Media LLC, 2018)

8. S. Ansari, A. Das, A. Bhattacharya, Resonant inductive wireless power transfer of two-coil system with class-E resonant high frequency inverter, in *2019 6th International Conference on Signal Processing and Integrated Networks (SPIN)* (2019)

9. T. Qiao, X. Yang, X. Lai, H. Tang, Modeling and analysis of multi-coil magnetic resonance wireless power transfer systems, in *2018 IEEE PELS Workshop on Emerging Technologies: Wireless Power Transfer (Wow)*, Montréal, QC (2018), pp. 1–6

10. M. Nisshagen, E. Sjöstrand, Wireless power transfer usingresonant inductive coupling, design and implementation of an IPT system with one meterair gap in the region between near-range and mid-range. Master thesis, Department of Energy and Environment, Chalmers University of Technology, Gothenburg, Sweden (2017)

11. S. Chopra, P. Bauer, Analysis and design considerations for a contactless power transfer system, in *2011 IEEE 33rd International Telecommunications Energy Conference (INTELEC)*, Amsterdam (2011), pp. 1–6

12. A.P. Sample, D.T. Meyer, J.R. Smith, Analysis, experimental results, and range adaptation of magnetically coupled resonators for wireless power transfer. IEEE Trans. Industr. Electron. **58**(2), 544–554 (2011)

13. R. Ajey Kumar, H.R. Gayathri, R. Bette Gowda, B. Yashwanth, WiTricity: wireless power transfer by non-radiative method. Int. J. Eng. Trends Technol. (IJETT) **11**(6), 290–295 (2014). ISSN:2231–5381. www.ijettjournal.org. Published by seventh sense research group

14. M.A. Hassan, A. Elzawawi, Wireless power transfer through inductive coupling, in *Recent Advances in Circuits* (2015). ISBN: 978–1–61804–319–1

15. Y. Burali, C. Patil, Wireless electricity transmission based on electromagnetic and resonance magnetic coupling. Int. J. Comput. Eng. Res. **3**, 48–51 (2012)

Model Predictive Control Based Inverter to Enhance the Performance Under Grid Frequency Variations

T. Ratna Rahul[ID]

1 Introduction

Distributed Generator (DG) is the heart of any microgrid application. Overall, the distributed generators gained their prominence owing to eco-friendly, socio-economic benefits, and technical advancements. Photovoltaic (PV)-based DGs have been drawing significant attention both from the industrial and academic perspectives. Despite high capital cost and low efficiency, the PVs are considered as one of the alternate sources of energy generation owing to their developments and potential benefits awarded during long run and increase in the durability of solar panels. The control strategies must be effective for the DG to enhance their performance. So, for this effect, the control of inverter plays a vital role and the transient performance as well as dynamic performance should be given good importance.

The control of inverter when it is synchronized with the grid in terms of controlling the active and reactive powers poses a challenging task [1, 2]. Controlling methods are developed for the DGs in injecting the powers of both active as well as/reactive demanded by the grid thereby stabilizing the system with the maintenance of voltage and frequency. Different control methodologies are proposed in literature for the real as well as reactive powers. The control method [2, 3] employed decouples the current in three phases to a synchronous reference frame (d-q). In general, mostly the current strategies employed use the conventional synchronous frame of reference where the low pass filters are essential and the tuning of those filters are also essential in improving the transient as well as steady-state responses of the system. The tuning of this filters also adds up in addition with the controllers developed for control of real as well as reactive powers.

T. Ratna Rahul (✉)
Ramaiah Institute of Technology, Bengaluru, India
e-mail: ratna.rahul1988@gmail.com

© The Author(s), under exclusive license to Springer Nature Singapore Pte Ltd. 2021
J. Kumar and P. Jena (eds.), *Recent Advances in Power Electronics and Drives*, Lecture Notes in Electrical Engineering 707,
https://doi.org/10.1007/978-981-15-8586-9_12

Mostly the controllers employ two control loops, one for the generation of reference currents based on the real as well as reactive powers as required by the grid and second being the switching pulse generation for the control of inverter. The controllers generally utilizing the two loops are conventional PI controllers, PR controllers, sliding mode control, fractional-order controllers, and predictive based current control. Further the controllers need the generation of reference currents as demanded by the grid. Theoretical-based results are presented related to model predictive control [4, 5]. Therefore, the control strategy needs the generation of reference current which further depends on the synchronization of the grid frequency which further uses the phase-locked loops (PLL) [3]. Hence the control of active and reactive powers is becoming further a complex in the tuning of PLL for the control. Hence, the control of power with frequency is quite essential and a challenging task.

In this context, the paper proposes a single-stage model predictive controller which is free of cumbersome tuning procedures, purely mathematically developed and is adaptable to the changes in the frequency of the grid and very quick in delivering real as well as reactive powers necessitated to grid.

2 System Modeling

The schematic of the three-phase grid-connected DG is presented in Fig. 1. The output voltage of DG is V_{Pcc} and current I_g are given to the MPC-based space vector modulation (SVM). The methodology of the process of grid synchronization is adopted from [6]. The powers transferred from the DG to grid are labeled as P_g, Q_g. The complete schematic diagram starting from the three-phase DG untill the point of common coupling (PCC) in a single line form can be represented as in (1).

$$V_i = I_f R_f + L_f \frac{dI_f}{dt} + V_{Pcc} \tag{1}$$

$$I_f = I_c + I_g \tag{2}$$

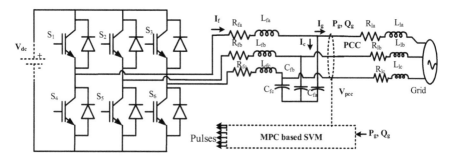

Fig. 1 Schematic of MPC-based SVM applied to three-phase DG

Differentiating (2) results in

$$\left.\begin{aligned}\frac{dI_f}{dt} &= \frac{dI_c}{dt} + \frac{dI_g}{dt}\\ I_c &= C_f\frac{dV_{Pcc}}{dt}\end{aligned}\right\} \tag{3}$$

The currents and voltages are transformed into α-β domain and can be represented as $V_{P\alpha}, V_{P\beta}, I_{0\alpha}, I_{0\beta}$

$$\left.\begin{aligned}V_{Pcc\alpha} &= V_m \cos\omega t; \quad \frac{dV_{Pcc\alpha}}{dt} = -\omega V_{Pcc\beta}\\ V_{Pcc\beta} &= V_m \sin\omega t; \quad \frac{dV_{Pcc\beta}}{dt} = \omega V_{Pcc\alpha}\end{aligned}\right\} \tag{4}$$

$$\left.\begin{aligned}I_{c\alpha} &= -C_f\omega V_{Pcc\beta}; \quad \frac{dI_{c\alpha}}{dt} = -\omega^2 C_f V_{Pcc\alpha}\\ I_{c\beta} &= C_f\omega V_{Pcc\alpha}; \quad \frac{dI_{c\beta}}{dt} = -\omega^2 C_f V_{Pcc\beta}\end{aligned}\right\} \tag{5}$$

Generally, the apparent power is represented as

$$\left.\begin{aligned}S &= V I_0^*\\ P + jQ &= (V_{Pcc\alpha} + jV_{Pcc\beta})(I_{g\alpha} - jI_{g\beta})\\ p &= V_{Pcc\alpha}I_{g\alpha} + V_{Pcc\beta}I_{g\beta}\\ Q &= V_{Pcc\beta}I_{g\alpha} - V_{Pcc\alpha}I_{g\beta}\end{aligned}\right\} \tag{6}$$

Differentiating P and Q of (6) results in

$$\left.\begin{aligned}\frac{dP}{dt} &= \left(\frac{dV_{Pcc\alpha}}{dt}I_{g\alpha} + \frac{dI_{g\alpha}}{dt}V_{P\alpha} + \frac{dV_{Pcc\beta}}{dt}I_{g\beta} + \frac{dI_{g\beta}}{dt}V_{Pcc\beta}\right)\\ \frac{dQ}{dt} &= \left(\frac{dV_{Pcc\beta}}{dt}I_{g\alpha} + \frac{dI_{g\alpha}}{dt}V_{Pcc\beta} - \frac{dV_{Pcc\alpha}}{dt}I_{g\beta} - \frac{dI_{g\beta}}{dt}V_{Pcc\alpha}\right)\end{aligned}\right\} \tag{7}$$

Evaluation of (7) requires $\dot{I}_{g\alpha}$ and $\dot{I}_{g\beta}$, differentiating $I_{g\alpha}$ and $I_{g\beta}$ results in (8)

$$\left.\begin{aligned}\dot{I}_{g\alpha} &= \dot{I}_{f\alpha} - \dot{I}_{c\alpha}\\ \dot{I}_{g\beta} &= \dot{I}_{f\beta} - \dot{I}_{c\beta}\end{aligned}\right\} \tag{8}$$

Substituting (1), (3), (4), (5), in (8) and on simplification results in (9).

$$\left.\begin{array}{l} \dot{I}_{g\alpha} = \dfrac{V_{i\alpha} + V_{Pcc\alpha}(\omega^2 L_f C_f - 1) - \omega R_f C_f V_{Pcc\beta} - I_{g\alpha} R_f}{L_f} \\[2mm] \dot{I}_{g\alpha} = \dfrac{V_{i\alpha} + V_{Pcc\beta}(\omega^2 L_f C_f - 1) - \omega R_f C_f V_{Pcc\alpha} - I_{g\beta} R_f}{L_f} \end{array}\right\} \tag{9}$$

Substituting (9) in (7) and writing the powers in state-space representation as

$$\begin{bmatrix} \dot{P} \\ \dot{Q} \end{bmatrix} = \frac{1}{L_f}\begin{pmatrix} -R_f & -\omega L_f \\ \omega L_f & -R_f \end{pmatrix}\begin{bmatrix} P \\ Q \end{bmatrix} + \frac{1}{L_f}\begin{pmatrix} V_{Pcc\alpha} & V_{Pcc\beta} \\ V_{Pcc\beta} & -V_{Pcc\alpha} \end{pmatrix}\begin{bmatrix} V_{i\alpha} \\ V_{i\beta} \end{bmatrix} + \frac{1}{L_f}*$$

$$\begin{bmatrix} (\omega^2 L_f C_f - 1)V_{Pcc\alpha} - \omega R_f C_f V_{Pcc\beta} & (\omega^2 L_f C_f - 1)V_{Pcc\beta} - \omega R_f C_f V_{Pcc\alpha} \\ -(\omega^2 L_f C_f - 1)V_{Pcc\beta} + \omega R_f C_f V_{Pcc\alpha} & -(\omega^2 L_f C_f - 1)V_{Pcc\alpha} - \omega R_f C_f V_{Pcc\beta} \end{bmatrix} * \begin{bmatrix} V_{Pcc\alpha} \\ V_{Pcc\beta} \end{bmatrix} \tag{10}$$

where $\dot{P} = \dfrac{dP}{dt}, \dot{Q} = \dfrac{dQ}{dt}$

Considering powers as state variables, V_i as input, (10) can be written as

$$\dot{X}_l = A_l X_l + B_l V_w + C_l V_{Pccl} \tag{11}$$

where $V_w = \begin{bmatrix} V_{i\alpha} \\ V_{i\beta} \end{bmatrix}$; $V_{Pccl} = \begin{bmatrix} V_{Pcc\alpha} \\ V_{Pcc\beta} \end{bmatrix}$

$$C_l = \frac{1}{L_f}\begin{bmatrix} (\omega^2 L_f C_f - 1)V_{Pcc\alpha} - \omega R_f C_f V_{Pcc\beta} & (\omega^2 L_f C_f - 1)V_{Pcc\beta} - \omega R_f C_f V_{Pcc\alpha} \\ -(\omega^2 L_f C_f - 1)V_{Pcc\beta} + \omega R_f C_f V_{Pcc\alpha} & -(\omega^2 L_f C_f - 1)V_{P\alpha} - \omega R_f C_f V_{Pcc\beta} \end{bmatrix}$$

$$A_l = \frac{1}{L_f}\begin{pmatrix} -R_f & -\omega L_f \\ \omega L_f & -R_f \end{pmatrix}; B_l = \frac{1}{L_f}\begin{pmatrix} V_{Pcc\alpha} & V_{Pcc\beta} \\ V_{Pcc\beta} & -V_{Pcc\alpha} \end{pmatrix}$$

Discretizing (11) and simplifying results as

$$X_l(k+1) = X_l(k)A_{ld} + B_d V_w + C_d V_{Pccl} \tag{12}$$

where $A_{ld} = e^{A_l T_s}$; $B_d = B_l T_s$; $C_d = C_l T_s$;

The objective function for the grid-connected DG is the active and reactive power control and is considered as

$$g = \sqrt{\left((P_l(k+1) - P_{ref})^2 + (Q_l(k+1) - Q_{ref})^2\right)} \tag{13}$$

The objective function is evaluated based on the switching states of three-phase grid-connected DG using space vector modulation and the minimum value of the evaluated g at different switching states is selected and given to the DG (Fig. 2).

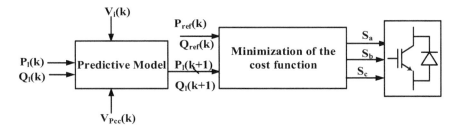

Fig. 2 Model predictive control implemented to grid-connected DG

$$g_{sl} = \min(g_l), \ for \ l = 1, 2...8 \tag{14}$$

3 Results and Discussion

In order to validate the performance of the developed control mechanism for three-phase grid-connected DG, the simulation results are performed in Matlab/Simulink. The parameters considered for simulation are presented in Table 1.

For the validation of the performance of the developed controller, the controller is tested with different cases. Initially, the DG is supplying 10 kW of power to the grid. At 1 s, an additional amount of 5 kW is demanded by the grid. It can be seen from Fig. 3, with the capacity of the DG possessing, the controller is able to supply the additional 5 kW with almost 1 cycle (20 ms) without any error or fluctuations in powers. Further, it is also tested whether the controller is able to track the powers when the DG is checked for the decrease in the active and reactive powers at a different power factor. At 1 s, the active and reactive powers are changed from 15 kW and 7.5 kVAr (0.9 pf) to 8 kW and 5 kVAr (0.85 pf) which can be seen in Figs. 3 and 4, respectively. It can be observed that there are no fluctuations or oscillations in both active and reactive powers and the controller is very quick to track the changes.

Table 1 Parameters for simulation

Parameters	Values
V_{dc}	750 V
V_{Pcc}	(415)V (RMS)
V_{ref}	415 V
f_p	(49.5–50) Hzz
R_f, L_f	0.15 Ω, 15 mH z
C_f	50 μF
P_g	(8–15) kW
Q_g	(5–7.5) kVAr

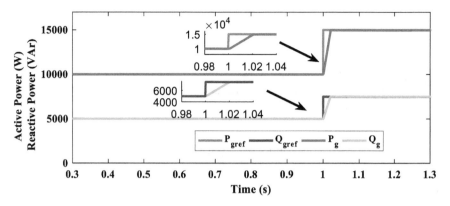

Fig. 3 Change of reference powers at 0.9 pf demanded by grid

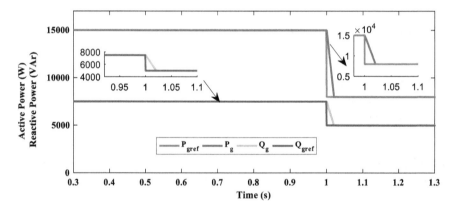

Fig. 4 Change of reference powers at 0.85 pf demanded by grid

The controller is checked with the variation of grid frequency whether during the transition in the frequency any power oscillations of high magnitude are present or not by using a programmable voltage source at 1 s. The variation of active and reactive powers with the change of grid frequency (50–49.5 Hz) is presented in Fig. 5. The variation is very low in terms of 10 W and 5 VAr, respectively. The magnitudes of the power ripples present are very low (0.1%). The current THD is presented in Fig. 6, which is 0.98%, well below the current THD as per IEEE 1547 standards. It can be clearly observed with the results that the developed control methodology is excellent in tracing powers demanded by the grid with low THDs and the controller is free of tuning procedures employed in both the controller as well as the tuning of low pass filters. Finally, the controller is very quick and good in adapting to the variations in the grid frequency (50–49.5 Hz).

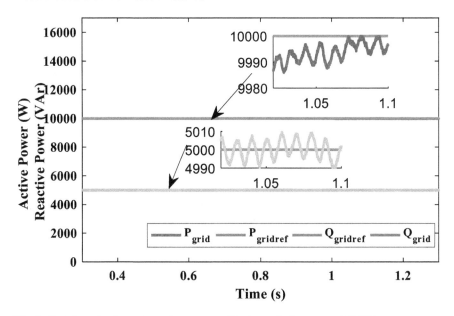

Fig. 5 Variation of active and reactive powers with change in frequency at 49.5 Hz

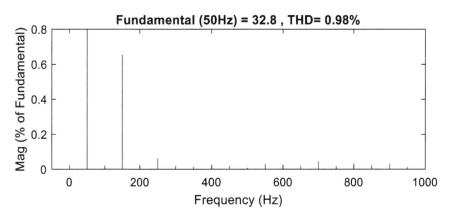

Fig. 6 THD of current waveform

4 Conclusion

The paper presented a single-stage model predictive control for three-phase grid-connected DG. Results indicate that the developed control methodology is quick in tracing the required real as well as reactive powers necessitated by the grid and the control method is quick in adapting to the subtle variations in frequency.

References

1. C. Ying-Chun, K. Yu-Lung, C. Hung-Shiang, et al., A novel single-switch resonant power converter for renewable energy generation applications. IEEE Trans. Ind. Appl. **50**(2), 1322–1330 (2014)
2. N. Altin, S. Ozdemir, H. Komurcugil, I. Sefa, Sliding-mode control in natural frame with reduced number of sensors for three-phase grid-tied LCL-interfaced inverters. IEEE Trans. Ind. Electron. **66**(4), 2903–2913 (2019)
3. B. Xianwen, Z. Fang, T. Yuan et al., Simplified feedback linearization control of three phase photovoltaic inverter with an LCL filter. IEEE Trans. Power Electron. **28**(6), 2739–2752 (2013)
4. A.W. Nasir, I. Kasireddy, A.K. Singh, Performance evaluation of model predictive control for a water bath temperature system based on its integer and non-integer order models. Int. J. Pure Appl. Math. **116**(10), 397–410 (2017)
5. T. Liu, Ch. Xia, T. Shi, Robust model predictive current control of grid-connected converter without alternating current voltage sensors. IET Power Electron. **7**(12), 2934–2944 (2014)
6. P.C. Sekhar, R.R. Tupakula, Model predictive controller for single phase distributed generator with seamless transition between grid and off-grid modes. IET Gener., Transm. Distrib. **13**(10), 1829–1837 (2019)

Stabilization of Boost Converter Connected to Photovoltaic Source Using PID Controller

Divesh Kumar ⓘ

1 Introduction

To increase the proficiency of PV systems, fuel cell and other DC power systems, the conversion losses can be minimized by using the DC voltage directly in place of converting it into AC and again back into DC for many applications [1]. To deal with the different voltage requirements, efficient multilevel DC–DC boost converter are widely used for the industrial as well as domestic applications [2, 3].

Many researchers proposed various control techniques to compensate the non-minimum phase behavior [4, 5]. An open loop multi-source input topology is proposed and this converter can be used to integrate the solar panels. This topology has advantages of low stress across switches. The size of the converter is small due to high-frequency operation. The main problem with this topology is that as the number of stage increases, the size of the capacitor increases and it will make it bulky [6–8]. Boost converters are widely used in HVDC transmission, non-conventional energy system like solar or wind turbine, fuel cell, dc grid, electric vehicle and space automation, etc. [9–11]. To obtain high gain in converters, the use of transformers is not practicable as it produces non-linearites [12, 13].

If load is changed the output parameters also changes. To regulate output voltage constant, the duty cycle should be controlled. Efficient DC–DC converters are required for PV system. Shadow effect reduces the generated output voltage and affects the solar system. It can be eliminated by using a suitable controller [14, 15]. The components used in DC–DC converter do not operate ideally so produce some power loss and affect the efficiency and accuracy of the boost converter. Power loss due to internal components of boost converter should be compensated. Conventional converter usually operates at high duty cycle to get higher gain from the converter.

D. Kumar (✉)
Department of ECE, GLA University, Mathura, Uttar Pradesh, India
e-mail: kambojdivesh@gmail.com

© The Author(s), under exclusive license to Springer Nature Singapore Pte Ltd. 2021   131
J. Kumar and P. Jena (eds.), *Recent Advances in Power Electronics and Drives*, Lecture Notes in Electrical Engineering 707,
https://doi.org/10.1007/978-981-15-8586-9_13

At high rate, inductor may get saturates. A low duty cycle is used to operate boost converter. Dependency of the conventional energy resources is decreasing day by day. Electric vehicle needs various supplies for charging purpose so multilevel boost converters are widely used [16, 17].

1.1 Boost Converter

Figure 1 shows a circuit diagram of conventional boost converter. The increase in output voltage gives its name as boost or step-up. The function of the capacitor at the resultant side is to make the resultant voltage ripple free and an inductor is added to the input side to make the current corresponding to the voltage set ripple free [1].

Switch-on State. When switch is closed, it will connect the anode of the diode to the –ive terminal and diode will be in reverse bias. Output stage will be isolated from the input side and inductor current will continue to flow and energy will be supplied from the source. Figure 2 shows the equivalent circuit for the state when switch is closed.

when switch is closed for period (d Ts),

$$V_s - \frac{dI_L}{dt} = 0 \tag{1}$$

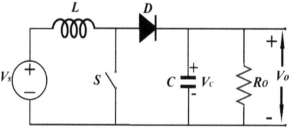

Fig. 1 Boost converter

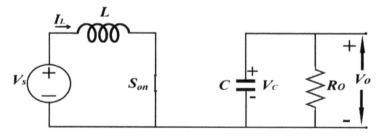

Fig. 2 Switch-on state

Fig. 3 Switch-off state of boost converter

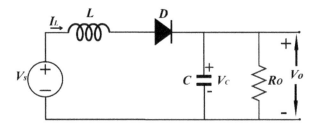

$$\frac{V_c}{R} + \frac{CdV_c}{dt} = 0 \tag{2}$$

Equations (1) and (2) can be represented in state-space model as

$$\begin{bmatrix} \frac{dI_L}{dt} \\ \frac{dV_c}{dt} \end{bmatrix} = \begin{bmatrix} 0 & 0 \\ 0 & -\frac{1}{RC} \end{bmatrix} \begin{bmatrix} I_L \\ V_C \end{bmatrix} + \begin{bmatrix} \frac{1}{L} \\ 0 \end{bmatrix} V_S \tag{3}$$

$$V_o = \begin{bmatrix} 0 & 1 \end{bmatrix} \begin{bmatrix} I_L \\ V_C \end{bmatrix} \tag{4}$$

Switch-off State. If the switch is off, energy is delivered from the inductor as well as the input source. The output capacitance maintains the voltage across the load constant. The equivalent circuit for switch-off state is shown in Fig. 3.

For switch-off period $((1 - d) Ts)$,

$$V_S - V_C - L\frac{dI_L}{dt} = 0 \tag{5}$$

$$I_L - \frac{V_c}{R} - \frac{CdV_c}{dt} = 0 \tag{6}$$

From Eqs. (5) and (6)

$$\begin{bmatrix} \frac{dI_L}{dt} \\ \frac{dV_c}{dt} \end{bmatrix} = \begin{bmatrix} 0 & -\frac{1}{L} \\ \frac{1}{C} & -\frac{1}{RC} \end{bmatrix} \begin{bmatrix} I_L \\ V_C \end{bmatrix} + \begin{bmatrix} \frac{1}{L} \\ 0 \end{bmatrix} V_S \tag{7}$$

$$V_o = \begin{bmatrix} 0 & 1 \end{bmatrix} \begin{bmatrix} I_L \\ V_C \end{bmatrix} \tag{8}$$

1.2 Control Strategy of DC–DC Power Converter

Output of boost converters is to be regulated within a specified tolerance limit. The converter resultant V_o is compared with V_{oref} and the error goes to the controller. As the control action, duty cycle is varied by the controller so that output voltage will remain constant.

State-Space Averaging Technique. State equation and output equation are given by Eqs. (9) and (10), respectively.

$$\dot{x} = [A_1 d + A_2(1 - d)]x + [B_1 d + B_2(1 - d)]V_S \tag{9}$$

$$V_o = [C_1 d + C_2(1 - d)]x \tag{10}$$

where $A_1 = \begin{bmatrix} 0 & 0 \\ 0 & -\frac{1}{RC} \end{bmatrix}$, $B_1 = \begin{bmatrix} \frac{1}{L} \\ 0 \end{bmatrix}$, $C_1 = [0\ 1]$, $D_1 = [0]$ is obtained from the switch-on state Eqs. (3) and (4) and $A_2 = \begin{bmatrix} 0 & -\frac{1}{L} \\ \frac{1}{C} & -\frac{1}{RC} \end{bmatrix}$, $B_2 = \begin{bmatrix} \frac{1}{L} \\ 0 \end{bmatrix}$, $C_2 = [0\ 1]$, $D_2 = [0]$ is obtained from switch-off state Eqs. (7) and (8).

Here current through the inductor and output voltage are considered to be two states of the system. The state-space averaging model is designed by averaging the on-state and off-state equations and is expressed as

$$A = A_1 d + A_2(1 - d) \tag{11}$$

$$B = B_1 d + B_2(1 - d) \tag{12}$$

$$C = C_1 d + C_2(1 - d) \tag{13}$$

Table 1 shows the converter data for analysis. High value of source voltage or output voltage results in rise of the ripple current. A low value of inductor improves the load transient while large value reduces the ripple current. The value of the capacitor is decided based on ripple voltage. Transfer function with respect to input

Table 1 Design parameters

Design parameter	Values
Source voltage (V_S)	12 V
Desired output voltage (V_{oref})	24 V
Inductance (L)	150 μH
Capacitance (C)	470 μF
Load resistance (R_O)	50 Ω
Switching frequency (f_S)	20 kHz

and duty cycle are given by as follows.

$$\frac{V_o(s)}{V_{in}(s)} = C(SI - A)^{-1}B + D \tag{14}$$

$$\frac{V_o(s)}{d(s)} = C(SI - A)^{-1}[(A_1 - A_2)x + (B_1 - B_2)V_{in}] + (C_1 - C_2)x \tag{15}$$

By using the Eqs. (9)–(13) and the design parameter from Table 1, the transfer function are obtained as in Eqs. (16) and (17).

$$\frac{V_o(s)}{V_{in}(s)} = \frac{7.092 \times 10^6}{s^2 + 44.33s + 3.546 \times 10^6} \tag{16}$$

$$\frac{V_o(s)}{d(s)} = \frac{-2128s^3 + 1.701 \times 10^8 s^2 - 0.0006695s + 6.036 \times 10^{14}}{s^4 + 88.65s^3 + 7.094 \times 10^6 s^2 + 3.144 \times 10^8 s + 1.257 \times 10^{13}} \tag{17}$$

Design of PID Controller

The PID controller is to be tuned for the compensation of input voltage fluctuation and load variation. For tuning of PID controller, the values of ultimate gain (K_C) and ultimate time period (T_U) are obtained from the Root Locus plot and then gain parameters are calculated by using Zeigler–Nichols method. The block diagram of the Boost Converter connected to photovoltaic source using PID controller and the corresponding Simulink model are shown in Fig. 4 and Fig. 5, respectively.

The Root Locus plot is shown in Fig. 6. From this plot, the values of critical gain (K_C) and critical time period (T_U) is found to be $K_C = 0.0209$ and $T_U = 0.0024$ s, respectively. By Zeigler–Nichols table, the controller parameters are obtained as

$K_p = 0.01045$

$K_i = 8.7$

$K_d = 0.000003135$

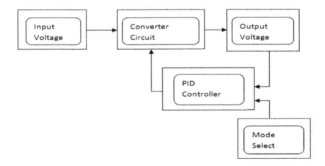

Fig. 4 Block diagram of the boost converter connected to photovoltaic source and PID controller

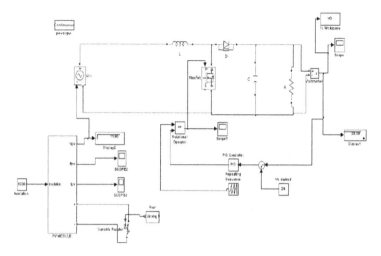

Fig. 5 Simulink model of boost converter connected to photovoltaic source and PID controller

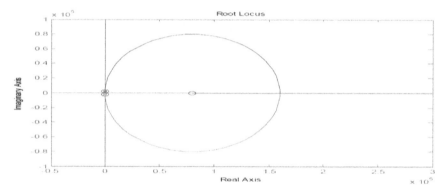

Fig. 6 Root locus plot

2 Results

2.1 Step Response

Open loop step response of the converter is shown in Fig. 7 and closed loop step response of the boost converter is shown in Fig. 8. Open loop system response is sluggish and closed loop response with PID controller is smooth, and settling time is also reduced to a very significant value.

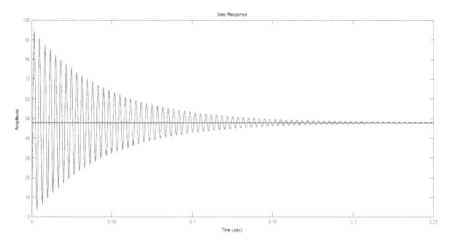

Fig. 7 Step response of the open loop system

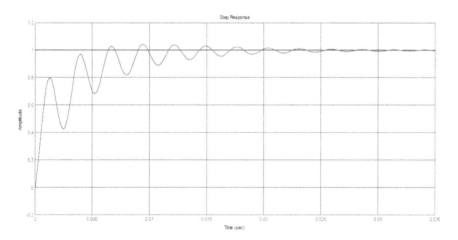

Fig. 8 Step response of the closed loop system

2.2 Performance Comparison of Time-Domain Specifications

Table 2 shows the comparison between the performance specification of boost converter without controller and with controller.

From the above observation, it is observed that by proper tuning of PID controller, the peak overshoot in output voltage and output voltage settling time reduces significantly.

Table 2 Performance specification

Time-domain specification	Boost converter without controller	Boost converter with controller
Rise time (s)	0.0056	0.0034
Settling time (s)	0.175	0.012
Peak overshoot (%)	96	5

Table 3 Output voltage with varying insolation

Solar irradiance (W (peak)/m^2)	Output voltage
1200	24.08
1000	24.12
800	24.05
600	24.09
400	24.14

2.3 Output Voltage

The Solar PV module voltage depends upon the solar irradiance. This voltage is fed to the boost converter as the input voltage. As irradiance varies the input voltage of the boost converter varies and it results in the variation in the output voltage of boost converter. By using a proper controller, the output voltage of the converter remains constant as solar irradiance varies. Table 3 shows the output voltage for different insolation levels.

Figure 9 shows the controlled output voltage. From Table 3 and Fig. 9, it is analyzed that the output voltage remains almost same with varying insolation levels from 1200 to 400 W (peak)/m^2.

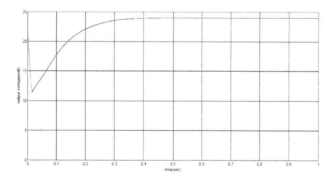

Fig. 9 Output voltage response

3 Conclusion

The efficiency of PV module is very poor and the output voltage depends on insolation and temperature. So there is a need to increase the DC output voltage level as well as to regulate the output voltage of the boost converter. The effect of PID controller on output voltage of boost converter connected to solar photovoltaic source is studied. It is found that the use of PID controller improves the performance of integrated system significantly. The problem of insolation variation is also compensated by the controller. Root locus plot is used to calculate the controller constants for tuning by Ziegler–Nichols method. Output voltage of the proposed converter remains almost constant (24 V) with changing insolation levels. Therefore, this high degree of regulation can be achieved by using PID controller for voltage regulation of solar PV based boost converter system.

References

1. D. Kumar, D. Kalra, D. Kumar, Design and analysis of a multilevel DC–DC boost converter, in *International Conference on Intelligent Computing and Smart Communication 2019. Algorithms for Intelligent Systems* ed. by G. Singh Tomar, N. Chaudhari, J. Barbosa, M. Aghwariya (Springer, Singapore, 2020), pp. 1065–1075
2. S. Kouro, M. Malinowski, K. Gopakumar, J. Pou, L. Franquelo, B. Wu, J. Rodriguez, M. Perez, J. Leon, Recent advances and industrial applications of multilevel converters. IEEE Trans. Ind. Electron. **57**(8), 2553–2580 (2010)
3. J. Candelaria, J.-D. Park, VSC-HVDC system protection: a review of current methods, in *Proceedngs of IEEE Power Systems Conference and Expo* (2011), pp. 1–7. C.Y. Chan, Simplified parallel-damped passivity-based controllers for DC–DC power converters. Automatica, **44**, 2977–2980 (2008)
4. Y.I. Son, I.H. Kim, Complementary PID controller to passivity-based nonlinear control of boost converters with inductor resistance. IEEE Trans. Control Syst. Technol. **20** (2012)
5. V.A.K. Prabhala, P. Fajri, V.S.P. Gouribhatla, B.P. Baddipadiga, M. Ferdowsi, A DC–DC converter with high voltage gain and two input boost stages. IEEE Trans. Power Electron. **31**(6), 4206–4215 (2016)
6. A. Alateeq, Y. Almalaq, M. Matin, A switched-inductor model for a non-isolated multilevel boost converter, in *Power Symposium (NAPS) 2017 North American* (2017), pp. 1–5
7. J.C. Rosas-Caro, J.C. Mayo-Maldonado, J.E. Valdez-Resendiz, A. Valderrabano-Gonzalez, The resonant DC–DC multilevel boost converter, in *International Conference on Electronics, Communications and Computers (CONIELECOMP)* (Cholula, 2018), 145–151
8. J.C. Rosas-Caro, J.M. Ramirez, F.Z. Peng, A. Valderrabano, A DC–DC multilevel boost converter. IET Power Electron. **3**, 129–137 (2010)
9. B. Axelrod, Y. Berkovich, A. Shenkman, G. Galon, Diode-capacitor voltage multipliers combined with boost-converters: topologies and characteristics. IET Power Electron. **5**, 873–884 (2012)
10. F.H. Dupont, C. Rech, R. Gules, J.R. Pinheru, Reduced-order model and control approach for the boost converter with voltage multiplier cell. IEEE Trans. Power Electron. **28**, 3395–3404 (2013)
11. A.S. Musale, B.T. Deshmukh, Three level DC–DC boost converter for high conversion ratio, in *2016 International Conference on Electrical, Electronics, and Optimization Techniques (ICEEOT)* (Chennai, 2016), pp. 643–647

12. A. Amir, A. Amir, H.S. Che, A. Elkhateb, N.A. Rahim, Comparative analysis of high voltage gain DC–DC converter topologies for photovoltaic systems. Renew. Energy **136**, 1147–1163 (2019)
13. G. Buticchi, L.F. Costa, M. Liserre, Multi-port DC/DC converter for the electrical power distribution system of the more electric aircraft. Math. Comput. Simul. **158**, 387–402 (2019)
14. J.C. Rosas-Caro, J.M. Rendón Ramírez, P.M. Garcia-Vite, Novel DC–DC multilevel boost converter, in *2008 IEEE Power Electronics Specialists Conference* (2008), pp. 2146–2151
15. I.M. Pop-Calimanu et al., A new hybrid inductor-based boost DC–DC converter suitable for applications in photovoltaic systems. Energies **12**(2), 252 (2019)
16. J. Biela, U. Badstuebner, J.W. Kolar, Impact of power density maximization on efficiency of DC–DC converter systems. IEEE Trans. Power Electron. **24**(1), 288–300 (2009)
17. J.C. Mayo-Maldonado et al., Modelling and control of a DC–DC multilevel boost converter. IET Power Electron. **4**(6), 693–700 (2011)

A New Model of Zero Energy Air Cooler: An Cost and Energy Efficient Device in Exploit

K. Akash and A. Sheik Abdullah ⓘ

1 Introduction

This mechanism of proposed structure has been provided with a cylindrical opening over the neck and the specified cork region. It has been proposed that when the area size decreases then the pressure inside it increases and then it cools the air to the specified funnel area.

- The bottles are said to be in a state of hydrated with the help of Metal Oxide Framework, which is capable of holding water at a microscopic level [1] (pore diameter being 6 Å).
- It has been provides with an state to reduce the interior temperature to a level of 10 °C, which then assigned with the air conditioning model.

The principle behind the working of the cooler are simple physics and engineering concepts it's easy to set up and self-sustaining model and hence needs not much of human interference. It is eco-friendly and cost effective. The measurement of the cost effectiveness is depicted in Fig. 1.

K. Akash
Department of Mechanical Engineering, Thiagarajar College of Engineering, Madurai, Tamil Nadu, India
e-mail: kaakash11c@gmail.com

A. Sheik Abdullah (✉)
Department of Information Technology, Thiagarajar College of Engineering, Madurai, Tamil Nadu, India
e-mail: aa.sheikabdullah@gmail.com

J. Kumar and P. Jena (eds.), *Recent Advances in Power Electronics and Drives*, Lecture Notes in Electrical Engineering 707,
https://doi.org/10.1007/978-981-15-8586-9_14

Fig. 1 Cost effectiveness of
proposed cooler

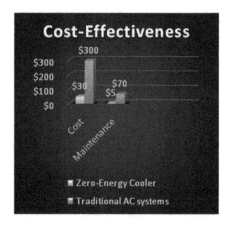

2 Objectives of the Proposed Work

The proposed work mainly focuses on the cost effective model to be provided to the common people with more efficient and adorable platform. The following are the key objectives of the model:

- Zero energy
- More convenient mode of installation
- Cost effectiveness
- Environment friendly
- Reusability
- Self-sustaining (zero carbon effect)
- Subsequent paragraphs, however, are indented.

The product completely focuses on the zero level of carbon effect in which the amount of the level of carbon dioxide to the atmosphere and the model is completely framed with a self-sustaining level of design.

Among the overall 2.8 billion populations in the more specific regions of the world there are about 8% who owns air conditioners when compared to that of the 80% of the people living in US and in Japan. The contact to the cooling in most of the regions of the world can be no longer considered a luxury and is becoming a necessarily important thing [2]. The workflow of the process is depicted in Fig. 2.

3 Experimental Setup

The following things are required to implement this project:

- Used 1–2 L PET Bottles (Say X*X ft. Then you require X^2 bottles)

Fig. 2 Process workflow

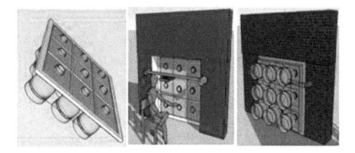

Fig. 3 Experimental design

- Sheet Made of Metal or Wood or Any preferred material (Based on the size of your window).
- Metal Oxide Framework for improved cooling purposes rendered CAD model of zero energy air cooler. The experimental design is mentioned in Fig. 3.

4 Solution Testing and Implementation

The idea of our project is to make the proper use of engineering principles in helping the society and integrate it with technology. So basically, there are three main engineering concepts applied in the prototype and as they follow:

- MOF (metal organic framework) and its psychometric analysis
- Evaporative Cooling
- Joule's law.

Only two levels of headings should be numbered. Lower level headings remain unnumbered; they are formatted as run-in headings.

About MOF, or generally known as metal organic framework which has been here since 2000s but due to cost factors haven't been used, Now with development and synthesis of various structures have enabled us to improve their application various fields like Adsorption, Gas storage, Conductivity etc. [3].We felt that utilization of MOF would be the game changer in the game. By further investigating researchers found a new compound MOF also known metal organic framework which could

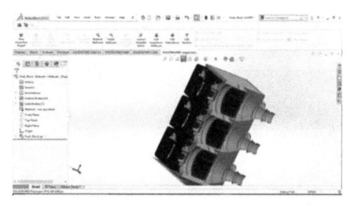

Fig. 4 CAD rendering of the product

hold water molecules (Wonderful scientific inventions are arrived at by mimicking the nature) [4]. The product rendering is given in Fig. 4.

Now the all new aluminum based Metal Oxide Framework has opened up the scope for new areas of application. The idea we intend to propose is based on the application of its ability to hold water molecules as a cluster of 8 molecules in one frame (i.e. used already for CO_2 and Methane absorption processes with help of zirconium based MOF-801 framework which can hold upto 20% of its own weight and the thus absorbed gases can be transported easily in large quantities as in case of methane) [5].

The MOF-303 based aluminum metal oxide composed framework. Then by adapting aluminum instead of using zirconium as a metal and thereby using water instead of the composed organic solvents. Hence the newly adhered structure is based on the XHH topology which is then built by Al (OH) (–COO) 2 which is formed as the secondary building unit. The implementation levels is depicted in Fig. 5.

The entire structure of MOF-303 is modeled into one-dimensional pores with a facilitation of 6A diameter [1]. The model is provided with a large water capacity framework of about: 0.48 gg-1. The isotherm is specifically suited for water absorption with a high performance material. The inflection of the isotherm has been made

Fig. 5 Implementation levels

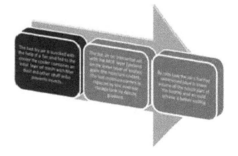

at a point of about 0.15, with a plateau of about 0.3, thereby providing a good level of temperature as observed [6].

Solid waste management has become one of major crisis. So in our product we tend to reduce and reuse solid waste. Reusable water (PET) and other plastic bottles of same or different sizes are used here. So used bottles instead of destroying the soil, it can be used in this prototype which will reduce the soil contamination effectively and better usage of the used bottles. Also recycled plastic may be used [7].

Integrating science and society is what engineering is about. Here we use joule's second law where a compressed gas regardless its volume is released into a free space, thus it cools down. So when we use this, we get that hot air coming from outside is made to pass through the larger end of the bottle and compress it towards the end of the nozzle and when the slightly compressed air is let out it cools down rapidly as we get cool air. This product is could be an effective alternative to traditional air conditioner. No eco-friendly air cooler (air conditioner) is available at this rate and totally different from any other in the market [8].

5 Experimental Results

For every HVAC system psychometric analysis will provide an accurately designed system requirements.

It proceeds as follows:

- Determination of the Outlet conditions
- Rate of air flow over for a unit area
- Determine return air condition entering the air-handling unit.
- Helps us predict the time consumption and other parameters such as humidity and temperature [9].

Based on a simulated environment with optimal inlet conditions the following calculations were made. The cross section view is depicted in Fig. 6.

Rate of loss of water from evap. system: ω_{water} = To be Determined.

The cycle is continuous and rapid also from X-ray diffraction test it is confirmed that there is no loss in property even after multiple cycles as given in Fig. 7. The parameter setting for rate of loss of water and determination of replacement is depicted in Tables 1 and 2.

Fundamentally, the "TG" of TG/DTA is very similar to standard thermo gravimetric analysis (TGA). A TG/DTA measures the change in sample weight as a function of temperature (and/or time) under controlled gas atmosphere and temperature. Graphing the percent weight change over a programmed temperature range enables the study of physical or chemical processes that have caused the sample to lose or gain weight [10].

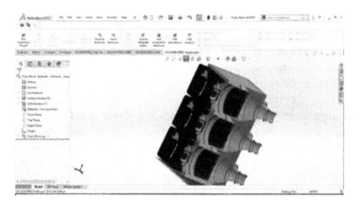

Fig. 6 Cross-sectional view of the model

Fig. 7 One hundred fifty
cycles of RH swing cycling
of scaled-up activated
MOF303 at 25 °C in a TGA

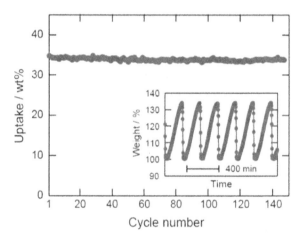

Table 1 Determination of rate of loss of water

Temperature at inlet T_1	41 °C
Temperature at outlet (Expected) T_2	25 °C
Specific humidity of dry air at 40 °C (ω_1)	0.009 kg/kg
Final specific humidity at 25 °C (ω_2)	0.017 kg/kg
Sub wet bulb temperature	18.3 °C
Specific volume (MOF)	0.3670 cm³/g
$d(m_{air})/dt = \rho \, d(V)/dt$	
$d(m_{water})/dt = d(m_{air})/dt \, (\omega_2 - \omega_1)$	3.699 m³/min = 0.0698 kg/s
Rate of loss of water from system in a single cycle	$0.0698 \times (0.017 - 0.009) = 5.512 \times 10^{-4}$ kg/s
Rate of loss of water from evap. system: ω_{water}	5.512×10^{-4} kg/s

Table 2 Determination of replacement of water per cycle

ρMOF	1.1277
Specific volume (MOF)	367 m^3/kg
Volume of water lost	0.00552 m^3/s
Sorption rate up to saturation	2.112 × 10^{-4} kg/s
Therefore time taken for re-absorption of lost water	2 s

6 Conclusion

We obtained about 55% of correlation of results on comparison of the actual performance of the model with the theoretical values. Which is indeed common cause of the ideal considerations of the outlet in theoretical sections such as a relative humidity of 100% is never attainable. It was also observed that with optimal input condition at laboratory condition (*a very low humidity at inlet.*) shows that the proposed model has a higher value of effectiveness at arid regions as in Fig. 8. From Thermo-gravimetric analysis it was clear that the MOF-303 retention property isn't lost even after multiple cycles suggesting that it can be economical compared to other cooling systems.

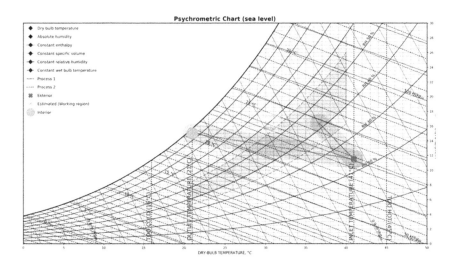

Fig. 8 Psychrometric plot of working conditions

7 Significant Measures and Expectations

7.1 Electricity Insufficiency

Electrical energy consideration is one of the most significant crises for providing it to the economy. A balance between electricity production and consumption is a major threat in this world which is to be focused with care. In this animated world that sometimes animates us, (i.e.) we could live without food and water but we will never be able to live without the electricity since it rules the world. It became a great stress factor now-a-days. Relieving ourselves from the stress is possible when we are shifting towards an eco-friendly Zero Energy Air cooler. Its impact will be tremendous in such a way that we will also be able to comfort us with a cooler without electricity.

7.2 Solid Waste Supervision

Waste management is considered to be one of the most important features with the increasing economical changes. Every year the production of waste has been increasing rapidly by many of the organizations which are about 11.2 billion tones among the worldwide. The process of reusability of waste materials and its compositions has to be studied and they have to be reused to any of the extent into usable proportion. Among the collected wastes worldwide it has been composed of about 5% of greenhouse gas emission every year. The waste collected may be composed of about some of the hazardous materials which are considered to be one of the major challenges to the faster growing economy. The waste pollutes and contaminates the water level and also the fertility of the soil. This has also to be considered for the reversal of waste into actionable materials. The statistical report is given in Fig. 9. The pollutants dissolved in the water and the soil may create various infectious diseases to the common community and cause severe damage to the ecosystem.

Fig. 9 Statistical report for waste management

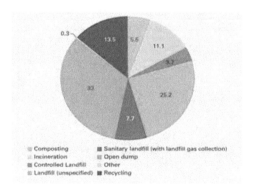

The best gatherable solution is the minimization of waste and its reusability. The reuse of the plastic bottles can be done while deploying the zero energy air cooler. We could do solid management far better since there is a lion's share for the plastic bottles in the solid waste management. At certain circumstances where the waste materials if can't be avoided then the process of recycling and reusing it should be considered as the third option in the waste management system.

8   Future Scope

Using zero percentage of any sources of electricity makes us pure and free from global warming that includes the green house effects which finally contributes a solution to the climate action. As the society is concerned with the environmental impacts and its considerations the generation of eco-friendly environment is most needed. Therefore, this prototype is expected to have massive impact on all regions especially those people in rural areas. Earlier our ancestors used clever ways to beat the heat like cavemen used caves to protect from heat; Chinese used hand held fans in second centuries like Middle East used underground tunnels to draw heat out and people who used to live under the mud built houses.

The Major Problem with Mud Buildings is that the construction requires the removal of top level of the soil which is important for Agriculture. The alternative solution out there would be zero energy air cooler.

One main disadvantage is that nowadays, the rural areas don't have these mud houses everywhere since it is not that much strong and safe to live in and heat is uncontrollable in summer. This problem can be overcome by using this product as it is cost wise beneficiary as well as eco-friendly.

Another main problem is that for people working in small scale factories. The places where they work are hotter. They all cannot afford to buy an average air conditioner. The Zero energy Air Cooler can reduce this heat emission inside factories to considerable levels [8]. A person who do horticulture often need temperature lower than they actually have. A prototype like this could come in handy to those people. A various number of real time applications are present here.

Utilizing the environment friendly air conditioning model will definitely reduce the electricity in different ways of action when mounting this particular model into action. Also, in turn it provides enough cooling mechanism for the environment with cost effective benefits. When we reduce the amount of energy, we reduce the risk of harmful emissions of gases that affect the ozone. This way more and more we will get to know the eco-friendly ways to save our environment [11].

CFC's (Chloro Fluro Carbon) has a more life time when they land up in the atmosphere (Stratosphere) where they are in a stage to destroy the ozone layer, thereby the protection of the environment is destroyed with effect from harmful UV rays. The CFCs consequences showing the harmfulness of those chemical compound present in it were doing much to the ozone layer.

References

1. World Health Organization, Health topics: health impact assessment (2019)
2. Y. He, R. Krishna, B. Chen, Metal–organic frameworks with potential for energy-efficient adsorptive separation of light hydrocarbons (2012)
3. S. Xiang, Y. He, Z. Zhang et al., Microporous metal-organic framework with potential for carbon dioxide capture at ambient conditions. Nat. Commun. **3**, 954 (2012)
4. F. Fathieh, M.J. Kalmutzki, E.A. Kapustin, P.J. Waller, J. Yang, O.M. Yaghi, Practical water production from desert air. Sci. Adv. **4**(6), eaatb3198 (2018)
5. F. Fathieh, M.J. Kalmutzki, E.A. Kapustin, P.J. Waller, J. Yang, O.M. Yaghi, Practical water production from desert air. Sci. Adv. (2018)
6. N. Hanikel, M.S. Prévot, F. Fathieh, E.A. Kapustin, H. Lyu, H. Wang, N.J. Diercks, T. Grant Glover, O.M. Yaghi. Rapid cycling and exceptional yield in a metal-organic framework water harvester. ACS Central Sci. **5**, 1699–1706 (2019)
7. S. Cui, M. Qin, A. Marandi et al., Metal-organic frameworks as advanced moisture sorbents for energy-efficient high temperature cooling. Sci. Rep. **8**, 15284 (2018). https://doi.org/10.1038/s41598-018-33704-4
8. B. Gido, E. Friedler, D.M. Broday, Assessment of atmospheric moisture harvesting by direct cooling. Atmos. Res. **182**, 156–162 (2016)
9. A. Sheik Abdullah, S. Selvakumar, A.M. Abirami, An introduction to data analytics: its types and its applications, in *Handbook of Research on Advanced Data Mining Techniques and Applications for Business Intelligence* (IGI Global, 2017). ISBN: 978-1-5225-2031-3, eISBN: 978-1-5225-2032-0. https://doi.org/10.4018/978-1-5225-2031-3.ch001
10. K. Akash, Zero energy air cooler (2018), https://www.iwma.in/yeswin/senior-3.pdf
11. K. Keerthy, A. Sheik Abdullah, S. Chandran, Analysis of ground water quality using statistical techniques: a case study of Madurai City, in *Artificial Intelligence and Machine Learning Applications in Civil, Mechanical, and Industrial Engineering*. Advances in Computational Intelligence and Robotics (ACIR) (2019). ISSN: 2327–0411 EISSN: 2327–042X(150619–102530)

Sliding Mode Controller Design for Wind Energy System to Enhance Power Profile and Stability

Abhishek Saraswat, Mohsin Kamal Siddiqui, and Sheetla Prasad

1 Introduction

Wind energy is a source that is pollution-free and inexhaustible towards the environment. Thus, the wind energy conversion system is one of the inherent sources for the future. In the near future, many countries have decided their goals for additional depletion of CO_2 emission and phase-out of fossil fuels, incremented consumption of renewable energy. At present, the most significant and effective source of renewable energy is wind energy. In the past decade, the first vertical and horizontal axis-based windmills were developed and used to convert wind energy into electricity. Several developments are adapted by researchers to enhance overall stability of wind system components [1]. Due to dependency on the geographical areas, a control scheme is designed to fulfil two main tasks; first one is to maximizing the power output from the wind against atmosphere and other one is to minimizing the fatigue of the turbine [2]. In [3], authors focused on the stability of the system, its reliability and how the system could be safe against the unpredictive nature of wind. Thus, wind energy system became a vast research on overall system stability, linearization and wind power fluctuations due to unknown nature of wind speed, etc., interest among the researchers.

Due to unknown nature of wind speed, a variable speed wind turbine is designed and played a key role to reduce the effect of wind speed uncertainties. The wind turbine is controlled and operated in the specific range of the wind speeds. Beyond specific limits, the turbine will stop to protect the generator and turbine. However, the operational zones of wind energy system are classified into three regions. The first zone is a low wind speed section where the turbine stops and their connections

A. Saraswat · M. K. Siddiqui · S. Prasad (✉)
School of Electrical, Electronics and Communication Engineering, Galgotias University, Greater Noida, India
e-mail: sheetla.prasad@galgotiasuniversity.edu.in

© The Author(s), under exclusive license to Springer Nature Singapore Pte Ltd. 2021
J. Kumar and P. Jena (eds.), *Recent Advances in Power Electronics and Drives*, Lecture Notes in Electrical Engineering 707,
https://doi.org/10.1007/978-981-15-8586-9_15

break from the grid in case of preventing any damages. The second zone is moderate wind speed section that only committed/restricted to the cut-in wind speed at the point where the turbine starts to produce power and achieved rated power as higher as possible at the rated speed. In the high wind speed zone, there is a finite amount of turbine power and due to this, turbine and generators are lightly loaded. It must be noticed that, for defending the turbine from mechanical high loading, it is to be shut down as soon as it reaches the cut out wind speed. Even though it is possible that the wind turbine speed is fixed or variable, it is certainly possible to achieve the extracted speed to be maximized only with variable speed wind turbines as discussed in [4]. As unknown uncertainty in wind speed, there may be so many variations in the wind power output. Thus, it is necessary to design a controller so that the wind system operates from fixed speed to variable speed [5]. In addition to battery storage system, a coordinated controller is designed to reduce wind power fluctuations due to wind speed variation as given in [6]. To minimize wind power fluctuations, a model predictive controller (MPC) is developed for wind energy system in [2, 6–9]. The sliding mode controller (SMC) is invariant against fluctuations in the system [10, 11]. A second-order SMC is implemented to control the torque and illustrated high maximum power point tracking (MPPT) against fluctuations in the torque [12]. Wind energy system power output is used to regulate frequency with a coordinated controller as discussed in [10, 13]. To reduce complexity in control scheme design process, a state space representation is demonstrated to evaluate the synchronous and induction generator-based wind turbines stability through its improved controllability and observability [3, 5, 14].

On other hand, the modern wind power plants are interfaced with the power electronics convertor and required to operate according to grid codes. The wind power plant acts closely as much as possible to the conventional power plants, which allows us the wide range of power output control [8]. As the infiltration of wind turbine is continuously increasing in the grid, there is a reduction in the power grid inertia that is extracted from the traditional power plant generators. The converter of rotor side controls the output power and voltage of generator that is measured at the grid terminals while the grid side converter helps in keeping the DC link voltage at constant level by generating and absorbing active power from the grid [15]. It noticed that the existence of high infiltration of wind power might result in the depletion of grid virtual inertia of the power systems [10, 16, 17]. While designing the structure of wind turbine, the most difficult issue is to construct the control system which is capable of tolerating the fluctuating of the structural load. Hence, researcher faced multiple challenges in control aspect due to fluctuating nature of wind energy that continuously makes changes within very less period of time.

However, the SMC technique is a more efficient to control various industrial applications against disturbances. In this study, the wind power fluctuations are minimized using first-order sliding mode control strategy in coordination with pitch angle mechanism even in the presence of wind speed variation. The optimal wind power and corresponding generator speed are obtained using "*fminsearch*" function in the MATLAB. The power error and generator speed error are calculated and applied to sliding mode controller and pitch angle control mechanism, respectively. The controller changes system damping ratio by use of non-linear function through LMIs "*mincx*" optimization tools, and pitch angle control mechanism varies pitch angle to maintain constant tip speed ratio. Thus, a variable system damping characteristics are obtained using nonlinear function and switching plane against wind uncertainty. Hence, wind power remains invariance against wind turbulence thrust change and ensures asymptotic convergence.

Remaining segments of the present study is arranged as wind energy system is represented in state space dynamics in Sect. 2, trailed by the proposed SMC structure for mentioned wind energy system under Sect. 3 followed by detailed closed-loop stability that is analyzed in Sect. 3. The simulated result analysis of the proposed controller is done in Sect. 4. Conclusion lastly drained from the illustrated work done is given in Sect. 5.

2 Wind Energy Description

As there are many uncertainties regarding the speed of wind, it may be in high speed and as well as in very low within a very short time interval. To control blade angle with minimization of wind power fluctuation simultaneously, the whole wind system is linearized for small perturbation in wind system. The turbine consists of three blades with variable speed and variable pitch angle control facilities. The generated turbine torque drives induction generator rotor and converts into electricity. Hence, the generated turbine torque is considered as equivalent wind output power in this study. The basic equations of wind energy system are described as follows:

$$\lambda_s = \frac{\omega_t r}{v} \tag{1}$$

where the terms λ_s, ω_t, v and r are tip speed ratio, turbine angular speed (rad\s), wind speed (m/s) and turbine radius (m), respectively.

The estimated power curve of 3.6 MW [10] is considered here and described as follows:

$$c_p(\lambda_s, \theta) = \sum_{i=0}^{4} \sum_{k=0}^{4} \sigma_{ik} \theta^i \lambda_s^k \tag{2}$$

where term θ is pitch angle (degree) and power coefficients σ_{ik} is given in [10].

The approximated turbine mechanical torque can be represented using power curve and given as follows:

$$\Gamma_m = 0.5\pi \rho r^3 c_t(\lambda_s, \theta) v^2 \tag{3}$$

where variable term $c_p(\lambda_s, \theta) = \frac{c_t(\lambda_s, \theta)}{\lambda_s}$ and wind power is calculated as

$$P_g = \omega_t \Gamma_m \tag{4}$$

The mechanical toque is obtained in shaft from turbine blades at angular speed ω_t. The turbine shaft and generator shaft is mechanically coupled and balanced using gearbox teeth ratio. The gearbox teeth ratio is considered as unity and balanced shaft torque can be represented as Γ_{tg} [10]. The generator angular speed ω_g varies according to generator breaking electromotive torque Γ_g^{br}. The turbine and generator shaft dynamics are indicated as two-mass equations and written as

$$2H_t \dot{\omega}_t = \Gamma_m - B_t \omega_t - B_{tg}(\omega_t - \omega_g) - \Gamma_{tg} \tag{5}$$

$$\dot{\Gamma}_{tg} = K_{tg}(\omega_t - \omega_g) \tag{6}$$

$$2H_g \dot{\omega}_g = \Gamma_{tg} - B_g \omega_g + B_{tg}(\omega_t - \omega_g) - \Gamma_{tg}^{br} \tag{7}$$

where terms H_t, H_g, B_t, B_g are inertia and damping coefficients of turbine and generator, respectively. The term B_{tg} is coupled damping coefficient. The pitch angle servo control dynamic with its time constant T_θ and estimated optimal generator angular speed ω_g^{op} is written as

$$\dot{\theta} = -\frac{1}{T_\theta} \theta + \frac{1}{T_\theta}(\omega_g^{op} - \omega_g) \tag{8}$$

The complete wind energy system dynamic is represented in state space equivalent equations as

$$\dot{x} = Ax + B\Gamma_g^{br} + F\Gamma_m$$

$$y = C x \tag{9}$$

The matrices order is given appropriate system dimensions and Eq. (9) is transferred into regular form [8] and given as

$$\begin{bmatrix} \dot{z}_1 \\ \dot{z}_2 \end{bmatrix} = \begin{bmatrix} A_{11} & A_{12} \\ A_{21} & A_{22} \end{bmatrix} \begin{bmatrix} z_1 \\ z_2 \end{bmatrix} + \begin{bmatrix} 0 \\ B_2 \end{bmatrix} \Gamma_g^{br} + \begin{bmatrix} D_{w1} \\ 0 \end{bmatrix} \Gamma_m$$

$$y = [C_1 C_2]z \tag{10}$$

Table 1 System variable nomenclature

Variable descriptions	Variable descriptions
v, r: Wind speed in m/s and turbine radius in m	ω_t, ω_g: Wind turbine and generator angular speed (rad/s)
λ_s, θ: Tip speed ratio and pitch angle in degree	$c_t(\lambda_s, \theta), c_p(\lambda_s, \theta)$: Estimated power curve
Γ_m: Wind turbine mechanical torque in N m and power	P_g: Wind power in watt
H_t, H_g: Inertia of turbine and generator	B_t, B_g: Damping coefficient of turbine and generator
B_{tg}, K_{tg}: Coupling damping coefficients	$\Gamma_{tg}, \Gamma_g^{br}$: Shaft and generator EM breaking torque (N m)
T_θ: Pitch angle servo system time constant (s)	ω_g^{op}, P_g^{op}: Wind optimal generator speed and power
$s, \varphi(\tau)$: Switching plane and nonlinear function	K, P: Feedback gain and positive definite matrix

All matrices value and system variables are given in Appendix and Table 1, respectively. The proposed control strategy is discussed in the next section for wind energy system Eq. (10).

3 Proposed Control Strategy and Its Convergence

The effect of fluctuations/disturbances in system on SMC scheme is negligible. The selection of switching plane is highly desirable and it is represented as

$$s = \left[K - \varphi(\tau) A_{12}^T P \ \ 1 \right] \begin{bmatrix} z_1 \\ z_2 \end{bmatrix} \tag{11}$$

where constant term K is state feedback gain and designed at low system damping value. The term $\varphi(\tau) = -\gamma e^{-k\tau^2}$ is used here to change system damping value and as a result, it damped out wind system power fluctuation according to $\tau = P_g^{op} - P_g$. As per maximum power point, the pitch servo control changes blade orientation angle according to error signal $\omega_g^{op} - \omega_g$ and detailed schematic representation is shown in Fig. 1.

The SMC design process, i.e. reaching and sliding mode is applied on Eq. (11) and after simplification by using Eq. (10), it gives

$$\dot{z}_1 = (A_{11} - A_{12}s_1)z_1 + F_1\Gamma_{m1} \tag{12}$$

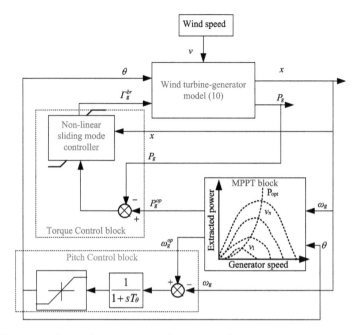

Fig. 1 The proposed control strategy schematic representation

where term $s_1 = K - \varphi(\tau)A_{12}^T P$ and vary according to variation in $\varphi(\tau)$. Thus, dynamics of Eq. (13) depends on $\varphi(\tau)$ value in terms of $\tau = P_g^{op} - P_g$. Whenever term τ is zero, i.e. nonlinear function $\varphi(\tau)$ is zero and it represents maximum point on power curve. Then, Eq. (12) is written as

$$\dot{z}_1 = (A_{11} - A_{12}K)z_1 + F_1\Gamma_{m1} \tag{13}$$

Otherwise, term τ is non-zero, i.e. nonlinear function $\varphi(\tau)$ may reach its maximum value γ. Then, Eq. (12) is written as

$$\dot{z}_1 = (A_{11} - A_{12}\gamma)z_1 + F_1\Gamma_{m1} \tag{14}$$

Let Eq. (13) variable final state feed gain may be represented as

$$K_f = K + \gamma A_{12}^T P \tag{15}$$

The Eq. (15) is represented as

$$A_{12}^T P - \frac{K_f - K}{\gamma} = N, \|N\| \leq \mu \tag{16}$$

Thus, dynamic of wind energy system is described by Eqs. (13) and (16) with proper selection of K and converted into linear matrix inequalities (LMIs) optimization framework using Schur complements [11]. Select P in such way that following LMIs hold simultaneously.

$$P > 0$$
$$\left[(A_{11} - A_{12}K)^T P + P(A_{11} - A_{12}K)\right] < 0$$
$$\begin{bmatrix} \mu I & N \\ N^T & \mu I \end{bmatrix} > 0 \qquad (17)$$

Theorem 1 *Due to simple design process of SMC and its fast convergence, the control law is developed with suitable selection of switching plane and nonlinear function [10, 11] as given below*

$$\Gamma_g^{br} = -B_2^{-1}\left(s^T Az - \frac{d\varphi(\tau)}{dt} A_{12}^T P s_1 + \overline{\kappa}s + \rho sign(s)\right) \qquad (18)$$

where variable terms $s, \overline{\kappa}, \rho, \varphi(\tau), P$ *are switching plane, positive constant* ($0 < \overline{\kappa} < 1$)*, boundary limit constant* $\rho \geq \|v\|$*, nonlinear function and positive definite matrix, respectively.*

Proof To obtain convergence condition, consider Lyapunov's function for the Eq. (10) as

$$V = 0.5 s^T s \qquad (19)$$

Differentiate the above equation and after substitution from Eq. (12), it gives

$$\dot{V} = s^T\left[s^T Az - \frac{d\varphi(\tau)}{dt} A_{12}^T P s_1 + (s_1 F_1 + F_2)\tau + B_2 \Gamma_g^{br}\right]$$

From Eq. (18), it is written as

$$\dot{V} \leq -s^T \rho \, sign\,(s) - \varsigma_{min}(\varepsilon)\|s\|^2 + \|s\|(\|s_1\|\|F_1\| + \|F_2\|)\rho$$

Then, $\dot{V} < -\chi|s|$ for some $\chi > 0$ and $\|s\| > \frac{1}{\varsigma_{min}(\varepsilon)}[(\|s_1\|\|F_1\| + \|F_2\|)\rho]$. The control strategy converges asymptotically. This completes the proof.

4   Results and Discussions

The SMC strategy is applied on wind energy system and demonstrated its performance through MATLAB© simulation. The proposed control strategy aims to minimize fluctuations in wind power output against unpredictable nature of speed of wind
and applied on three blades 3.6 MW energy systems [10]. The wind energy system
matrices and its parameters are given in appendix. Three simulation scenarios are
considered to show effectiveness of the proposed control scheme and described as

1. *Constant wind speed profile throughout time.*
2. *Step increasing change variation in wind speed.*
3. *Step decreasing change variation in wind speed.*
4. *Comparative analysis of the proposed method with integral SMC and traditional
   PI controller.*

The wind speed patterns are generated for 40 s time frame window with full
turbulent using the *"Class-A Kaimal"* turbulence spectra. The average wind speed is
10 m/s [18, 19] with tip speed ratio limit $2 < \lambda_s < 13$ and maximum power curve
$c_p^{max} = 0.4288 pu$. Thus, detailed discussion with justification is illustrated in the
following sections.

4.1   Constant Wind Speed Profile Throughout Time

The wind speed pattern with constant average wind speed 10 m/s is considered as
given in Fig. 2a and simulated wind system dynamics with the proposed strategy are
shown in Fig. 2 and Fig. 3, respectively. From said Fig. 2, it is seen that the power
curve is obtained at constant tip speed ratio through controlled generator torque
against variation in wind speed. The corresponding turbine and generator speeds,
pitch angle effort and effective wind power output are given in Fig. 3. It is evident
that the proposed control strategy performed well in presence of wind speed. The
obtained wind power fluctuation has minimum oscillations and enhances power grid
reserve capacity. Thus, SMC strategy requires less virtual inertia in the traditional
power network and as a result power system has less frequency oscillations. However,
the proposed method is enhanced its application.

4.2   Step Increasing Change Variation in Wind Speed

The wind speed pattern with step change in wind speed average value is used as
given in Fig. 4a in this subsection and its dynamic responses are shown in Figs. 4 and
5. The wind speed average is increased in steps from 8 to 14 m/s as given in Fig. 6a.
The tip speed ratio response, estimated power curve and controlled generator torque

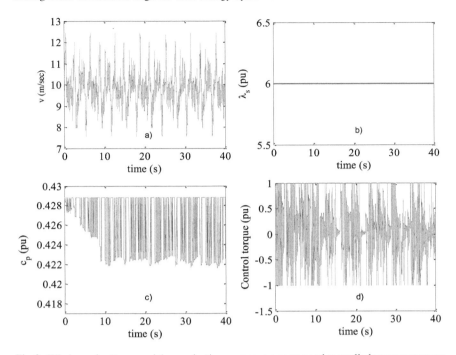

Fig. 2 Wind speed patterns; and tip speed ratio, power curve energy and controlled torque responses

responses are represented in Fig. 4, respectively. Similarly, turbine and generator speeds, pitch angle and wind output power responses are also shown in Fig. 5. It is evident that SMC strategy is protected wind energy system even in presence of high wind turbulence force. The proposed controller and pitch angle control scheme both are operated smoothly and reduce oscillation in wind power with reduction in estimated power curve. Thus, the proposed control scheme increased wind system stability and its application in variable speed geographical area.

4.3 Step Decreasing Change Variation in Wind Speed

The wind speed pattern with step change in wind speed average value is considered as given in Fig. 6a in this subsection and its dynamic responses are shown in Figs. 6 and 7. The wind speed average is decreased in steps from 14 to 8 m/s as given in Fig. 6a. The tip speed ratio response, estimated power curve and controlled generator torque responses are represented in Fig. 6, respectively. Similarly, turbine and generator speeds, pitch angle and wind output power responses are also shown in Fig. 7. It is seen that initially a high wind turbulence thrust is given on wind blades and the proposed control strategy is capable to protect as well as reduce wind power fluctuations with reduced estimated power curve. The pitch angle servo mechanism

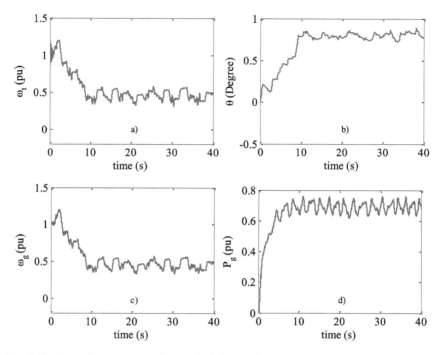

Fig. 3 Turbine and generator angular speed, pitch angle and wind output power responses

control approach issues command to change orientation of blade face smoothly. Thus, the proposed controller is capable to protect and minimize wind power oscillation and enhanced overall system stability even in presence of wind uncertainties.

However, SMC is able to control wind energy system against wind speed unpredictable nature uncertainty through variable closed-loop damping ratio phenomenon. This is achieved through nonlinear function and switching surface with proper use of LMIs optimization "*mincx*" MATLAB tools. The control generator torque signal gives marginal chattering effect in presence of wind speed uncertainties. Thus, SMC-based control structure enhances its application.

4.4 Comparative Analysis of the Proposed Method with Integral SMC and Traditional PI Controller

The proposed controller, integral SMC [20] and traditional proportional-integral (PI) controller ($k_p = 1.34, k_i = 1.564$)-based wind energy system responses are compared in presence of wind speed pattern as shown in Fig. 2a. The turbine speed, control signal, power curve and wind output power are shown in Fig. 8. The integral SMC responses have oscillations due to lack of coordination between pitch angle

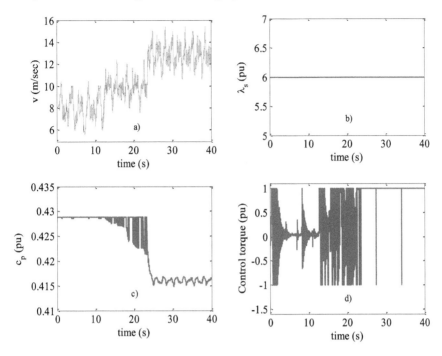

Fig. 4 Wind speed patterns; and tip speed ratio, power curve energy and controlled torque responses

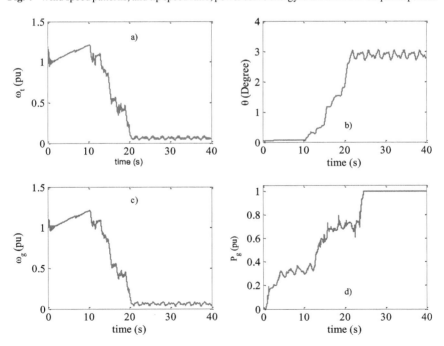

Fig. 5 Turbine and generator angular speed, pitch angle and wind output power responses

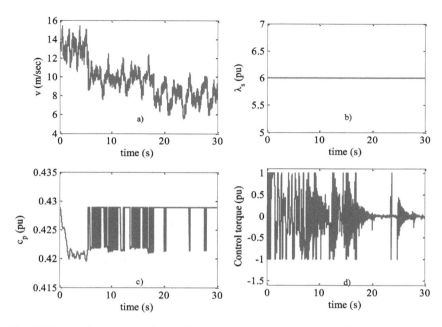

Fig. 6 Wind speed patterns; and tip speed ratio, power curve energy and controlled torque responses

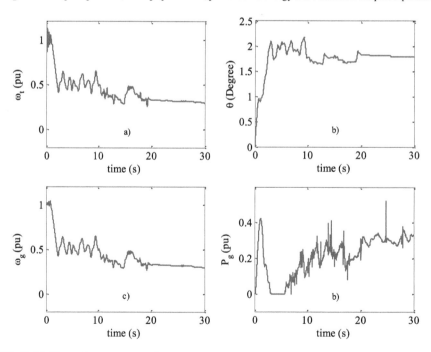

Fig. 7 Turbine and generator angular speed, pitch angle and wind output power responses

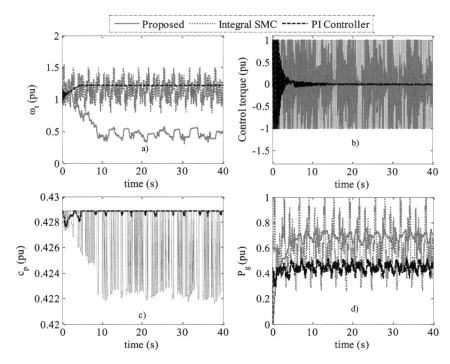

Fig. 8 A comparative analysis of the proposed controller responses with integral SMC [20] and PI controller

control mechanism and integral SMC while the proposed controller has comparatively less oscillations in the system responses. Thus, the proposed controller is robust and effectiveness compared to integral SMC [20] and traditional PI controller.

5 Conclusion

To reduce power oscillation, a SMC-based control scheme was proposed in this study for variable wind speed energy system. The control scheme asymptotic convergence strategy was proved using the Lyapunov theorem. A variable system damping property was achieved through nonlinear function and switching plane. The variable matrix P was obtained using LMIs optimization function "*mincx*" and finally, state feedback gain was calculated. The optimal power and corresponding generator speed are obtained using "*fminsearch*" function in MATLAB and applied to SMC and pitch angle control scheme, respectively. The nonlinear function and pitch angle are varied according to minimize power oscillations at constant tip speed ratio. In addition, SMC and pitch angle control schemes were able to minimize power oscillations and protect wind energy system against wind speed uncertainty. From simulation response analysis, the proposed control strategy is capable to reduce power oscillations as well as enhanced closed-loop stability against wind unknown nature uncertainty and also compared with existing techniques. In future, SMC-based controller will be analyzed considering double-fed induction generator wind energy system.

Appendix: Wind Energy System Matrices and Its Parameters

$$A = \begin{bmatrix} -\frac{(B_t + B_{tR})}{2H_R} & -\frac{1}{2H_t} & \frac{B_{tR}}{2H_t} \\ -K_{tg} & 0 & -K_{tg} \\ \frac{B_{tR}}{2H_R} & \frac{1}{2H_R} & \frac{(B_t + B_{tR})}{2H_R} \end{bmatrix}, B = \begin{bmatrix} 0 \\ 0 \\ -\frac{1}{2H_R} \end{bmatrix}, F = \begin{bmatrix} 0 \\ 0 \\ \frac{1}{2H_t} \end{bmatrix},$$

$$C = \begin{bmatrix} 1 & 0 & 0 \\ 0 & 1 & 0 \end{bmatrix}, x = \begin{bmatrix} \omega_t & \omega_r & \Gamma_{tg} \end{bmatrix}^T$$

Rated capacity: 3.6 MW; number of blades: 3; blade diameter: 104 m; wind speed (variable): 8–13 m/s; cut-in wind speed: 5 m/s; cut-out wind speed: 27 m/s; mechanical shaft system (on 3.6 MW base): $H^t = 2.49$ s; $H_g = 0.9$ s; $D_t = D_g = 0$; $D_{tg} = 1.5$; $K_{tg} = 296.7$ pu. $\lambda^{opt} = 6.212$ pu; $\omega_{base} = 1.335$ pu; $C_p^{out} = 0.4288$ pu; $\omega_r^{max} = 1.33$ pu; $\omega_r^{min} = 0.7$ pu; $C_{p\_max} = 0.4288$; $C_{p\_min} = 0.288$ $\alpha_{i,j}$ coefficients of C_p are given in Table 4-7 in [10].

References

1. J.K. Kaldellis, D. Zafirakis, The wind energy revolution: a short review of a long history. Renew. Energy **36**, 1887–1901 (2011)
2. A. Jain, G. Schildbach, L. Fagiano, M. Morari, On the design and tuning of linear model predictive control for wind turbines. Renew. Energy **80**, 664–673 (2015)
3. H. Bassi, Y. Mobarak, State-space modeling and performance analysis of variable-speed wind turbine based on a model predictive control approach. Eng. Technol. Appl. Sci. Res. **7**(2), 1436–1443 (2017)
4. M.A. Abdullah, A.H.M. Yatim, C.W. Tan, R. Saidur, A review of maximum power point tracking algorithms for wind energy systems. Renew. Sustain. Energy Rev. **16**, 3220–3227 (2012)
5. C.E. Ugalde-Loo, J.B. Ekanayake, N. Jenkins, State-space modeling of wind turbine generators for power system studies. IEEE Trans. Ind. Appl. **49**(1), 223–232 (2013)
6. M. Khalid, A.V. Savkin, Model predictive control for wind power generation smoothing with controlled battery storage, in *Joint 48th IEEE Conference on Decision and Control and 28th Chinese Control Conference* (2009), pp. 16–18
7. M. Mirzaei, M. Soltani, N.K. Poulsen, H.H. Niemann, An MPC approach to individual pitch control of wind turbines using uncertain LIDAR measurements, in *2013 European Control Conference (ECC)* (2013), pp. 490–495
8. T.G. Hovgaard, S. Boyd, J.B. Jørgensen, Model predictive control for wind power gradients. Wind Energy (Wiley) **18**(6), 1–12 (2013)

9. X. Liu, Y. Zhang, K.Y. Lee, Coordinated distributed MPC for load frequency control of power system with wind farms. IEEE Trans. Industr. Electron. **64**(6), 5140–5150 (2017)
10. T.H. Mohamed, J. Morel, H. Bevrani, T Hiyama Model predictive based load frequency control design concerning wind turbines. Int. J. Electr. Power Energy Syst. **43**, 859–867 (2012)
11. S. Prasad, S. Purwar, N. Kishor, Non-linear sliding mode control for frequency regulation with variable speed wind turbine systems. Int. J. Electr. Power Energy Syst. **107**, 19–33 (2019)
12. S. Prasad, S. Purwar, N. Kishor, Non-linear sliding mode load frequency control in multi-area power system. Control Eng. Practice **61**, 81–92 (2017)
13. B. Yang, Y. Tao, H. Shu, J. Dong, L. Jiang, Robust sliding-mode control of wind energy conversion systems for optimal power extraction via nonlinear perturbation observers. Appl. Energy **210**, 711–723 (2018)
14. B. Yang, T. Yu, H. Shu, Y. Zhang, J. Chen, Y. Sang, L. Jiang, Passivity-based sliding-mode control design for optimal power extraction of a PMSG based variable speed wind turbine. Renew. Energy **119**, 577–589 (2018)
15. B.H. Chowdhury, H.T. Ma, Frequency regulation with wind power plants, in *2008 IEEE Power and Energy Society General Meeting—Conversion and Delivery of Electrical Energy in the 21st Century* (2008), pp. 1–5
16. C.E. Ugalde-Loo, J.B. Ekanayake, State-space modelling of variable-speed wind turbines: a systematic approach, in *IEEE International Conference on Sustainable Energy Technologies (ICSET)* (2010), pp. 1–6
17. A. Zertek, G. Verbic, M. Pantos, Optimized control approach for frequency-control contribution of variable speed wind turbines. IET Renew. Power Gener. **6**, 17–23 (2012)
18. R.G. de Almeida, J.A. Peças Lopes, Participation of doubly fed induction wind generators in system frequency regulation. IEEE Trans. Power Syst. **22**(3), 944–950 (2007)
19. J. Mann, Wind field simulation. Probab. Eng. Mech. **13**(4), 269–282 (1998)
20. D. Qian, S. Tong, H. Liu, X. Liu, Load frequency control by neural-network-based integral sliding mode for nonlinear power systems with wind turbines. Neurocomputing **173**(3), 875–885 (2016)

Planning of Multi-renewable DG with Distribution Network Reconfiguration Based on MOFF Approach Using PSO Technique

Balmukund Kumar, Bikash Kumar Saw, and Aashish Kumar Bohre

1 Introduction

Distribution system is an important portion of the power system, which undergoes an enormous quantity of power losses, voltage deviation with reliability (RL) issue, and it needs to be reduced by means of different heuristic optimization techniques. Dispersed generation and reconfiguration is a kind of solution, efficient assignment of DGs, and appropriate reconfiguration with tie switches can enhance voltage profile along with system reliability, which offers adequate loss mitigation. In recent ages, deep studies have been carried out to discover numerous added impacts of distribution system such as reliability, protection, and stability. Dispersed generation is the production of electricity near to the consumption place, i.e., load end in the range of 1–50 MW [1, 2]. DG placement in distribution system network can diminish numerous difficulties to a great degree, for example, reduction in power losses, improvement in voltage profiles, drop in load demand, power supply to the consumer with improved quality, cost minimization at peak operation, enhanced safety, strength and reliability of grid, reduced pollutant emission, and so on. In [1], by using mixed-integer nonlinear programming optimization (MINLPO) method, the optimal planning of DGs are found and sited in distribution systems/networks to decrease loss and to improve profile of system voltage. Teng et al. [3] and Díaz et al. [4] presented load flow analysis which is used in the present work. Many researchers have stated

B. Kumar · B. K. Saw (✉) · A. K. Bohre
Department of Electrical Engineering, National Institute of Technology, Durgapur-09, Durgapur, WB, India
e-mail: bks.18ee1105@phd.nitdgp.ac.in

B. Kumar
e-mail: bmkumar14301@gmail.com

A. K. Bohre
e-mail: aashishkumar.bohre@ee.nitdgp.ac.in

that different optimizing techniques [5–11] are available for the determination of optimal planning of renewable DGs which can maintain the performance, operational, economical, and reliability issues. The Genetic Algorithm optimization technique is given in [5–7], while the particle swarm optimization technique is given in [8–10] are used for distribution system planning. In [11], it is discussed that what kind of commonly used technologies are available to harvest the energy by using DGs near to the consumption place and their different types of module sizes. Prakash et al. [12] presented a comparable study for optimal placement of DGs in the 69-bus IEEE Radial Distribution Network (IEEE RDS) between PSO and bat algorithm (BA). Sedighizadeh et al. [13] presented a work based on reconfiguration using minimization of objectives with loss and reliability indices. In [14], depending on the injected power DGs are characterized as four types. Reddy et al. [17] proposed a work, namely, reconfiguration of IEEE 69-bus RDS using PSO. The DG planning with reconfiguration is a complex target to achieve with a fitness function which is having multi-objective function which constitutes nonlinear constraints. In the presented work, the PSO technique is used with the aim to evaluate the optimum planning of multi-renewable DGs with reconfiguration. For this work, type-1 DG has been considered [10, 14]. In the used methodology, the radial property of the distribution system network is always maintained for load flow study (LFS). BIBC and BCBV based backward–forward sweep method [3, 4] is used for load flow study, which is more superior to the basic load flow method for radial systems. This method has been further employed to form a multiple objectives based fitness function that is coded in MATLAB/MatPower tool [15]. The proposed work is organized in different sections as follows: Section-I contains introduction, methodology is given in Section-II, In Section-III optimization technique is explained, Section-IV is about result and discussion, and Section-V is about conclusion. The different cases of the proposed work are as follows:

- Cases-1: Base Case system
- Cases-2: The System including reconfiguration
- Cases-3: The System including only single DG
- Cases-4: The System including only multi-DG
- Cases-5: System with reconfiguration and single DG
- Cases-6: System with reconfiguration and multi-DG.

2   The MOFF-Based Methodology

The Multiple Objective Based Fitness Function Formulation.

In this section, advanced multi-objective fitness function (MOFF) is introduced, which consists of distinct system performance and reliability assessment parameter. This novel fitness function-based problem is utilized for the optimum allocation of multiple renewable DGs with reconfiguration via PSO technique for different cases.

$$MOFF = c_1 \times APLI + c_2 \times VolDI + c_3 \times RPLI + c_4 \times RLI \qquad (1)$$

where c_1, c_2, c_3, and c_4, having the values 0.40, 0.20, 0.25, and 0.15, respectively, are weight factors which define how much weight has been given to each variable system indices based on priority basis [10]. APLI, VolDI, RPLI, and RLI are active power losses indices, voltage deviation indices, reactive power losses indices, and reliability indices, respectively. Within the fitness function priorities have been given to all the important factors that will make the distribution system more efficient and reliable. In the said work with the proposed methodology, active power losses (APL), reactive power losses (RPL) of system are reduced as shown in Table 1, although voltage profile and the system reliability has improved as presented in Table 2. For load flow study, BIBC and BCBV based backward–forward sweep method [3, 4] is considered which is more superior to basic load flow method.

- The BIBC and BCBV based backward–forward sweep method to update the voltage and current of each buses and branches which are calculated as follows: -
- Load current at each bus is given by

$$LC(i)^k = \left(\frac{PL(i) + QL(i)}{V(i)^k} \right) \qquad (2)$$

Table 1 Losses of the 69 bus RDS for all cases

Cases	Losses		
	P Loss	Q Loss	S Loss
Case-1	0.225	0.10216	0.247107
Case-2	0.067788	0.082142	0.106501
Case-3	0.083483	0.040513	0.092794
Case-4	0.070233	0.03234	0.077321
Case-5	0.039669	0.03568	0.053354
Case-6	0.028828	0.030911	0.042268

Table 2 Reliability indices of the IEEE 69 bus RDS for all cases

Cases	Reliability Indices				RL in %
	SAIFI	SAIDI	CAIDI	AENS	
Case-1	0.89024	0.53414	0.6	0.71671	85.29
Case-2	0.74094	0.44456	0.6	0.63513	87.01
Case-3	0.42983	0.2579	0.6	0.52597	90.27
Case-4	0.38794	0.23276	0.6	0.41892	91.2
Case-5	0.32359	0.19415	0.6	0.37731	91.97
Case-6	0.27864	0.16719	0.6	0.33704	92.58

- Branch current is given as

$$BC = [BIBC][LC] \tag{3}$$

- Updated voltage at each bus is calculated by,

$$V_m = V_n - BC(n-1) \times Z_{nm} \tag{4}$$

- Voltage deviation at each bus is given by

$$[\Delta V] = [BCBV][BC] \tag{5}$$

- Apparent power loss is calculated as

$$S = \sum_{j=1}^{NB} [\Delta V][BC]^* \tag{6}$$

- P_l and Q_l are given as

$$P_l = Real(S) \tag{7}$$

$$Q_l = Imag(S) \tag{8}$$

- Fitness function indices calculation by using the below formulae: -
- Active power losses index is given by

$$APLI(\alpha) = \frac{P_l(case - \alpha)}{P_l(case - 1)} \tag{9}$$

- Reactive power losses indices are obtained as

$$RPLI(\alpha) = \frac{Q_l(case - \alpha)}{Q_l(case - 1)} \tag{10}$$

- Voltage deviation index is calculated by

$$VolDI = max\left(\frac{\Delta V(n)}{v_{ref}}\right) \tag{11}$$

- Calculation of reliability index is considered based on the other system reliability parameters SAIFI, SAIDI, CAIDI, etc. which are given as

$$SAIFI = \frac{\sum_{z=1}^{N} \gamma_z * MVA_z}{\sum_{z=1}^{N} MVA_z} \tag{12}$$

$$SAIDI = \frac{\sum_{z=1}^{N} UN_z * MVA_z}{\sum_{z=1}^{N} MVA_z} \qquad (13)$$

$$CAIDI = \frac{SAIDI}{SAIFI} \qquad (14)$$

$$Reliability = 1 - \frac{AENS}{P_D} \qquad (15)$$

where i = 2 to N, N denote total number of bus, k = Iteration, j = 1 to NB, NB denote total number of Branches, LC = Load Current at each bus, BC = Branch Current of each branch, P_l and Q_l Active and Reactive Load on each bus, V = Bus voltage, m and n denotes 'To Bus' and 'From Bus', α = case number (2–6), n = 2 to N, γ_z = failurerate, UN_z = unavailability, MVA_z isApparentpowerdemandoneachbus, MW_Z = Activepowerload/demandoneachbus, S-Load = Total connected apparent power load, AENS = Average-energy-not-supplied, P_D = Total active power load/demand.

3   Optimization Technique

The first-ever PSO technique was stated in 1995 by Eberhart and Kennedy [9]. It is a population-based optimization technique [8, 9]. In PSO, in its search area every swarm follows a stated inertia and velocity by associated iterations. Based on the swarm last local best understanding and the past best understanding in its neighborhood the speed and way of the velocity are in tune. The characteristics of the swarm are to fly in the direction of an auspicious zone in a specified search area. PSO counts each individual's movement in complete exploration area through a specific stated velocity which updates based on the swarm's local-movement understanding and by the colleagues movement understanding [8–10]. In Fig. 1, the PSO technique algorithm flowchart has shown. In PSO the population of particles are initialized randomly and the corresponding updates in the particles' position are totally based upon the local best and neighbor best experience [10]. At every iteration updated new values of the velocity are the deciding factor for updated new value of the population. Swarm updated population is the sum of population of previous iteration and the velocity of current iteration in the next iteration. Suppose total population is P and respective velocity is v. The velocity and population for ith iteration is v $(i + 1)$ and X $(i + 1)$, respectively, similarly v and X are the velocity and position of previous iterative process. These two equations are given in order to update the velocity and position for i + 1th iterative process as [6–8].

$$v(i + 1) = w(i) \times v(i) + c_1 \times rand \times (X_{locbest} - X)$$
$$+ c_2 \times rand \times \left((X_{gbest}) - X\right) \qquad (16)$$

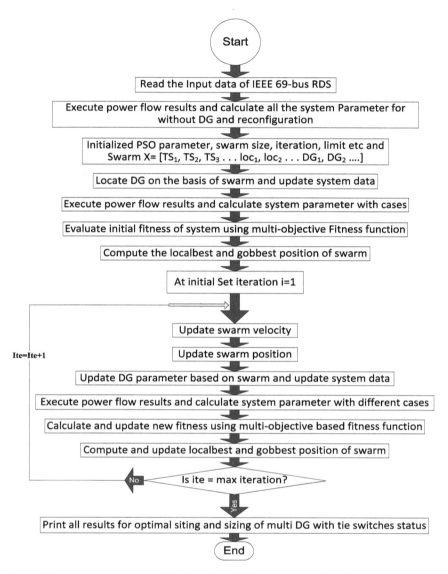

Fig. 1 Flowchart to find optimum DG allocation with Reconfiguration by PSO

$$X(i+1) = X(i) + v(i+1) \tag{17}$$

where P (i = 1, 2, 3....P) is number of population, $X_{localbest}$ and X_{gbest} are the local and global best populations, respectively.

v = Velocity, X = Swarm population for PSO.

i = ith iteration, w = inertia, rand = random number between 0 and 1, respectively, and c_1, c_2 are the constriction factors and these are positive constant numbers [6–8].

The flowchart to determine optimal planning with reconfiguration in IEEE 69-bus RDS is presented in Fig. 1. Five switches (TS$_1$, TS$_2$, TS$_3$, TS$_4$, TS$_5$,) three DG locations (Loc$_1$, Loc$_2$, Loc$_3$) and sizing of DGs are considered in the flowchart; these are converted and represented as swarm or (particle).

4 Results and Discussion

The presented work is implemented with MATLAB/MatPower tool and for the analysis of the obtained result, standard 69 bus IEEE-RDS [16] is considered as shown in Fig. 2 to verify the proposed methodology. In case-1, base case is considered without reconfiguration and DG, and obtained results are active power loss (APL) as 0.225 p.u., reactive power loss (RPL) as 0.10216 p.u. and maximum voltage deviation (VolD) of the system as 0.19081 p.u. with respect to the reference voltage. While in case-2, the standard IEEE 69 bus RDS with only reconfiguration is considered by maintaining the radiality of the system, the observed results are APL as 0.067788 p.u., RPL as 0.082142 p.u. and maximum VolD of system as 0.13423 p.u., respectively, with respect to base case as shown in Tables 1, 2, and 3. In the same fashion, single DG has been implemented by using PSO technique in Case-3, where APL, RPL, and VolD is reduced by 68%, 68.34% and 37.59%, respectively, as shown in below three tables with respect to Case-1. In Case-4, multi-DG has been considered by using PSO technique where got the APL reduction 69.87%, RPL reduction 19.59%, VolD decrease by 29.65%, and hence the reliability is increased to 91.20% as shown in Table 2 and Fig. 12 in comparison with the base case.

Thereafter, after reconfiguration with DGs placement has implemented then better reduction in all indices has noticed, and therefore two more cases, namely, Case-5 and 6 are considered. In Case-5, reconfiguration with single DG by using PSO

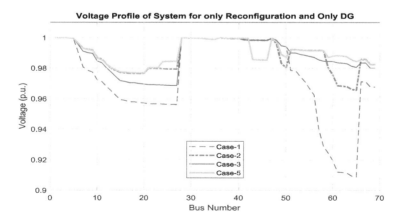

Fig. 2 Enhanced voltage profile of the 69 bus RDS for Case-1, 2, 3, 5

Table 3 Fitness function of the IEEE 69 bus RDS for all cases

Cases	Fitness function indices				Fitness function
	APLI	VolDI	RPLI	RLI	
Case-1	1	0.173463	1	1	0.83469
Case-2	0.30128	0.12202	0.80401	0.88309	0.47838
Case-3	0.37103	0.11926	0.39655	0.66167	0.37065
Case-4	0.31215	0.10919	0.31655	0.59816	0.31556
Case-5	0.17631	0.11183	0.34924	0.54542	0.26201
Case-6	0.12812	0.10576	0.30256	0.5045	0.22371

technique is implemented, then the APL is reduced by 82.36%, RPL is reduced by 65.07%, and VolD is also decreased by 35.53% as compared to Case-1. Multi-DG with reconfiguration by using PSO technique has been implemented in case-6 and considerable reduction is observed in the multi-objective function where APL is reduced by 87.18%, RPL is reduced by 69.74%, and VolD is decreased by 39.02%, hence the reliability is increased up to 92.58% as given in Table 2 and presented in Fig. 12 as bar chart compared with the basic case.

After analyzing the obtained result, it can be concluded that the optimum allocation of multi-renewable DGs including reconfiguration using PSO Technique is an effective approach to minimize MOFF and its minimized data is presented in Table 3. Through this proposed work we achieved our target with 87.18% APL reduction and improved reliability of 92.58% along with better quality voltage profile as shown in Tables 1, 2 and Figs. 1, 2, 3, 4, 5, 6, 7, 8, 9, 10, 11, 12.

In the comparison table, literature [1, 14] presented optimal DG placement using MINLP Technique and PSO Technique of the IEEE 69-bus RDS, respectively, whereas in [17] reconfiguration using PSO for the same has been carried. For the

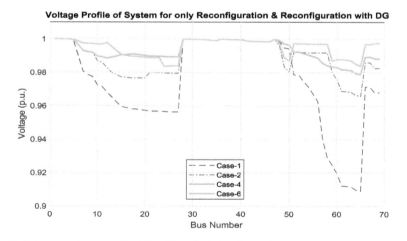

Fig. 3 Enhanced voltage profile of the 69 bus RDS for Case-1, 2, 4, 6

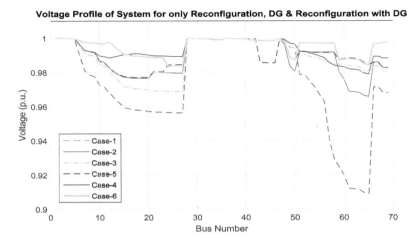

Fig. 4 IEEE 69 bus RDS improved voltage profile of the for all cases

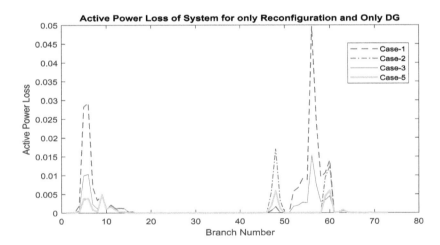

Fig. 5 EEE 69 bus RDS active power loss for Case-1, 2, 3, 5

said cases in Table 4 the reduced active power losses are 69.53, 69.89, and 99.82 kW but in the proposed work of this paper the active power loss is reduced to 28.82 kW and also voltage profile is upgraded with reference to base case as given in Table 1.

5 Conclusion

The optimum allocation or planning (siting and sizing) of DGs including reconfiguration based on a multiple objective-function (MOFF), which consist of numerous

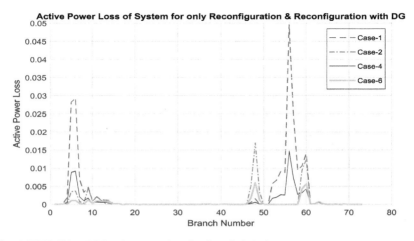

Fig. 6 IEEE 69 bus RDS active power loss for Case-1, 2, 4, 6

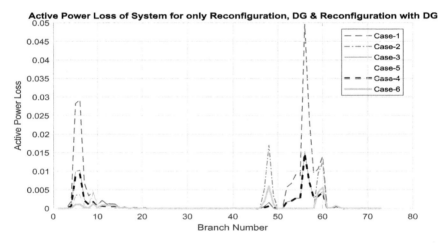

Fig. 7 Active Power Loss of the 69 bus of RDS for all cases

network parameters using PSO is evaluated on IEEE 69-bus RDS. Finally, network reconfiguration with multi-DG is implemented using PSO technique with MOFF. In the observed outcomes, APL, RPL, and VolD are decreased by 87.18, 69.74, and 39.02% as shown in Table 1. Correspondingly the overall reliability is enhanced up to 92.58% along with improved voltage profile as illustrated in result analysis with respect to the first case, i.e., with the base case. The outcomes are analyzed for different cases and also compared with the results of finally implemented case, i.e., reconfiguration with multi-DG (Case-6) with the existing work as illustrated in Table 4. It is concluded that the presented approach is effective for the decrease in apparent power. Therefore, the power losses (active and reactive) and the voltage

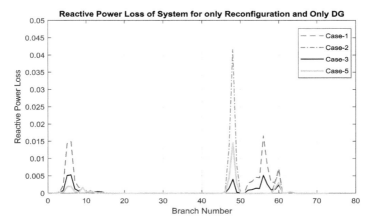

Fig. 8 System reactive power loss for Case-1, 2, 3, 5

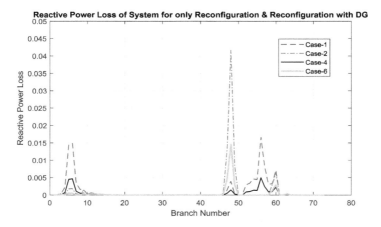

Fig. 9 system reactive power loss for Case-1, 2, 4, 6

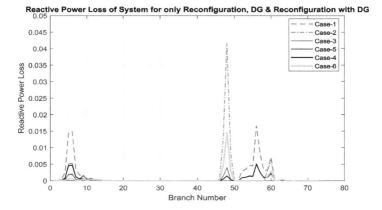

Fig. 10 Reactive power loss of the IEES 69 bus RDS for all cases

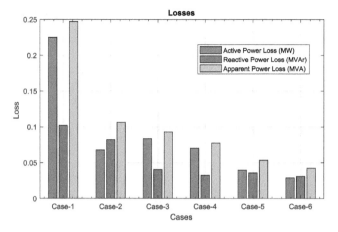

Fig. 11 Losses of the IEES 69 bus RDS for all cases

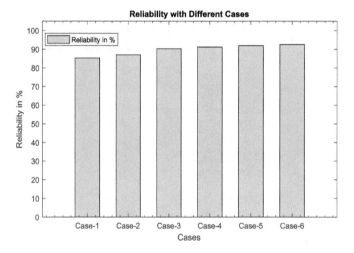

Fig. 12 Reliability of the 69 bus RDS for all Cases

Table 4 Comparative analysis of 69 bus RDS existing work with the proposed work

Existing/Proposed work	optimization technique	power loss (kW)	Voltage deviation
Alam et al. [1]	MINLP	69.53	–
Prakash et al. [14]	PSO	69.89	–
Reddy et al. [17]	PSO	99.62	0.1572
Proposed work	PSO	28.82	0.1163

deviation reduce significantly. Moreover, the system voltage profile and reliability are enhanced.

References

1. A. Alam, et al., Power loss minimization in a radial distribution system with distributed generation, in *2018 International Conference on Power, Energy, Control and Transmission Systems (ICPECTS)* (IEEE, 2018)
2. T. Ackermann, G. Andersson, L. Söder, Distributed generation: a definition. Electr. Power Syst. Res. **57**(3), 195–204 (2001)
3. J.-H. Teng, A direct approach for distribution system load flow solutions. IEEE Trans. Power Deliv. **18**(3), 882–887 (2003)
4. G. Díaz, J. Gómez-Aleixandre, J. Coto, Direct backward/forward sweep algorithm for solving load power flows in AC droop-regulated microgrids. IEEE Trans. Smart Grid **7**(5), 2208–2217 (2015)
5. J.H. Holland, *Adaptation in Natural and Artificial Systems Ann Arbor,* vol. 1 (The University of Michigan Press, 1975), p. 975
6. D.E. Goldberg, *Genetic Algorithms in Search, Optimization and Machine Learning* (Addism1—Wesley, Reading, MA, 1989)
7. R.L. Haupt, S. Ellen Haupt, *Practical Genetic Algorithms* (2004)
8. J. Kennedy, R. Eberhart, Particle swarm optimization (PSO). In *Proceedings of IEEE International Conference on Neural Networks* (Perth, Australia, 1995)
9. R. Eberhart, J. Kennedy, A new optimizer using particle swarm theory, MHS'95, in *Proceedings of the Sixth Int Symposium on Micro Machine and Human Science* (IEEE, 1995)
10. A.K. Bohre, G. Agnihotri, M. Dubey, Optimal sizing and sitting of DG with load models using soft computing techniques in practical distribution system. IET Gener., Transm. Distrib. **10**(11), 2606–2621 (2016)
11. L.I. Dulău, M. Abrudean, D. Bică, Distributed generation technologies and optimization. Procedia Tech. **12**, 687–692 (2014)
12. R. Prakash, B.C. Sujatha,. Optimal placement and sizing of DG for power loss minimization and VSI improvement using bat algorithm, in *2016 National Power Systems Conference (NPSC)* (IEEE, 2016)
13. M. Sedighizadeh, M. Esmaili, M.M. Mahmoodi, Reconfiguration of distribution systems to improve reliability and reduce power losses using imperialist competitive algorithm. Iran. J. Electr. Electron. Eng. **13**(3), 287–302 (2017)
14. D.B. Prakash, C. Lakshminarayana, Multiple DG placements in distribution system for power loss reduction using PSO algorithm. Procedia Tech. **25**, 785–792 (2016)
15. R.D. Zimmerman, C.E. Murillo-Sanchez, Matpower 4.1' (2011). https://www.pserc.cornell.edu//matpower/
16. A. Swarnkar, N. Gupta, K.R. Niazi, A novel codification for meta-heuristic technique used in distribution network reconfiguration. Electr. Power Syst. Res. **81**(7), 1619–1626 (2011)
17. A.S. Reddy, M.D. Reddy, Optimization of network reconfiguration by using particle swarm optimization. *2016 IEEE 1st International Conference on Power Electronics, Intelligent Control and Energy Systems (ICPEICES).* (IEEE, 2016)

Application of SVC and STATCOM for Wind Integrated Power System

Vikas Kumar Tiwari and Atma Ram Gupta⑩

1 Introduction

Renewable energy resources (RES) are the most valuable asset to the power sector. Therefore, their integration to the main grid is also a topic of research and development. But integration of these resources have several drawbacks, hence wind farms are no exception. Wind integrated system faces various shortcomings related to power quality and dynamic performance of the wind turbine. Power quality characteristics like flicker, fault ride through, low voltage ride through (LVRT), voltage dips/swells, and high voltage ride through (HVRT), phase imbalance or low power factor are dominant influence of power utilities and distribution systems [1]. Power quality issues are increasing day by day as RES are coming in picture, which impact power grid stability and also lead to compromise the reliability of power supply [2, 3].

Various researches have been done to solve wind farm power quality issues and grid-connection challenges which ensure grid stability, and trust in employing non-conventional sources [4]. Increasing demand for power and shortcomings of conventional resources have impact in proportional increase in renewable energy demand. Now, there are various FACTS configurations available which can solve these problems, but researches have shown that SVC and STATCOM are one of the best available options [5]. STATCOM has its own advantage which makes it a most suitable FACTS device on application point of view. It reduces size and has a quick response which is a basic and necessary requirement for on-field installation [6]. It can separate out the undesirable harmonic component present in the system, and therefore source current and voltage has zero phase difference between them.

V. K. Tiwari · A. R. Gupta (✉)
Electrical Engineering Department, NIT Kurukshetra, Kurukshetra 136119, India
e-mail: argupta@nitkkr.ac.in

V. K. Tiwari
e-mail: vk95tiwari@gmail.com

© The Author(s), under exclusive license to Springer Nature Singapore Pte Ltd. 2021   181
J. Kumar and P. Jena (eds.), *Recent Advances in Power Electronics and Drives*, Lecture Notes in Electrical Engineering 707,
https://doi.org/10.1007/978-981-15-8586-9_17

Utilization factor also increases by overcoming different power quality problems. Fast Fourier Transform (FFT) analysis with and without controller is done, and it is seen that Total Harmonic Distortion (THD) can be almost eliminated from the voltage and current waveform [7].

In the beginning of this paper, a brief introduction of the dominant power quality issues is given. In the next section, FACTS devices implementation is presented as a solution of the problem with the main focus of the SVC and STATCOM. Basics of SVC and STACOM design and layout have been explained for brief information. Further, V-I characteristics as well as applicability of these device in context of wind farm have been explained. A detailed comparison is also done on applicability of these devices in wind integrated system. Because of the several advantages, STATCOM gets an edge over SVC which is also reflected by the researchers' interest in it. In the next section different areas of research in SVC and STATCOM related to the problems in wind farms have been explained which covers the major areas where these devices have found best suitability. In the further section, some additional benefits of these devices in the field of large wind integration as well as offshore wind has been discussed. In the last section, precise conclusion has been given on suitability of these devices in wind farms.

2   Power Quality Issues in Wind Integrated System

Wind being non-uniform over geographical locations and time period brings fluctuation in the system. Due to variation in the speed, mechanical power varies which results in the non-uniform torque and subsequently variable electrical power. It causes not only power quality problems but also transient problems occur in the system. That is why they are equally important for the research. Power quality parameters which are affected most are voltage dips, sag and swell and harmonic contents.

The type of wind generator is important in evaluating their power quality characteristics. Wind generators which utilize power converters are more sensitive to power quality issues and hence proper monitoring is required. The uplift in load demand tries to reduce the speed of the turbines at the generation plant. It causes sag in voltage along with frequency variation which is affected more severely [8].

Technological advancement in the power electronic switches being used causes significant reduction in the size of the power system. Use of electronic converter is most common in almost all control schemes of FACTS devices. As a result, current distortion limit is always under the radar. Reactive power imbalance is also a frequent phenomenon in these cases.

3 Implementation of FACTS Device

FACTS devices are used to enhance the quality of power with adequate reactive power support along with a controller. A basic layout of the connection topology of these devices is explained in Fig. 1 which explains the position of all components in the system for proper operation. The FACTS devices used mostly in the wind generation system are SVC, STATCOM, and Unified Power Quality Controller (UPQC). Different FACTS devices have different priority for use. For example, to reduce THD and to improve voltage profile during the fault, STATCOM is preferred. Voltage ride through capability of the system can be improved by this device after voltage collapse below the critical level [9].

Drastic improvement can be seen in the transient characteristics of the system along with grid stability is also improved using FACTS devices. Sudden voltage collapse occurring in the system can also be avoided with the help of these devices.

3.1 Compensation Through SVC and STATCOM

The basic design of the SVC is based on the variable equivalent shunt admittance as explained in Fig. 2. Variation of the firing angle of the thyristor results variable

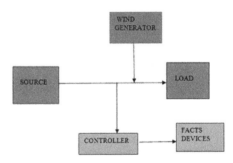

Fig. 1 FACTS device basic connection to grid integrated wind farm

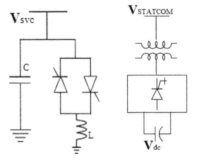

Fig. 2 Basic layout of SVC and STATCOM

reactance [10]. Line compensation done by the SVC controller modulates the net susceptance offered to the system [6].

Placement of SVC at optimal location increases the voltage stability of the system [11]. STATCOM works on the basic principle of voltage conversion from fixed DC to variable AC with the help of voltage source converter as explained in Fig. 2 [12]. Voltage level at STATCOM terminal and grid decides direction and nature of reactive power [3].

3.2 Control Schemes for SVC and STATCOM

There are various control schemes which are used in these devices like Clarke and Park transformation, Hysteresis controller, Space Vector Pulse Width Modulation (SPWM), etc. These techniques are used to produce control signals for controllers. Based on the reference voltage and actual voltage of the system switching signal is given to power electronics switch used in the VSC of the STATCOM. Thus reactive power compensation is carried out to maintain the voltage level above the threshold limit to avoid tripping [3].

Effect of Control Schemes: SPWM control techniques are mostly used for voltage sags and swell and reactive power control, whereas Clarke and Park transformation is used for instantaneous reactive power control which helps to minimize Total Harmonic Distortion (THD) and improve power factor. Hysteresis current controller is used in case of sudden decrease in the voltage of the system, like the case of sudden fault at the terminals of wind farm.

4 Comparative Analysis of SVC and STATCOM

4.1 In Terms of Characteristics

The area of application and compensation provided by FACTS device can be explained with the help of V-I characteristics. In Fig. 3, it is clear that maximum current injection of STATCOM for compensation can be made at a low voltage of around 20% of maximum voltage. Therefore, maximum reactive power can be transferred even at a lower magnitude of voltage [5]. Nature of graph is also linear which indicates that compensation provided by the STATCOM varies in proportion to voltage change. The maximum current compensated by the SVC varies in linear relation with voltage. But if we talk about maximum VAR compensation then it has a square relation with it.

Fig. 3 Voltage-current
characteristics of SVC and
STATCOM

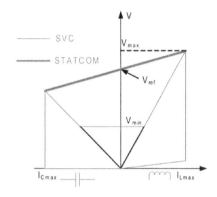

4.2   Application Point of View

After application of FACTS devices in the wind farm the voltage, reactive power, and rotor speed are constantly monitored. Thereafter, a comparison is made between before and after application of FACTS devices which shows considerable improvement. When all parameters are taken into account as described in Table 1 then it is clear that although few drawbacks are present in both the FACTS devices but if we take speed of operation and occupied area in list then the most practical comparison can be made. Because slow operation may cause tripping of the system which may incur more cost than designing cost of the devices.

Hence although STATCOM is somehow costlier than SVC, it is preferred in most of the cases due to its fast and superior operation [13]. Additionally hybrid combination of battery, Super Conducting Magnetic Energy Storage (SMES) etc., are possible in case of STATCOM. It can be seen from the researches also as most of the researches are done for the STATCOM. By controlling the d- and q-axis parameters of the VSC voltage of STATCOM required operation that can be carried out for compensation.

Table 1 Comparison of SVC and STATCOM in application point of view

Parameters	STATCOM	SVC
Speed of operation	Quicker operation	Slow
Cost-benefit analysis	Higher cost	Less costly
Losses	More	Less
Area occupied	Smaller in size	Occupies more area
Applicability	Wide area of applicability (Battery, SMES integration possible)	Lesser range of applicability
Compensation characteristics	Constant compensation	Reduction with drop in voltage

5 Effectiveness of SVC and STATCOM in Wind Integrated System

Unlike conventional power plant, wind power system has a different set of characteristics and grid code requirement; new standards are set and required for viable operation [14]. Study of FACTS devices has become a burning topic of research for the stability and voltage balance of the system [15]. Application of the wind power plant is not straight forward; technocrats have to keep in the mind about different adversities that have to be faced in case of worst-case scenarios like faults and voltage imbalance. Hence, FACTS devices come as a rescue for providing essential aid like reactive power support. In Table 2 suitability of FACTS device for different issues in wind integrated system has been explained. It is evident from it that in application point of view SVC and STATCOM have proved to be best among all devices.

5.1 Stability Point of View

For flexible and reliable power control SVC and STATCOM are used [16] to ensure proper loading and they also sustain power oscillation damping present in the system. In [17], stability for turbines with lower speed is discussed, but the level of converter used is two which is only not applicable in the case of higher voltage of operation. Hence, later studies [18] shows that with increasing level of the converter we can stabilize the system even for higher voltages. In [19], stability is improved with the help of STATCOM and dynamic breaking resistor. Ability of STATCOM to supply power control even in dynamic condition makes it one of the important FACTS devices [20]. Variation in the speed of the wind has an impact on the transient behavior of the fault, which is shown in [21].

5.2 Voltage Recovery and LVRT

The ability of the different types of wind turbine to recover from the low voltage occurring in the system due to unwanted sudden fault in the system defines the

Table 2 SVC and STATCOM effectiveness in different cases	List of problems in wind farms	Suitable FACTS device
	Voltage stability	STATCOM, SVC
	Power oscillation damping	STATCOM
	Fault ride through capability	STATCOM, SVC
	Harmonics	STATCOM
	Flicker mitigation	SVC

effectiveness of the FACTS device. In [22], it is shown that after the fault has been cleared, application of STATCOM in induction generator type wind turbine maintains the voltage above the threshold limit. If voltage falls below certain level then protection unit of the wind farm trips the generator.

In [23], STATCOM application for the case of rapid voltage change has been illustrated. Some better control schemes can increase the recovery rate of the system. Hence in [23], linear optimal control scheme has been replaced with the Proportional Integral (PI), Proportional Integral Derivative (PID) control schemes and a comparison has been done. Low voltage ride through capability has been accessed with the help of SVC and STATCOM. Conventional method of using rotor resistance control has some serious disadvantages in terms of power loss in the form of rotor copper loss. Hence use of SVC and STATCOM has been analyzed in the wind integrated system which does not have these drawbacks [14].

5.3 Applicability

A number of studies have been done in application point of view of SVC and STATCOM [17, 18, 22]. Integration of the wind farm in the grid comes up with several negative consequences which can be eliminated with the help of STATCOM [24].

In conventional system, switched capacitor was used for reactive power support of the system but it results in overvoltage in the system. This can be stopped with the help of STATCOM which is shown in [25]. In [26], it is shown that STATCOM is one of the fastest devices for reactive power support and stability which aid value to its applicability in the system.

5.4 Sub Synchronous Resonance (SSR) and Oscillations

In power system, network generators suffer electromechanical oscillation which may cause instability on the network. In most cases, these oscillations are local in nature which means they are subjected to generator. But in case of weak transmission network, they may get transferred to neighborhood generators. Most basic solution to them is Power System Stabilizer (PSS), but in case of inter-network oscillations FACTS devices are preferred. In [15], use of SVC and STATCOM to damp out oscillation present in the system is discussed and benefits of use of shunt controllers have been emphasized compared to traditional PSS.

For bulk transfer of power over a power system network series reactance of transmission line should be least. That is why series compensation is done but because of compensation subsynchronous resonance occurs in the system. In [27], remedy for this problem is given with the help of SVC and STATCOM through MATLAB/Simulink.

5.5 Flicker Mitigation

Voltage fluctuation observed at the load side by the consumer is normally referred as flicker. The variation in the load voltage is somehow related to wind turbine power fluctuation. Hence it becomes mandatory to rectify this problem near wind farms. Several researches have been done to relate this problem rectification with STATCOM and SVC [28–30].

5.6 Dynamic Performance

Transient behavior analysis of the wind integrated system is also equally important because any disturbance in form of fault or power oscillation has direct impact on the slip characteristics and rotor movement of generator. Real-time analysis using power system computer-aided design (PSCAD) has been done in [31]. In another paper [32], the dynamic performance of wind turbine has been tested for power oscillation and damping. Several cases of simulation have been done to explain the importance of voltage and power stability. Ability of shunt controller to restore the system for stable condition has been tested. It is found that STATCOM is the most suitable device for dynamic control of the wind farm. There are some limitations in STATCOM and Static Synchronous Series Compensator (SSSC) which tends to destabilize the system. Automatic gain controller has been modeled in [33] with the help of gate turn off (GTO) AC–DC converter. It takes care of inherent delay caused by the Phase-Locked Loop (PLL) in the typical controller and improves the dynamic performance of the system.

5.7 Active and Reactive Power Control

Reactive power transfer is possible in the basic STATCOM controller but for active power support of Energy Storage System (ESS) is necessary. Large penetration of wind energy in the grid causes poor power quality and instability. ESS combination results in smoother operation than conventional approach [34]. There are various control schemes related to the ESS. In conventional feedback control system, a few more addition has been done in [35]. In this method, state charge limits as well as charging and discharging rates have been considered. Depending on wind forecasting, power levels are chosen at which power is transferred.

In reactive power control, approach toward power management has been diverse and that is why new methods have started coming in market. Use of heuristic dynamic programming is introduced in [36] which uses neural network function to develop the neurocontroller. In case of any fault occurring in the system, STATCOM helps to control the voltage and current parameter in doubly fed induction generator circuit.

6 SVC and STATCOM in Large Wind Power Integration and Offshore Wind Farm

6.1 Integration in Weak Transmission Network

Normally wind farms are installed near the sub-transmission or distribution level where the grid is assumed to be weak and fragile. Therefore, instability in terms of voltage can be seen if we are planning large-scale integration. When plenty of wind power is transferred over a transmission network then it may congest the network. In weak transmission network, this problem becomes more severe. Hence, there should be some limit decided for a particular network. For the study of all the relevant power quality issues in the system, first of all a weak transmission network is proposed with a set of wind turbine. Thereafter, supervisory control and data acquisition (SCADA) analysis is done for pointing out major issues and a STATCOM controller connected to check the voltage fluctuations present for a shorter duration of time [37].

Two states of arts have been chosen for the study; one is SVC and other is STATCOM with energy storage system. Depending upon the X/R ratio, active and reactive power has been discussed. For larger X/R ratio, reactive power compensation is recommended while in case of low X/R ratio active power insertion seems beneficial for voltage support to the system. Different case studies have been chosen to showcase the effectiveness of SVC in application point of view in the practical field. Design of energy storage system along with SVC by ABB Company has been discussed for compensation of 600 kVar and 600 kW [38].

6.2 Application in Offshore Wind Farms

For offshore wind farm study combination of an offshore wind farm and Marine current farm is taken in [39] which uses modal control theory for frequency domain analysis. STATCOM along with PID damping controller is used to make the system stable in the adverse situation like faults and oscillation present in the system. Offshore wind turbines are located far away from the terrestrial place; hence a lot of cabling is required in the system which somehow increases the net capacitance of the system. Reactive power has been a problem for wind farms; this additional burden causes different set of norms for these types of wind generation systems.

There are two types in which compensation can be done in these systems named as substation compensation and wind turbine compensation schemes. In future, dedicated offshore turbines have to be developed so that the main emphasis can be on power transfer to the main grid [40].

7 Conclusions

If we look up at all FACTS devices having shunt control then SVC and STATCOM are most obvious choice to rectify the problems of wind farm. In wind integrated system where reactive power is one of the main concerns for voltage support, these device acts as the most appropriate solution. STATCOM is proved to be more beneficial compared to SVC in terms of speed of operation and occupied area, hence they are used in most of the cases. But it does not reduce the importance of SVC because if we look up for solution for flicker as well as frequency control then they are most economic for use. If we focus on the problems faced due to integration of the wind energy in the grid then we can really analyze the usefulness of these devices. After study of all the researches done in the field, like stability, low voltage ride through capability, flickers, dynamic performance, Sub Synchronous Resonance (SSR) and oscillation present in the system, it can be concluded that SVC and STATCOM are most genuine devices for mitigating these problems. They are able to solve almost all the problems present in the wind integrated system. Variation in the control scheme used in these devices has also led to smoother operation. By considering the factors destabilizing the system, elements can be added in the conventional control schemes for improvement in compensation schemes.

In the last case, integration of large wind system in the weak network system has been analyzed. Even with addition of wind energy into grid, some limitations are there in term of the congestion caused in the network. Hence, a threshold is set in case of the weaker transmission network to avoid any tripping of the network. Offshore wind farms have been discussed in addition to observing the wide area of application of SVC and STATCOM in the wind integrated system.

References

1. P. Pourbeik, R.J. Koessler, D.L. Dickmander, W. Wong, Integration of large wind farms into utility grids (Part 2 – Performance Issues), in *IEEE Power Engineering Society General Meeting Conference* (2003), pp. 1520–1525
2. A.L.T. Garcia, J.M.M. Ortega, E.R. Ramos, Reactive power compensation strategies for wind generation in distribution systems: a power quality approach. International Journal of Energy Technology and Policy **3**(3), 196–212 (2005)
3. V. Suresh Kumar, A.F. Zobaa, R. Dinesh Kannan, K. Kalaiselvi, Power quality and stability improvement in wind park system using STATCOM. Jordan J. Mech. Ind. Eng. **4**(1), 169–176 (2010)
4. K. Elkington, Modelling and control of doubly fed induction generators in power systems: towards understanding the impact of large wind parks on power system stability. Ph.D. thesis, KTH, Electric Power Systems, SE-100 44, Stockholm, Sweden (2009). ISBN: 978-91-7415-264-7
5. L. Xu, L. Yao, C. Sasse, Comparison of using SVC and STATCOM for wind farm integration, in *International Conference on Power System Technology* (2006), pp. 1–7
6. Y.L. Tan, Analysis of line compensation by shunt-connected FACTS controllers: a comparison between SVC and STATCOM. IEEE Power Eng. Rev. 57–58 (1999)

7. D. Srinivas, Power quality improvement in grid connected wind energy system using facts device and PID controller. IOSR J. Eng. **2**, 19–26 (2012). https://doi.org/10.9790/3021-021 111926
8. G. Elsady, Y.A. Mobarak, A.-R. Youssef, STATCOM for improved dynamic performance of wind farms in power grid, in *Proceedings of the 14th International Middle East Power Systems Conference (MEPCON'10)*, Cairo University, Egypt (2010)
9. C. Rahmann, H.J. Haubrich, A. Moser et al., Justified fault ride through requirements for wind turbines in power systems. IEEE Trans. Power Syst. **26**, 1555–1563 (2011)
10. U. Eminoglu, M.H. Hocaoglu, Effect of SVCs on transmission system voltage profile for different static load models, in *Universities Power Engineering Conference*, vol. 1 (2004), pp. 11–14
11. R. Panda, P.K. Satpathy, S. Paul, Application of SVC to mitigate voltage instability in a Wind system connected with grid. Int. J. Power Syst. Oper. Energy Manag. (IJPSOEM) **1**(1), 90–96 (2011)
12. H. Byung-Moon, L. Burn-Kyoo, J. Young-Soo, L. Kwang-Yeol, Application feasibility analysis of STATCOM for wind power system with induction generator. Trans. Korean Inst. Electr. Eng. **53**(6), 309–315 (2004)
13. D. Lijie, L. Yang, M. Yiqun, Comparison of high capacity SVC and STATCOM in real power grid, in *2010 International Conference on Intelligent Computation Technology and Automation*, Changsha (2010), pp. 993–997
14. M. Molinas, J.A. Suul, T.M. Undeland, Low voltage ride through of wind farms with cage generators: STATCOM versus SVC. IEEE Trans. Power Electron. **23**, 1104–1117 (2008)
15. N. Mithulananthan, C.A. Canizares, J. Reeve, G.J. Rogers, Comparison of PSS, SVC, and STATCOM controllers for damping power system oscillations. IEEE Trans. Power Syst. **18**(2), 786–792 (2003)
16. S.M. Muyeen, M.A. Mannan, M.H. Ali, R. Takahashi, T. Murata, J. Tamura, Stabilization of grid connected wind generator by STATCOM. Power Electron. Drives Syst. **2**(28-01) (2005)
17. S.M. Muyeen et al., Stabilization of grid connected wind generator by STATCOM, in *International Conference on Power Electronics and Drive Systems (IEEE PEDS 2005), Conference CDROM*, Malaysia (2005), pp. 1584–1589
18. S.M. Muyeen et al., Application of fuzzy logic control based STATCOM to stabilize wind turbine generator system, in *XVII International Conference on Electrical Machines (ICEM 2006)*, Paper No. 198, Chania, Crete Island, Greece (2006)
19. X. Wu, A. Arulampalam, C. Zhan, N. Jenkins, Application of a static reactive compensator (STATCOM) and a dynamic braking resistor (DBR) for the stability enhancement of a wind farm. Wind Eng. **27**(2), 93–106 (2003)
20. S.M. Muyeen, M.A. Mannan, M.H. Ali, R. Takahashi, T. Murata, J. Tamura, Stabilization of grid connected wind generator by STATCOM. IEEE Power Electron. Drives Syst. **2**, 28-01 (2005)
21. M.S. Elmoursi, A.M. Sharaf, Novel STATCOM controllers for voltage stabilization of standalone hybrid schemes. Int. J. Emerg. Electr. Power Syst. **7**(3), 1–27 (2006). Art 5
22. Z. Chen, F. Blaabjerg, Y Hu, Voltage recovery of dynamic slip control wind turbines with a STATCOM, in *International Power Electronic Conference (IPEC05)*, S29-5 (2005), pp. 1093–1100
23. A. Jain, K. Joshi, A. Behal, N. Mohan, Voltage regulation with STATCOMs: modeling, control and results. IEEE Trans. Power Del. **21**(2), 726–735 (2006)
24. S.-I. Jang, J.-H. Choi, I.-K. Park, H.-M. Hwang, D.-M. Choi, K.-H. Kim, N.S. Yoo, A study on the STATCOM application for efficient operations of wind farm **52**(5), 250–256 (2003)
25. B.-M. Han, B.-K. Lee, Y.-S. Jon, K.-Y. Lee, Application feasibility analysis of STATCOM for wind power system with induction generator **53**(6), 309–315 (2004)
26. Z. Saad-Saoud, M.L. Lisboa, J.B. Ekanayake, N. Jenkins, G. Strbac, Application of STATCOMs to wind farms. IEE Proc. Gener. Transm. Distrib. **145**(5), 511–516 (2014)
27. M.S. El-Moursi, B. Bak-Jensen, M.H. Abdel-Rahman, Novel STATCOM controller for mitigating SSR and damping power system oscillations in a series compensated wind park. IEEE Trans. Power Del. **25**(2), 429–441 (2010)

28. T. Sun, Z. Chen et al., Flicker mitigation of grid connected wind turbines using STATCOM, in *Proceedings of 2nd IEEE International Conference on Power Electronics, Machines and Drives*, PEMD04 (2004)
29. T. Sun, Z. Chen, F. Blaabjerg, Flicker mitigation of grid connected wind turbines using STATCOM, in *Conference on Power Electronics, Machine and Drives*, vol. 1 (2004), pp. 175–180
30. M.I. Marei, T.H.M. El-Fouly, E.F. El-Saadany, M.M.A. Salama, A flexible wind energy scheme for voltage compensation and flicker mitigation, in *IEEE Power Engineering Society General Meeting*, vol. 4 (2003), pp. 2497–2500
31. H. Gaztanaga, I. Etxeberria-Otadui, D. Ocnasu, B. Seddik, Real-time analysis of the transient response improvement of fixed speed wind farms by using a reduced-scale STATCOM prototype. IEEE Trans. Power Syst. **22**, 658–666 (2007). https://doi.org/10.1109/TPWRS.2007.895153
32. G. Elsady, Y. Mobarak, A.-R. Youssef, STATCOM for improved dynamic performance of wind farms in power grid (2010)
33. A.H. Norouzi, A.M. Sharaf, Two control schemes to enhance the dynamic performance of the STATCOM and SSSC. IEEE Trans. Power Del. **20**(1), 435–442 (2005)
34. Z. Yang, C. Shen, L. Zhang, M.L. Crow, S. Atcitty, Integration of a STATCOM and battery energy storage. IEEE Trans. Power Syst. **16**(2), 254–260 (2001)
35. S. Teleke, M.E. Baran, A.Q. Huang, S. Bhattacharya, L. Anderson, Control strategies for battery energy storage for wind farm dispatching. IEEE Trans. Energy Convers. **24**(3), 725–732 (2009)
36. W. Qiao, R.G. Harley, G.K. Venayagamoorthy, Coordinated reactive power control of a large wind farm and a STATCOM Using Heuristic Dynamic Programming. IEEE Trans. Energy Convers. **24**(2), 493–503 (2009)
37. C. Han, A. Huang, M. Baran, S. Bhattacharya, W. Litzenberger, L. Anderson, A. Johnson, A. Edris, STATCOM impact study on the integration of a large wind farm into a weak loop power system. IEEE Trans. Energy Convers. **23**, 226–233 (2008). 10.1109/TEC.2006.888031
38. R. Grünbaum, FACTS to facilitate AC grid integration of large scale wind integration (2012)
39. L. Wang, C. Hsiung, Dynamic stability improvement of an integrated grid-connected offshore wind farm and marine-current farm using a STATCOM. IEEE Trans. Power Syst. **26**(2), 690–698 (2011)
40. Y.A. Abdelaziz, A.M. El-Sharkawy, A.M. Attia, Effect of system topology on wind generators penetration maximization with TCSCs insertion. i-Manag. J. Power Syst. Eng. **5**(1), 1–9 (2017)

Optimization of Charge Transport Layer Thickness for Efficient Perovskite Solar Cell

Anjan Kumar, Roushan Kumar, and Sangeeta Singh

1 Introduction

Solar energy is one of the best renewable type of energy resource to fulfill the worldwide energy demands in the present situation. Sun-based energy together with wind, biomass, tidal, and geothermal energy are developing as substitute sources of energy for the energy crisis in our planet. Among all energy sources, Sun-based energy is the inexhaustible and clean kind of energy that provides us energy that is free from dangerous atmospheric gases and ozone-depleting substances caused by non-renewable energy sources [1]. Solar energy being one of the renewable, green, and unpolluted energy draws attention to the researchers worldwide due to its economical and vast accessibility.

Sun-based light is one of the sustainable power resource assets that can be the solution for continuous energy supply in the twenty-first century. To use Sun-based light more productively, low price, stable and well-organized Sunlight-based cells have been used for practical cases. So, photovoltaic technology is the optimum decision to come across the worldwide energy crisis because of its endless characteristics [14]. Sun-based cells are the optoelectronic gadgets that specifically transform daylight into electrical energy utilizing the best emerging material. Since many years, silicon-based solar cells are dominatingly utilized for changing Sunlight to energy [14]. On the other hand, single/multi-crystalline silicon and other thin film-based Sun-powered cells have constrained use in vast scale due to the production price and nature issues [14]. As a sustainable power source, Sun-based energy is one of the fastest approaches to illuminate the current energy crisis and environmental pollution [2]. In view of that, the analysis and development of high effectiveness and ease of

A. Kumar (✉) · R. Kumar
GLA University, Mathura, India
e-mail: anjan.kumar@gla.ac.in

A. Kumar · S. Singh
National Institute of Technology, Patna, India
e-mail: sangeeta.singh@nitp.ac.in

© The Author(s), under exclusive license to Springer Nature Singapore Pte Ltd. 2021   193
J. Kumar and P. Jena (eds.), *Recent Advances in Power Electronics and Drives*, Lecture Notes in Electrical Engineering 707,
https://doi.org/10.1007/978-981-15-8586-9_18

Sun-powered cells is the innovative establishment, in which, conventional crystalline silicon-based sunlight-based cells have been effectively commercialized irrespective of some restrictions. Thus, dye-based solar cells, quantum spot sunlight based cells (QDSCs) and perovskite Sun based cells (PSCs) have been exploited [10]. Till now, the most productive and reasonable way is the change of sunlight into power with the assistance of photovoltaic (PV) cells [12]. It has pulled in the consideration of world energy field because of the developing requests of sustainable power source and exhaustion of petroleum products. As far as minimal effort, light-absorbing capacity, outstanding carrier transport, great thermal stability, and also generally high photon to electric transformation effectiveness, the perovskite Sun-powered cells (PSCs) give economically and actually an elective course for photovoltaic devices which will satisfy the future energy crises [12]. As we are probably aware that Silicon (Si) Sun powered cells which produce Sun-based energy is as of now being used; however, these cells cannot produce so much amount of Sun-based energy because of its failure to consume energy not as much as its band gap, and thermalization of photons having energy more than the band gap which constrains the productivity up to 33%. If we talk about periodic table then we are mostly aware of Si but rather than Si, Ga, as also an element, which combines and forms 28% efficiency in single junction and more in multi-junction dope. Cd-Te is also a best compound for solar cell but not efficient. Tin and lead compound are very much useful for constructing light-harvesting devices, but we are not getting best efficiency as well as cheap solar cell. It is a best choice for a single junction solar cell having highest PCE up to 33.7% but single junction solar cells are not perfectly able to convert sunlight to electricity, so a multi-junction or tandem-based cell using organic–inorganic halide perovskite is used to overcome the efficiency problem. Perovskite was chosen because it is capable of being an efficient solar cell material despite being deposited at low temperature and requires low cost. Depending on the perovskite chosen, the efficiency limit can be increased from 23% which is currently present.

1.1 Perovskite Solar Cell

Perovskite-based Sun cells are cells that include a perovskite organized material as the active layer. This Sun-based cell was first revealed by Miyaska et al. in 2009, and now it has become so popular for practical use. The perovskite compound structure is mainly in the form of ABX_3 which is generally carbon–non-carbon-based compound with a lead or tin halide-based materials. So these material acts as a light-collecting active layer for perovskite Sun based cell. The most generally utilized perovskite material is methyl-ammonium lead tri-halide ($CH_3NH_3PbI_3$) where X is a halogen particle, and the band gap of methyl ammonium lead tri-halide is between 1.5 and 2.3 eV. According to halide structure present in perovskite solar cell, the Sun-powered cell proficiencies of a device utilizing these materials have increments from 3.8% in 2009 to 24.2 in 2019% [9]. Perovskite-based Sun cells have become the quickest developing sunlight-based innovation to date. Perovskite containing metal halides provides the best features that are valuable for Sun-powered cell applications. The materials required and the procedure of fabrication both takes place at a very low cost. Its high absorption coefficient and thickness around 500 nm makes it more effective to absorb the wide-ranging visible solar intensity spectrum. The above two features combined and provided a probability to make ease, high productivity, thin light weight, and adaptable Sun-oriented modules. Mixing of natural/inorganic perovskite materials (e.g., $CH_3NH_3PbI_3$) are the best safeguard materials for thin-film photovoltaic applications. The main reason for drastically incrementing power transformation efficiency of such devices is that perovskite materials have the vast majority of the properties required to be fulfilled, suitable direct band gap, great absorption factor, incredible carrier transport, and markable strength of defects [6]. Apart from ease and simplicity of formation, perovskite materials proposed a wide tune ability on synthesis and arrangement by genuinely changing the metal halide structure and also determined characteristic types [6]. Organic–inorganic perovskite are the best emerging photovoltaic materials [9]. The four-terminal tandem perovskite solar cell could lead to single or two layer cell in PCE. Lead halide perovskite, specifically methyl ammonium lead iodide ($CH_3NH_3PbI_3$), is presently the demandable and more usable material, shown with a crystal structure ABX_3, in which the A and B cations are coordinated with the X anions, and this class of perovskite provides noticeable photovoltaic performance records and started significant research prominence. This is mostly due to its ease of process ability and small price of materials. This paper mainly focuses on (a) different arrangement and possibility of perovskite material which can give better efficiency and low cost (b) all possibility of hole transport layer material (HTL) which affects the execution of perovskite material (c) similar different electron transport layer materials (ETL) (d) choosing of proper electrode (anode, cathode).

1.2 Perovskite Solar Cell Working Principle

When light enters from the glass which works as a substrate for the solar cell it allows passing it to the second layer of solar cell. The second layer of solar cell works as a transparent electrode which is nothing but different element doped oxides or generally we take fluorine doped metal oxide, which allows most of the sunlight to the perovskite material without proper loss of light intensity. The light intensity after passing from transparent electrode strikes on n-type material which is electron transport layer which works as a carrier for electron transport from perovskite layer where the light finally reaches the perovskite layer material which acts as absorber material; the function of this material is to absorb most of the light and generate most of the electron and hole. The electron is captured by the electron transport material layer and gap is captured by the positively charged transport layer material. So the flow of electron-hole pair generation takes place in proper way. On the top of hole transport layer material we evaporated the contact of gold or silver. So all the layer below perovskite material layer except glass works as an anode and the layer above the perovskite material layer works as a cathode for hole conduction. So electricity is generated with the help of these electrodes. With the proper doping of electron and hole material, increasing the conductivity of negatively charged and positively charged transport material and by adjusting the proper thickness and band gap of perovskite semiconductor material, the efficiency can be upgraded. Therefore, the role of negatively charged carrying material, hole transport material, and different possibility of perovskite material is very necessary to increase the sustainability and productivity of perovskite Sun-oriented cell.

2 Previously Published Perovskite Based Solar Cell

Despite the fact that the expense of silicon modules has reduced considerably, the expense of power created by photovoltaic cell is not going to decrease and it is higher than that of electricity produced by non-renewable energy sources. Therefore, the expansion handled by Photovoltaic Cell Company and their market is totally depending on government support, as it were, previously. To make the PV market more advance and effective, there is a direct need to build up a financially strong PV industry that can run their business without management support [5]. The push to bring down expenses has brought about the improvement of numerous new PV innovations dependent on different materials and minimal effort forms, for example, thin-film silicon Sun-powered cells [13] and color improved Sun-oriented cells (DSCs) [3]. By the way, the power conversion efficiencies of these Sun-based cells have not been sufficiently high for larger application use. As of late, perovskite Sun-based cells (PSCs) have pulled in wide attention in light of their high efficiencies directly accomplished in the research capability >20% [15], and there might be a solid possibility that a conversion efficiency of more than or close to 35% will be accomplished in the close future [4].

Today, perovskite Sun-oriented cells accomplished the efficiency of 24.2% with organic material as HTM [9]. The price of natural HTMs is a significant challenging issue and hence elective materials are necessary [14]. The fluorine doped hole transporting material with an adjusted energy level and a great crystal advancement temperature guarantees very effective and temperature-wise stable PSCs. The utilization of this material is to manufacture photovoltaic devices with 23.2% conversion proficiency with stable proficiency of 22.85% for little area ($0.094\,cm^2$) cells and 21.7% effectiveness for huge zone ($1\,cm^2$) cells. We likewise accomplish qualified proficiencies of 22.6% (little area cells, $0.094\,cm^2$) and 20.9% (larger area, $1\,cm^2$) The subsequent scheme demonstrates preferable thermal stability over the devices with spiro-OMeTAD, keeping up relatively 95% of its underlying implementation above $500\,h$ after warm tempering at $60\,°C$ [8].

3 Proposed Layered Structure of Perovskite-Based PV

In this perovskite structured Sun cell there are six layers and out of these three layers are main layers. These three layers mainly influences the performance of solar cell named as (1). Electron Transport Layer (2). Perovskite Layer (3). Hole Transport Layer. Apart from these layers, glass as a substrate layer followed by fluorine doped tin oxide act as an electrode. Silver or gold as a back contact layer.

As in Fig. 1, the sunlight falls on the glass which acts as a substrate. On top of glass layer, fluorine doped tin oxide is formed, which commonly acts as an electrode for this solar cell. The perovskite layer is packed in the middle of two essential layers which are generally the backbone of this solar cell structure. Below the perovskite layer electron transport layer exists which is important for the transportation of electron and also most of the electrons are collected here only. On the top of perovskite a hole transport material is present which is responsible for the carrying of hole generated from perovskite layer. The efficiency can be improved with the help of the following: (a) increasing the current flow capability of hole transport material by proper doping. (b) By taking suitable thickness of absorber as well as the band gap. With the proper thickness of positively charged carrying layer, negatively charged carrying layer, and perovskite layer, the productivity of the proposed perovskite Sun-based solar cell increases up to the markable value. Here titanium oxide based material is chosen as a negatively charged carrying layer because it has some special property like it is more electron enriched and avoid direct connection between the perovskite layer and FTO electrode. Positively charged layer also performs similar operation and it avoids the perovskite layer from the back contact such that maximum number of hole generated from the perovskite layer is collected on the only hole transport layer.

Fig. 1 Proposed perovskite structured solar cell

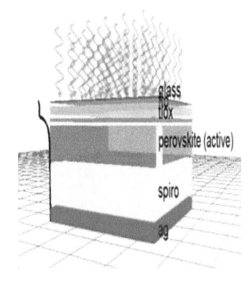

The perovskite layer thickness and their band gap affect mostly on the performance of cell because it acts as a light harvesting layer and it absorbs most of the light spectrum and produces maximum of negatively-positively charged pair.

4 Simulation Results of Proposed Solar Cell

Simulation is carried out in general purpose photovoltaic device model (gpvdm) software [11]. Electrical parameter of the main absorber layer is shown in Fig. 2.

4.1 Optimization of Charge Transport Layer Thickness

In order to optimize the charge transport layer thickness, HTL and ETL thickness is varied keeping perovskite layer thickness constant.

From Table 1, it is observed that at the constant thickness of perovskite layer ($3e^{-07}$m), if the thickness of negatively charged carrying layer and positively charged carrying layer changes then the efficiency, fill factor, short-circuit current, and open-circuit voltages also changes. From the table, it is found out that up to the thickness of $3 e^{-07}$m of negatively charged carrying layer and the thickness of $7e^{-07}$m of

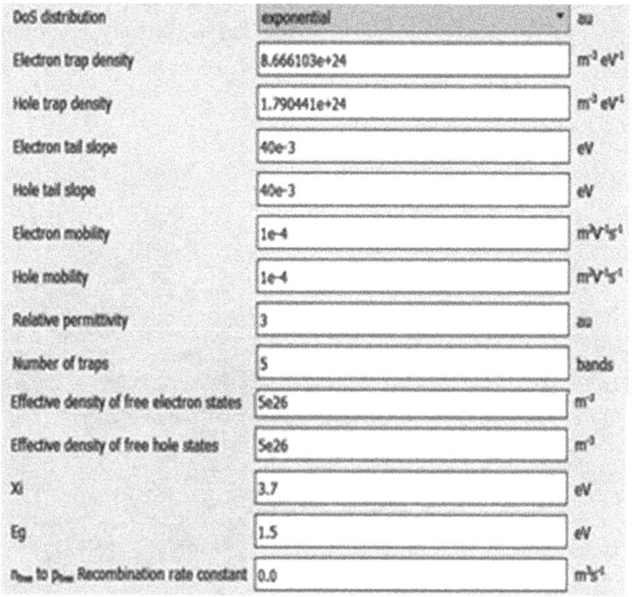

DoS distribution	exponential	au
Electron trap density	8.666103e+24	$m^3 eV^1$
Hole trap density	1.790441e+24	$m^3 eV^1$
Electron tail slope	40e-3	eV
Hole tail slope	40e-3	eV
Electron mobility	1e-4	$m^2V^1s^1$
Hole mobility	1e-4	$m^2V^1s^1$
Relative permittivity	3	au
Number of traps	5	bands
Effective density of free electron states	5e26	m^{-3}
Effective density of free hole states	5e26	m^{-3}
Xi	3.7	eV
Eg	1.5	eV
n_free to p_free Recombination rate constant	0.0	m^3s^1

Fig. 2 Electrical parameter of various layers

Table 1 Thickness of different layers

Thickness of ETL (m)	Thickness of HTL (m)	Efficiency of perovskite cell (%)
$6e^{-08}$	$2e^{-07}$	17.79
$5e^{-08}$	$2e^{-07}$	18.62
$2e^{-08}$	$5e^{-07}$	18.81
$1.5e^{-08}$	$5e^{-07}$	19.30
$3e^{-08}$	$2e^{-07}$	19.89
$2.5e^{-07}$	$4.5e^{-07}$	20.85
$4e^{-07}$	$6e^{-07}$	20.94
$3e^{-07}$	$7e^{-07}$	21.38
$4e^{-07}$	$7e^{-07}$	20.51
$5e^{-07}$	$7e^{-07}$	17.16
$5e^{-07}$	$2e^{-07}$	16.85
$6e^{-07}$	$2e^{-07}$	16.58
$1e^{-07}$	$7e^{-07}$	19.34
$2e^{-07}$	$8e^{-07}$	18.52

Table 2 Optimized thickness of different layers

Layer name	Thickness (m)	Layer type
Glass	$2.5e^{-08}$	Other
FTO	$1.2e^{-07}$	Contact
TiO_x	$3e^{-07}$	ETL
Peroyskite	$3e^{-07}$	Active-layer
Spiro	$7e^{-07}$	HTL
Ag	$2e^{-07}$	contact

hole transport layer the efficiency of solar cell is remarkable. But as soon as we increase the thickness of ETL and keeping HTL layer same then efficiency of solar cell reduces drastically, but when we decrease the thickness of ETL as thin as possible and increasing the thickness of HTL up to the certain thickness here mentioned as $7e^{-07}$ m, then the efficiency of perovskite cell increases. But after the threshold of $7e^{-07}$ m of the thickness of hole transport layer keeping the perovskite layer constant like $3e^{-07}$ m, the efficiency starts to decrease. So the role of thickness of perovskite layer plays a very important role to decide the thickness of ETL and HTL such that the efficiency of perovskite-based Sun cell increases and maintains its performance. The optimized thickness of different layers is shown in Table 2.

4.2 Value of Different Performance Parameters of Proposed Solar Cell at Optimized Charge Transport Layer Thickness

To get the desired result in terms of efficiency, fill factor, open-circuit voltage, and short-circuit current the maximum number of photons absorption show a vital character. The maximum number of photons absorption produces a maximum of negatively-positively charged pair which is very necessary for the best performance of the solar cell. Graph in Fig. 3a–d is generated when light is incident on proposed solar cell (Table 3).

To get optimized solar parameters, the absorber layer should set for optimum thickness. The thickness of the absorber layer varies from 0.15 to 0.56 μm. As the absorber layer thickness increases, the longer wavelength of sunlight will produce a large amount of negatively-positively charged pair generation. By reducing the thickness of the absorber layer, the depletion layer becomes very near or close to the back contact and more electrons will be accumulated on the back contact, so the chances of recombination can happen. So due to the less electron-hole pair generation the fill factor and efficiency reduces.

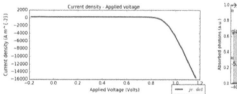 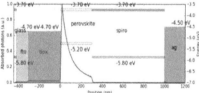

(a) Current Density V/S Applied Voltage Graph

(b) Photon Absorption V/S Position

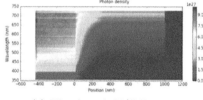 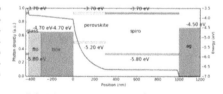

(c) Wavelength V/S Position

(d) Photon Density V/S Thickness

Fig. 3 Variation of different parameters when light is incident on proposed solar cell

Table 3 Performance parameter of proposed PSCs

Parameter	Value
Fill factor	83%
Power conversion efficiency	21.38%
Maximum power	213.84 W
V_{oc}	0.8455 V
Recombination time constant at V_{oc}	$4.94e^{-007}$ S
Recombination rate at V_{oc}	$9.02e^{+027}$ m^{-3} s^{-1}
Average carrier density at Pmax	$2.21e^{+021}$ m^{-3}
Recombination time constant	$7.89e^{-006}$ m^{-1}
Trapped electron at V_{oc}	$2.85e^{+020}$ m^{-3}
Trapped hole at V_{oc}	$1.34e^{+0.00}$ m^{-3}
Free electron at V_{oc}	$4.83e^{+031}$ m^{-3}
Free hole at V_{oc}	$4.961e^{+0.21}$ m^{-3}
J_{sc}	$-3.04e^{+002}$ A/m^2
Total carrier $((n + p)/2)$ at V_{oc}	$9.72e^{+0.11}$ m^{-3}

5 Conclusion

we conclude that the effect of thickness of both electron transport layer and hole transport layer shows a vital role to enhance the efficiency and performance of perovskite solar cell. The thickness between (200 and 350 nm) of ETL and the thickness between (300 and 700 nm) of HTL give an optimum result. As we increase the thickness of ETL by keeping the HTL and perovskite layer constant, the efficiency and the performance start to reduce, same for HTL up to the certain thickness of HTL and we get a desired result in terms of efficiency and performance. Apart from that threshold value of HTL, if we try to increase more the thickness of HTL and keeping perovskite layer as a constant then efficiency decreases. So a particular range of thickness of ETL and HTL is very important to maintain the performance and reproducibility of perovskite solar cell. The thickness of perovskite layer is much essential among all the consideration we choose, because the total amount of photons generated depends on the absorption of photons in the perovskite layer. More the thickness of perovskite layer, more will be the photons efficiency and so more number of electron-hole pair generation takes place. Therefore, the thickness of perovskite layer should be more as compared to the thickness of ETL and HTL to get the desired result. The bad gap of perovskite layer also should be in a desired range because it affects more on the performance of solar cell. As we increase the band gap of perovskite layer, the proficiency of solar cell increases. But up to the particular band gap (i.e., 3.1 eV) it increases and again when we increase the band gap then efficiency starts to decrease. So the best band gap of perovskite layer where the result is more valuable is 1.5 or 1.55 eV. At this bad gap, the proficiency of solar cell is recorded as the best efficiency.

References

1. M.K. Assadi, S. Bakhoda, R. Saidur, H. Hanaei, Recent progress in perovskite solar cells. Renew. Sustain. Energy Rev. **81**, 2812–2822 (2018)
2. J.L. Barnett, V.L. Cherrette, C.J. Hutcherson, M.C. So, Effects of solution-based fabrication conditions on morphology of lead halide perovskite thin film solar cells. Adv. Mater. Sci. Eng. **2016**, (2016)
3. F. Bella, C. Gerbaldi, C. Barolo, M. Grätzel, Aqueous dye-sensitized solar cells. Chem. Soc. Rev. **44**(11), 3431–3473 (2015)
4. D. Bi, W. Tress, M.I. Dar, P. Gao, J. Luo, C. Renevier, K. Schenk, A. Abate, F. Giordano, J.P.C. Baena et al., Efficient luminescent solar cells based on tailored mixed-cation perovskites. Sci. Adv. **2**(1), e1501170 (2016)
5. M. Cai, Y. Wu, H. Chen, X. Yang, Y. Qiang, L. Han, Cost-performance analysis of perovskite solar modules. Adv. Sci. **4**(1), 1600269 (2017)
6. Q. Chen, H. Zhou, Z. Hong, S. Luo, H.S. Duan, H.H. Wang, Y. Liu, G. Li, Y. Yang, Planar heterojunction perovskite solar cells via vapor-assisted solution process. J. Am. Chem. Soc. **136**(2), 622–625 (2014)
7. D. Forgács, L. Gil-Escrig, D. Pérez-Del-Rey, C. Momblona, J. Werner, B. Niesen, C. Ballif, M. Sessolo, H.J. Bolink, Efficient monolithic perovskite/perovskite tandem solar cells. Adv. Energy Mater. **7**(8), 1602121 (2017)
8. N.J. Jeon, H. Na, E.H. Jung, T.Y. Yang, Y.G. Lee, G. Kim, H.W. Shin, S.I. Seok, J. Lee, J. Seo, A fluorene-terminated hole-transporting material for highly efficient and stable perovskite solar cells. Nat. Energy **3**(8), 682–689 (2018)
9. A. Kumar, S. Singh, Numerical modeling of lead-free perovskite solar cell using inorganic charge transport materials. Mater. Today Proc. (2020)
10. F. Liu, Q. Li, Z. Li, Hole-transporting materials for perovskite solar cells. Asian J. Organic Chem. **7**(11), 2182–2200 (2018)
11. R.C. MacKenzie, Gpvdm user manual (2016)
12. K. Mahmood, A. Khalid, M.T. Mehran, Nanostructured ZNO electron transporting materials for hysteresis-free perovskite solar cells. Solar Energy **173**, 496–503 (2018)
13. J.B. Orhan, R. Monnard, E. Vallat-Sauvain, L. Fesquet, D. Romang, X. Multone, J.F. Boucher, J. Steinhauser, D. Dominé, J.P. Cardoso et al., Nano-textured superstrates for thin film silicon solar cells: status and industrial challenges. Solar Energy Mater. Solar Cells **140**, 344–350 (2015)
14. R. Rajeswari, M. Mrinalini, S. Prasanthkumar, L. Giribabu, Emerging of inorganic hole transporting materials for perovskite solar cells. Chem. Record **17**(7), 681–699 (2017)
15. W.S. Yang, J.H. Noh, N.J. Jeon, Y.C. Kim, S. Ryu, J. Seo, S.I. Seok, High-performance photovoltaic perovskite layers fabricated through intramolecular exchange. Science **348**(6240), 1234–1237 (2015)

A Comparative Study of Renewable Energy Resources in Distribution Network

P. V. N. R. Varun Teja, Vinay Kumar Jadoun, and Anshul Agarwal

1 Introduction

Now-a-days usage of renewable energy is increased, and many countries are establishing renewable energy targets for their electricity offer, as solar and wind tend to be a lot varying and unsure than typical sources, meeting these targets can involve changes to installation designing and operations. Grid integration is one that follows developing economical ways in which to deliver variable renewable energy (RE) to the grid is faced by many difficulties while integrating to grid [1].

Grid integration is generally divided into three general categories: capacity expansion, production cost, and power flow studies. As the above mentioned categories, in this paper, power flow studies are chosen because it is used to analyze the network of distributed network while integrating the renewable energy and emerging RE in the distribution network that makes network more complex which changes from radial to mesh network. Investigation of distribution power flow has turned into a difficult undertaking because of its unpredictability. Due to fluctuation in the power output of renewable energy voltage, system stability and reliability are concentrated areas. Energy storage system (ESS) is used to minimize the fluctuations of renewable energy. The main cause of reliability in the distribution network is power interruption between customers and Distribution Company. In demand side management, the production of renewable generation of energy is also increasing. In recent years,

P. V. N. R. Varun Teja (✉) · V. K. Jadoun
Manipal Institute of Technology, MAHE, Manipal, Karnataka, India
e-mail: pvnrvarun@gmail.com

V. K. Jadoun
e-mail: vjadounmnit@gmail.com

A. Agarwal
National Institute of Technology, Delhi, India
e-mail: anshul@nitdelhi.ac.in

© The Author(s), under exclusive license to Springer Nature Singapore Pte Ltd. 2021   203
J. Kumar and P. Jena (eds.), *Recent Advances in Power Electronics
and Drives*, Lecture Notes in Electrical Engineering 707,
https://doi.org/10.1007/978-981-15-8586-9_19

consumers are interested in the production of solar energy. It can be seen that most of the commercial buildings and educational institution are installing the photovoltaic (PV) system, and also they supply power to grid. Here, the main point is the initialization cost of PV system. It can be identified that the three challenges to understand the cost of PV system at high penetration [2].

To increase the renewable energy especially for PV system, government has implemented certain policies [3]. For example, in Germany, PV generation is one of the pillars of energy transition and there are reginal impacts and effects of distribution [4]. In olden days, there is no distribution generation (DG). After introducing renewable energy in distribution network DGs started existing. In recent years, there is development in active distribution system, and these must see its technical and economic aspects also [5]. Expansion of renewable energy in distribution network made way for smart grid and microgrid. Reduction of the energy losses and voltage deviation in smart grid are explained in [6]. Energy storage system is used for backup, i.e., while integrating renewable energy to the grid and maintaining the same power. At the present situation, in power distribution network what is the problem facing by ESS explained in [7]. For reduction in system power losses and voltage profile, an antlion optimization algorithm is used [8].

A microgrid is a compressed power grid that can work uninhibitedly or operate along with other grids. A microgrid can work on islanded or non-islanded mode. There are more advantages using renewable energy as it can reduce the CO_2 emission and also it reduces the usage of fossil fuels. Research work on electrical vehicles is also advanced and Tesla is the first company to manufacture the electrical cars in 2008 and similarly electrical trains, electric buses, etc.; by using these products you can reduce air pollution and sustain a green environment.

This paper discusses about the various load flows methods, energy storage system, distributed generation, introduction to microgrid and demand side management.

2 Distribution Power Flow Analysis

Generation of power is distributed through transmission lines to demand side (consumer) in such a way that the network is stable and reliable. By the development of technologies, power generation is also started in demand side and is rapidly increasing in the past few years. Renewable energy came into existence as there is increasing power in demand side and the government is also giving subsidy for those who are installing the PV panels on rooftop of houses, commercial buildings, etc. Renewable energy is being integrated to the main grid. After integrating the renewable energy (RE) to the distribution network, complexity comes to picture when we understand the power flow. There are different load flow methods such as Newton Raphson method (NR), Forward/Backward analysis, Sequential Numbering, Fuzzy Arithmetic, Fast Decoupled Power flow, and Linear Data structure. In this section, various load flow methods are discussed in detail. Network theory was used to solve the distribution network (DN), and to solve the load flow methods simple mathematical expressions have been used. Active and reactive power variables are used to solve the meshed transmission to reduce the computation burden. The decoupled method was introduced to fast convergence. It is better to avoid the NR method in modern DN. To illuminate the power flow legitimately, another calculation was suggested using only the bus branch details for estimation, repetitive decomposition (Lower–Upper) of the LU; it does not required for the Jacobin matrix substitution. In power flow, if you are considering the transformers, peculiarity issue emerges to defeats, and new Forward/Backward sweep technique was presented. This technique was improved using Kirchhoff Voltage law (KVL) for forward and Kirchhoff Current law (KCL) for backward method. New software was introduced for power flow analysis during both distribution and transmission. To avoid the sequential numbering, a new method was proposed, and it required initial node feeder of lateral. Mat power was proposed for steady-state operation. With the creation of an informative controller sector, another power flow calculation often uses fuzzy arithmetic and it was acquainted with unraveling the load flow issue. It neglects the sequential numbering and flat voltage at node. Space difficulty and time was further reduced by introducing new load flow based on linear data structure (LDS). Complex normalization was used to propose a new technique for better performance of the fast-decoupled method [9]. This technique cannot use when Forward/Backward sweep (FBS) method was used. A new technique was introduced by the combination of NR and FBS method. To avoid the voltage variation and power loss in distributed network a decision method was introduced where less losses and stable voltage are gained. Among all the load flow methods, FBS approach is ideally suited for analyzing load flow.

3 Distributed Generation

Generally, generation of electricity is produced by longer distances and it reaches to consumers via transmission and distribution. The conventional energy sources are centralized power like coal-based thermal power plant, hydro power plant, etc. By using the centralized method, huge investment cost, transmission losses, environmental impacts, etc., are involved. So, to avoid this problem, distributed generations are adopted. DGs can be described as a source of electricity directly linked to the distribution network or on the customer side. Figure 1 shows the centralized and decentralized (DG) system. Distributed generation is classified into three types: 1. Non-renewable energy DG source, 2. Renewable energy DG source, 3. Energy storage DG source [10]. In Distributed generation, difficulties are faced in technical and cost analysis. Optimum integration and preparation of DG into the DN can be accomplished. Improper planning of the distribution network affects the power quality and reliability of the network. Conventional technique and metaheuristic algorithms have been applied for Optimum DG preparation.

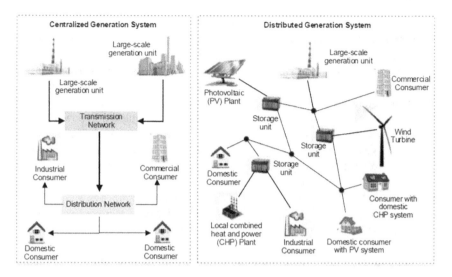

Fig. 1 Centralized and decentralized distribution network [12]

3.1 Conventional Technique

Conventional methods are given below:

Analytical technique

Analytical method conducts a computational energy DN study, and this results in a series of mathematical equations which can be used to formulate an objective function. In this method, it is easy to implement and guarantee DG planning solution convergence. These methods are adopted to improve the voltage profile and decrease the power loss. The author [11] suggested the analytical technique for optimal planning of distributed generation (OPDG), and he derived at some mathematical equations to determine the power loss minimization. Objectives for power loss minimization are power factor, optimal size, and location.

Exhaustive analysis

In this method, it has a single objective, i.e., to reduce the loss in the DG capability and placement in the network for the complete solution relative to space. A comprehensive assessment is computationally effective if the location of a single DG unit for a particular load generation situation is considered. It is a comprehensive assessment that was carried out using a multi-objective index. Multi-objective index is used to reduce the active and reactive power loss and voltage variation. The authors [12] used the multi-objective index to perform the exhaustive analysis to assess the minimization of power losses, active and reactive losses.

3.2 Metaheuristic Algorithms

These methods are mainly based on iteration, which analyzes applicant solutions by combining different ideas to regulate a subordinate heuristic approach. Some of the different algorithms are

Genetic Algorithm (GA)

This algorithm is a search algorithm based on the natural and genetic ideas. To improve the voltage stability, reduce the line losses and voltage variations, a new genetic algorithm is introduced in [12, 13] for the optimal DG. But here GA has the inconveniences of potential convergence of premature solutions and the numerical inefficiency associated with the repeated objective calculation.

Particle swarm optimization (PSO)

Particle swarm optimization [11] is driven primarily by bird flocks' social behavior. To minimize the power loss and to increase the voltages in distributed network are the objectives of the OPDG using the particle swarm method. Compared to genetic algorithm, it reduces the iterations. But the particle swarm method can rapidly merge and be trapped in local optima. Similarly, remaining methods [12] also have drawbacks and gets trapped by the local optima. Among all these methods, analytical method is suitable for the optimal planning for the distributed generation and it can achieve a better performance of power quality, security, improve voltage stability and reliability of the system. By adopting this method, it can achieve the objectives such as reduce the power losses, increase the voltage profile, and increase the cost savings of OPDG. To improve the power quality, multi-function distributed generation plays an important role in distributed network [13].

4 Energy Storage System

In the recent years, CO_2 emission has increased, so in order to reduce the CO_2 consumption, the world is looking at renewable energy and electrical vehicles. While integrating the renewable energy to distributed network, power fluctuations occur. So to avoid these fluctuations in distribution network, energy storage system introduces different technologies. These are battery energy storage system, pumped hydraulic storage, ultra-capacity energy storage system, flywheel energy storage system, super conducting magnetic energy storage, and compressed energy storage [6]. Storage system is classified into two types: 1. Storage medium and 2. Storage duration. Storage medium means which type of energy is stored like chemical, mechanical, and electric storage system [14]. Storage duration means discharge duration and it is also divided into two types: long-term and short-term technologies. Long term technologies mean huge storage system, its stores more amount of electricity. Short term means duration and capacity is less compared to long term and is suitable for improvement in power quality, frequency regulation. It has low energy density. Flywheel, super-conducting magnetic energy storage, and ultracapacitor come under short-term technologies [15]. For optimal size and placement (OSP) of the ESS we have different methods as shown below:

4.1 Analytical Method

In Analytical method, OSP of ESS is derived through the mathematical equations, algorithms, derivations, etc. Objective functions and predefined system constraints are repeatedly evaluated during the optimization process with a different set of parameters and the set of parameters including the position and efficiency of the ESS with sufficient output is selected as the optimal solution. The key goal of optimization after deployment of the Energy Storage Network was to optimize benefit. Determining the optimum capacity of ESS is by decreasing the investment and operating costs of BESS as well as the operating costs of microgrid. Microgrid are of two types, first is islanded mode and second is grid connected mode among them islanded has more flexibility in microgrid with increase in the large battery energy storage system capacity [15].

4.2 Mathematical Optimization

This method is based on the numerical methods to find the optimal solutions. The advantage of mathematical optimization is that an optimal solution is achieved. The allocation issue is modeled and solved for implementation in the form of numerical expression in ESS allocation. When linear functions are used by the mathematical representation, it is known as a LP model (linear programming) [15]. ESS sizing was developed using Mixed Integer Linear Programming, while in the optimization process a Monte Carlo simulation was used to compensate for random uncertainties and to assess the reliability of Microgrid.

4.3 Artificial Intelligence

Artificial intelligence (AI) does not require mathematical equations, numerical methods like analytical method, and mathematical model for OSP of ESS. The Artificial intelligence approach is to find a solution space. It replicates the behavior and process that are naturally present. AI does not have the optimal solution, but it satisfies the AI algorithm-based solution. The AI process is fast and less time consumption compared to the remaining methods. Finally, analytical and mathematical optimization requires more time to get an optimal solution, and in case of artificial intelligence it will give optimal solution with less complexity level. So, Artificial intelligence method is best compared to remaining method for optimization of placement and sizing of ESS. In ESS, hybrid meta-heuristic optimization approaches have been explained in [14].

5   Microgrid

Microgrid control innovations empower dynamic power of the board inside little frameworks with a different generation, and also load the executive's capacities. Microgrids are normally structured as frameworks with static limits and well-characterized grid connection point. System topology is frequently overlooked in microgrids since its electrical hugeness is constrained because of the closeness of resources and static nature. Microgrid can operate on islanded mode and it reduces the burden on main grid to protect the microgrid. Overcurrent relays are used, where overcurrent relays are unable to work, a new relay are used, i.e., signal processing technique. Contrary to this, renewable-based DG's elevated penetration into the present power systems can create fresh problems. The power flow from the substation to the customer is one-way. The major difficulties are growing Intolerance in the daily net load profiles (called the duck curve), extreme excess supply from DGs, and two-way energy flows. Overvoltage, line-overload problem is faced in the DN due to bidirectional flow of power; also system losses will increase. It may affect the adjacent buses [16]. Managing distribution and energy management systems are key enablers of microgrids that can play a significant role in reducing the adverse effects of DG penetration. High-resolution solar irradiance information was simulated using a virtual test bed and an LV network model taking into account the cloudiness and residential consumption profiles reflecting a day of each season. The severity of critical working circumstances was explored for daily operation in terms of voltage variation and line losses. Daily assessment is contrasted with the extreme case analysis. Similarly, in daily assessment, active power losses are much lower than in extreme cases. Voltage rises were found to be lower than extreme case assessments, but with longer periods of time. Likewise, power losses were smaller than the assessment of the worst scenario. The data observed in this article will contribute to device sizing research which is used in actual systems by showing the important instances which are more likely to occur during daily operation [16]. In latest years, huge development has done, and advance technology has been used. Researches are made more on interesting area like renewable energy which are integrated with microgrid. By this development, there is a chance that a greater number of microgrids are integrated with renewable energy and it rises a new problem, that is, fluctuations and uncertainty in the system. There are different methods to solve this problem, they are active distribution and virtual power plant methods. This method is able to manage and operate the multiple microgrids. Drawbacks of these methods are that they cannot fully

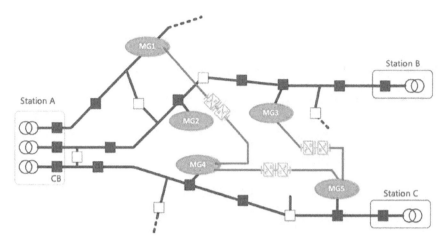

Fig. 2 Multiple microgrids are interconnected to the flexible power electronic device [17]

operate for different types of microgrid which is linked to the distribution network. So, another method can better operate the multiple grids with the help of power electronic device. Flexible interconnection device [17] helps to interconnect between the two different microgrid transformers. By using multiple microgrids, it reduces the installation capacity of energy storage system, power losses, and improves the efficiency of the system. It can improve the reliability of the system if multiple microgrids are planned. Figure 2 shows the interconnection of multiple microgrids with flexible interconnection device. Figure 2, shows three different transformers, stations, and 5 microgrids. It can be seen that different microgrids are connected to three flexible interconnection devices (FIDs), as shown in Fig. 2. MG1 is connected to MG4 through an FID and because of these they can easily support by each other. There are many advantages using this method [17].

5.1 Reliability

In distributed network, reliability also plays a major role for stability of the system. To improve the reliability in distributed network integrated with renewable energy, a distributed generation capacity is used to supply the power. Where the power is interrupted, energy storage system is also used for backup while fluctuations occur in the operation. The reliability of a consumer's power supply can be assessed by the effective interruptions on consumers and distributors under network failure circumstances. Alternate feeders for restoring supply, mesh grid, maintenance, etc., are the old ways of enhancing network reliability. ESS helps to minimize the fluctuation in distribution network while integrating with renewable energy and expanding the

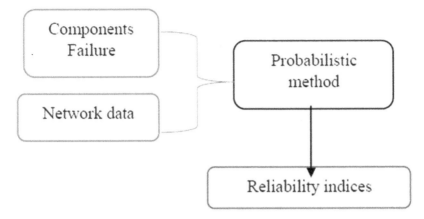

Fig. 3 I/O in traditional techniques of reliability [18]

generation. There are various techniques available for determining the reliability of the different network technologies.

In conventional distribution network, there are two types of reliability techniques, i.e., Analytical method and Monte Carlo simulation. Monte Carlo simulates the network component samples that fail to calculate the probability distribution of reliability. To calculate the average value of reliability, mathematical expression is used in Analytical method. These methods are used based on requirement analysis. In modern distribution, Markov model is used for distributed generation for reliability. Figure 3 shows a summary of the required I/O (input/output) data measured using conventional distribution network reliability assessment techniques.

5.2 Islanded Operation

Without the presence of electrical grid but the continuous supply of power to a location with the help of renewable energy is known as islanded mode. In recent years, renewable energy usage increased. These can improve the reliability in islanded mode with perfect planning. Generally, Monte Carlo simulation is used to assess the reliability of dispatchable and non-dispatchable distributed generation. Simulation takes more time to assess the fault and reliability of the system, so an alternative method can be chosen, i.e., analytical method because it does not consider the fault occurrence. For the wind DG, three states of Markov model is used for the average load of islanded location during fault condition. The three states are (a) Up Markov model (b) Down Markov model and (c) Derated Markov model. To avoid fluctuations in renewable energy, Markov models are also an alternative solution.

5.3   Grid Connected

DG enhances the efficiency of distribution systems in grid-connected working mode by decreasing the loading of machinery and allowing load transfers from neighboring feeders experiencing failures. In the presence of DG, the transfer capacity between feeders must be quantified in order to estimate the reliability impact. Analytical method is used to calculate the transfer capacity by a set of load concentrations in the network (low, medium, and high), and the analytical method is then incorporated into the Monte Carlo simulation to decrease computational time. The joint probability technique does not include time-dependent load and generation fluctuations during the outage. A case study was also present in [19], among all the methods Analytical method is suitable to improve the reliability of the system.

6   Demand Side Management

Generally, demand side means the consumer side. In previous years, the demand side management (DSM) helps to achieve the reduction in load and to reduce the usage of electricity; the government has also introduced many policies, schemes, awareness programs, etc. Mostly, the electricity is generated from coal or fossil fuels. As we know that fossil fuels are available to some extent only, the usage of the energy production is increased day by day. So, for alternatives, scientists look over to renewable energyies like solar, wind, biomass, hydro, etc., which are available free of cost and these will not be runout in future too. The government also encourages the people by providing policies, subsidies, etc., for conservation of energy. In recent years, usage of renewable energy is also increasing by the usage of solar water heaters, installation of PV, solar streetlight, solar cookers, etc.

DSM affords the energy system with improved reliability by decreasing general demand through energy efficiency and lowering peak demand through dispatchable programs. It also decreases the price of transmission and distribution from a supply-side resource. Energy Efficiency–the focus is on decreasing general power usage and high demand over several years as well. Peak Load Management–the focus is on continuously decreasing peak demand over a season. Demand Response–the focus is on reducing peak demand for a few days during the year for short periods of time. As Demand Side Management measures have not only helped utilities to decrease peak requirements for electricity, but also to reduce the huge investment in networks of generation, transmission, and distribution. Active distribution management is also known as the DSM with adequate coordination of DGs, voltage regulators, shunt capacitors, and other devices in the network enhance high safety and improve efficiency. In active distribution, there are different uncertainty techniques that are classified as probabilistic technique, Stochastic optimization, robust, possibilistic technique, hybrid probabilistic-possibilistic technique. Among these methods, one

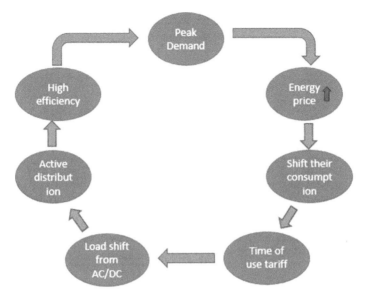

Fig. 4 Cycle of DSM—[21]

method is adopted after detailed analysis of each method [20]. For recent development in active distribution, we have to look into its economic and technical operation [4]. To reduce the investment and reliability cost, maximize the electrical vehicles charging service capability in joint planning with active distribution network. There are two models to solve the abovementioned problems; hence, a detailed analysis of each method is given in [21]. Figure 4 shows the cycle of demand side management (DSM), when peak demand increases energy price increases then consumers shift their consumption period to off peak time and this introduces the new time of use tariff (TOU). This method improves the efficiency of the system and reduction in peak demand.

7 Conclusion

An extensive literature survey has been conducted on grid integration problems and the techniques used to mitigate them currently. The increase in the complexity of the main flow network after grid integration is the major problem. A feasible technique to mitigate this problem is forward/backward sweep analysis. Proper placement of the distributed generation, energy storage system, fluctuations and system uncertainties are induced due to the addition of more Microgrids. These are the several factors to be noted while solving the grid integration problems. The methods existing in the literature to solve such problems can be classified as Conventional and Metaheuristic techniques. However, the trapping of local by the global optima while using the

Metaheuristic problems make them inferior to the Conventional techniques. These Metaheuristic methods for load flow can be further investigated for obtaining more optimized solutions.

References

1. S.R. Simon Sinsel, R.L. Riemke, V.H. Hoffmann, Challenges and solution technologies for the integration of variable renewable energy sources—a review. Renew. Energy, Elsevier **145**, 2271–2285 (2020)
2. A.W. Horowitz Kelsey, B. Palmintier, B. Mather, P. Denholm, Distribution system costs associated with the deployment of photovoltaic systems. Renew. Sustain. Energy Rev. Elsevier **90**, 420–433 (2018)
3. P. Neetzow, R. Mendelevitch, S. Siddiqui, Modeling coordination between renewables and grid: policies to mitigate distribution grid constraints using residential PV-battery systems. Energy Policy, Elsevier **132**, 1017–1033 (2019)
4. J. Többen, Analysis regional net impacts and social distribution effects of promoting renewable energies in Germany. Ecol. Econ., Elsevier **135**, 195–208 (2017)
5. M. Jabbari Ghadi, S. Ghavidel, A. Rajabi, A. Azizivahed, L. Li, J. Zhang, A review on economic and technical operation of active distribution systems. Renew. Sustain. Energy Rev., Elsevier **104**, 38–53 (2019)
6. H. Fallahzadeh-Abarghouei, S. Hasanvand, A. Nikoobakht, M. Doostizadeh, Decentralized and hierarchical voltage management of renewable energy resources in distribution smart grid. Electr. Power Energy Syst., Elsevier **100**, 117–128 (2018)
7. H. Sabooria, R. Hemmati, S.M. Sadegh Ghiasi, S. Dehghan, Energy storage planning in electric power distribution networks—a state of-the-art review. Renew. Sustain. Energy Rev., Elsevier **79**, 1108–1121 (2017)
8. P.D. Prasad Reddy, V.C. Veera Reddy, T. Gowri Manohar, Ant Lion optimization algorithm for optimal sizing of renewable energy resources for loss reduction in distribution systems. J. Electr. Syst. Inf. Technol., Elsevier **5**, 663–680 (2018)
9. B. Murugananthan, R. Gnanadass, N. P. Padhy, Challenges with renewable energy sources and storage in practical distribution systems. Renew. Sustain. Energy Rev., Elsevier **73**, 125–134 (2017)
10. H.A. Zubo Rana, G. Mokryani, R. Haile-Selassie, A. Jamshid, T. Niknam, P. Pillai, Operation and planning of distribution networks with integration of renewable distributed generators considering uncertainties: a review. Renew. Sustain. Energy Rev., Elsevier. **72**, 1177–1198 (2017)
11. H. HassanzadehFard, A. Jalilian, Optimal sizing and location of renewable energy-based DG units in distribution systems considering load growth. Electr. Power Energy Syst., Elsevier **101**, 356–370 (2018)
12. A. Ehsan, Q. Yang, Optimal integration and planning of renewable distributed generation in the power distribution networks: a review of analytical techniques. Appl. Energy, Elsevier **210**, 44–59 (2018)
13. Y. Naderi, S.H. Hosseini, S.G. Zadeh, B. Mohammadi-Ivatloo, J.C. Vasquez, J.M. Guerrero, An overview of power quality enhancement techniques applied to distributed generation in electrical distribution networks. Renew. Sustain. Energy Rev., Elsevier **93**, 201–214 (2018)
14. K.C. Das, O. Bass, G. Kothapalli, S. Mahmoud Thair, D. Habibi, Overview of energy storage systems in distribution networks: placement, sizing, operation, and power quality. Renew. Sustain. Energy Rev., Elsevier **91**, 1205–1230 (2018)
15. L.A. Wong, V.K. Ramachandaramurthy, P. Taylor, J.B. Ekanayake, S.L. Walker, S. Padmanaban, Review on the optimal placement, sizing and control of an energy storage. J. Energy Storage, Elsevier **21**, 489–504

16. M.A. Zehir, A. Batman, M.A. Sonmez, A. Font, D. Tsiamitros, D. Stimoniaris, T. Kollatou, M. Bagriyanik, A. Ozdemir, E. Dialynas, Impacts of microgrids with renewables on secondary distribution networks. Appl. Energy, Elsevier **201**, 308–319 (2017)
17. Y. Yang, W. Pei, Q. Huo, J. Sun, F. Xu, Coordinated planning method of multiple micro-grids and distribution network with flexible interconnection. Appl. Energy, Elsevier **228**, 2361–2374 (2018)
18. A. Escalera, B. Hayes, M. Prodanovi, A survey of reliability assessment techniques for modern distribution Networks. Renew. Sustain. Energy Rev., Elsevier **91**, 344–357 (2018)
19. J. Többen, Analysis regional net impacts and social distribution effects of promoting renewable energies in Germany. Ecol. Econ., Elsevier **135**, 195–208
20. A. Ehsana, Q. Yang, State-of-the-art techniques for modelling of uncertainties in active distribution network planning: a review. Appl. Energy, Elsevier **239**, 1509–1523 (2019)
21. S. Wang, F. Luo, Z.Y. Dong, G. Ranzi, Joint planning of active distribution networks considering renewable power uncertainty. Electri. Power Energy Syst. Elsevier **110**, 696–704 (2019)

Modelling and Control of Power Flow from Auxiliary Source in Hybrid Electric Vehicle

Kanchan Yadav and Sanjay Kumar Maurya

1 Introduction

HEV consist mainly of the electric motor, the energy storage and the controller to control the power flow. The power which is supplied to the motor is controlled by the controller, and thus indirectly it controls the forward and reverse power flow in the vehicle. The demand for these bidirectional controllers is raised as using alternative or auxiliary source for supply in HEV is gaining popularity [1]. For HEV an auxiliary source along with the main source is used for the energy storage and provides proper power management in the vehicle and thus indirectly improves the efficiency.

The schematic diagram shows the flow of energy in the system in Fig. 1. The power is distributed between the engine and the battery. To improve the energy efficiency of the vehicle, the energy recovery system plays an important role. When the regenerative braking takes place the mechanical energy which is recovered is stored in the auxiliary source [2]. When the acceleration/deceleration or the braking conditions exist, the electric motor will draw high current [3]. During acceleration the sensor voltage refers to the accelerate pedal and power is supplied with both the main, as well as auxiliary source. During breaking the sensor voltage refers to the brake pedal and with the help of converter the energy flows into the auxiliary source and thus restored.

K. Yadav · S. K. Maurya (✉)
Department of Electrical Engineering, GLA University, Mathura, India
e-mail: sanjay.maurya@gla.ac.in

K. Yadav
e-mail: kanchan.yadav_mt18@gla.ac.in

© The Author(s), under exclusive license to Springer Nature Singapore Pte Ltd. 2021   217
J. Kumar and P. Jena (eds.), *Recent Advances in Power Electronics and Drives*, Lecture Notes in Electrical Engineering 707,
https://doi.org/10.1007/978-981-15-8586-9_20

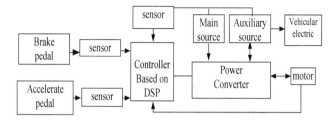

Fig. 1 Schematic diagram of energy recovery system

1.1 Power Flow in HEV

The power flow process is generally different for different configurations. Here the parallel configuration is taken and the schematic diagram in Fig. 2, will show the power flow path for the vehicle. For this parallel HEV the propulsion power will be supplied by the controller which converter the electrical energy to the required mechanical energy. For this configuration the power is supplied whether by the ICE or the electric motor or both can be used to supply the propulsion power [1]. When braking occurs, the energy gained is converted into electrical energy with the help of electric motor working as a generator and then this energy is stored in the energy storage system (ESS).

ESS is the very vital part as it directly has the effect on the efficiency of the vehicle. The batteries used as ESS must have high energy density, long cycle life; low internal resistance, etc. For the charging purpose battery must have high energy density. The hybrid containing battery and ultra-capacitor is going to be the attractive option for the vehicles due to efficient accumulation of charge carrying capacity. Other than hybrid storage, the batteries of Li-ion chemistries are widely coming as the good option to use in the HEV, as well as for pure EV [4]. Li-ion batteries come as a prompting solution as they have high energy, as well as power density. Thus, their modelling also becomes very important in order to predict the batteries *SOC* (state of charge) and SOH (state of health).

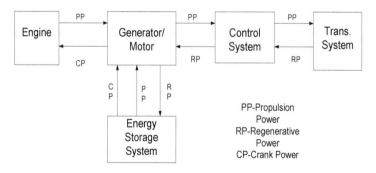

Fig. 2 The schematic diagram of power flow in HEV

For the actual driving condition that is start-stop, regenerative braking and the shifting of power flow, i.e. operating range of main and auxiliary source, the transmission system of HEV should be able to perform these functions [5].

The traction motors have to perform two different functions that are

- Providing constant torque for the normal mode having rated speed, and decreased torque in the extended mode for the above rated speed range.
- To capture the energy that is generated during regenerative braking, i.e. they can operate as a generator to convert this mechanical energy into electrical energy.

The driving in the urban condition encounters frequent stop and start of the vehicle. So to improve the fuel economy, control strategies used, and the regenerative braking system both play a very crucial role. During cruising, the electric drive used as a generator and charges the battery, while on climbing the grade, both the ICE and the motor will supply the energy [6]. Whenever brake pedal executes the control, signal will switch off the ICE power and the negative torque generated by the motor, thus making it operate as a generator thus utilizing the braking energy to recharge the battery.

1.2 Bidirectional Converter

In an electric vehicle, the power electronic circuits play an important role which includes DC-AC inverters and DC-DC converters. For supplying high- power electric motor DC-AC inverter is used and for supplying low power, low voltage loads DC-DC converters are used [7]. The advanced topologies of bidirectional high-power DC-DC converters have been developed for being applied in hybrid vehicles. The paper considers a model consisting of a non-isolated bidirectional dc-dc converter of half-bridge topology. The bidirectional dc-dc converter is used for battery charging, regenerative braking, and backup power requirements [3] (Fig. 3).

In buck mode, as it is the step down condition, the low voltage side will get the energy, thus power supplied to battery. For this case, the motor acts as generator and supply the energy generated during regenerative braking to the battery. During boost

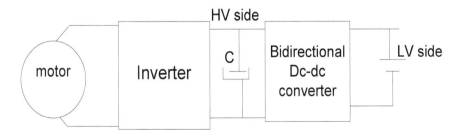

Fig. 3 Block diagram of HEV with bidirectional converter

mode, as the demand will be stepped up so, the power will be supplied from the battery. Thus in this mode, the power is supplied by both the main source and the auxiliary source [3].

The power flow from low voltage end that is from battery to high voltage end and thus it is referred as boost operation. When regenerative braking occurs the power flows in reverse direction that is to the low voltage side and thus used to recharge the battery and this mode referred as buck mode [8]. The bidirectional dc-dc converters constituting energy storage systems becomes very prompting for power related systems, thus it improves efficiency, as well as performance of the system. A drive cycle constitutes of acceleration, braking, and constant speed operation of a vehicle. In the acceleration, power demand increase while in braking power demands goes to zero, and the kinetic energy of vehicle can be stored in the battery. The bidirectional converter allows the power flow in both directions, and hence it is used to manage the power flow from a battery in the charging and discharging process. Here the converter will manage the DC constant voltage and the charging/discharging process will change the constant voltage of battery.

The power supplied during various operations are

(a) Only ICE—the power supplied by the main source, i.e. ICE only.
(b) ICE + EM—When energy requirement is more (during acceleration or at high speed) the motor starts supplying power parallelly with the ICE.
(c) ICE + Battery charging—when power requirement is less; the excess energy available is used to charge the batteries. It occurs during regenerative braking or cruising (when the propulsion power of ICE is higher than required).

The non-isolated bidirectional dc-dc converters used for charging/discharging the batteries are simple in structure, highly efficient, cost-effective and highly reliable. To maximize the efficiency of charging and discharging of the battery, voltage is varied by controlling the converter output which works in buck and boost mode [9].

The bidirectional converter used will transfer the energy to the battery when required and also recovers the energy during regenerative braking and makes the battery to charge. A bidirectional power converter is needed to regulate the charging and discharging, as well as power flow in the battery.

Here in this paper, a battery is used as energy storage with bidirectional dc-dc converter so that it can regulate charging and discharging mode. The converter control is done with the use of PID controller. When additional power is required, the energy storage will supply the load and during the regenerative braking the energy produced is stored back to the battery. The converter has two modes that include buck and boost.

2 Battery Characterization and Modelling

The Li-ion battery is used for the simulation. For the safe charging/discharging process the accurate *SOC* (state of charge) estimation is necessary. The correct *SOC* estimation shows the remaining driving range and the power requirement. Important parameters discussed for the ESS [4] are

- Battery Pack Capacity (Amp-hr)
- Peak Charge/Discharge Power (KW)
- Continuous Charge/Discharge Power (KW)
- Max. Operating Voltage (V)
- Min. Operating Voltage (V)
- Max. Self-Discharge Rate (Whr/day)
- Allowable Operating Range of *SOC* (%)
- Battery Cycle Life
- Estimation Accuracy of *SOC*.

2.1 Battery Modelling

Here a battery model is presented to meet all requirements having non-linear characteristics during charging/discharging, as well as their dependence on state of charge of battery. The elements for characterization are self-discharge resistance (Rs) which is function of open circuit voltage, charge resistance (Rc) and discharge resistance (Rd) which are different for charging/discharging. Due to battery internal chemistry, overcharge resistance (Rco) and over discharge resistance (Rdo) increases significantly. Battery capacity (Ah) is modelled varying with *SOC* [10] (Fig. 4).

From the given expressions the charging/discharging characteristics of LI-ion battery can be deduced [11].

$$V_{ch} = E_0 - K\frac{Q}{Q^1 + 0.1Q}i^* - K\frac{Q}{Q^1 - Q}Q^1 + A.e^{-BQ^1} - Ri$$

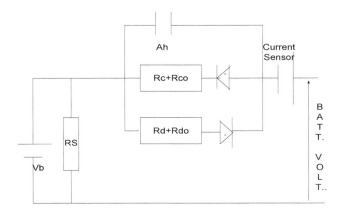

Fig. 4 Battery model with charging/discharging resistance

$$V_{disch} = E_0 - K\frac{Q}{Q-Q^1}i^* - K\frac{Q}{Q-Q^1}Q^1 + A.e^{-BQ^1} - Ri \qquad (1)$$

where, Vch = charging battery voltage (V); $Vdisch$ = discharging battery voltage (V); E_0 = constant voltage (V); i = current (A); K = polarization constant (Ah − 1), R = Internal resistance (Ω); Q^1 = extracted capacity (Ah); Q = maximum capacity (Ah); A = exponential zone voltage (V); B = exponential zone time constant (Ah) − 1, i^* = low frequency current dynamics (A).

Battery *SOC* is the capacity of charge that battery is able to supply. When battery is fully charged the battery *SOC* is 1 and when completely empty *SOC* is 0. For optimum operation the battery *SOC* will have to lie between 20 and 80% [12].

SOC of the battery is given by the equation-

$$SOC = 100\{1 - \frac{1}{Q}\int_0^t i(t)\mathrm{d}t\} \qquad (2)$$

where,
$i(t)$ is the current flowing through the battery.

Efficiency of Battery System- For a complete round trip of charging/discharging cycle, the efficiency is defined as the ratio defined as energy available to the energy supplied and during the cycle it is subjected to the charge balance, i.e. capacity in Amp-Hr [13]. In other words, we can say that the state of charge in the battery should be the same during the start and at the end of the cycle. Thus, the battery system efficiency can be calculated as

$$\eta_{bat}.(T, SOC) = \frac{\int_{ti}^{tf} V_t(t)I_{disch}(t)\mathrm{d}t}{\int_{ti}^{tf} V_t(t)I_{ch}(t)\mathrm{d}t} \times 100\% \qquad (3)$$

where ηbat is the battery system efficiency at the given temperature and *SOC*, ti is the start time, tf is the end time, $V_t(t)$ is cell terminal voltage, $I_{ch}(t)$ is the charging current, and $I_{disch}(t)$ is the discharging current.

Assumptions

- The battery *SOC* cannot go below 20% for the safety factor.
- For the charging and discharging some specific period of time is taken
- The daily mileage of vehicle would be logarithmic normal distribution
- The battery charging process must be continued up to the level when *SOC* reached to 90%
- Uniform distribution is taken for the staring activity of charging and discharging.

2.2 *Li-Ion Battery Parameters and Characteristics*

For any mid-size vehicle, taking a drive cycle of few miles, the battery capacity in KWh must be decided. As the battery charging is done in two modes namely charge depleting and charge sustaining. During CD mode the power is given by the battery, and once the battery level is below specified, it comes to the charge sustaining mode. In CS mode the vehicle will get power from both the ICE and the battery and the power distribution depends on the energy management system. During CS mode, the battery supplies power for short durations like climbing the grade, etc. and during braking it captures the braking energy. The CS mode occurs in a Li-ion battery when its *SOC* level reaches in range of 20–30%. The major limitation in Li-ion battery is the safety concern during overcharging of the battery. If a cell gets charged more than specified, the active material reduces and comes to the unstable point. Thus, a special electric circuit is very necessary to provide the protection of the battery from getting overcharged or over discharged.

The equations used for battery charging/discharging power will be the function of *SOC* and is given as [14]

$$Pcb(k) = -\frac{\eta_{bats}\mu_{c}^2\text{max} - \mu_{oc}(k)\mu_{c}\text{max}}{ri}\eta_{batp}$$

$$Pdb(k) = -\frac{\eta_{bats}\mu_{b}(k)^2 + \mu_{oc}(k)\mu_{c}\text{min}}{ri}\eta_{batp} \tag{4}$$

where,

μ_b is the battery voltage, μ_{oc} is the battery's' open circuit voltage, r_i is the batteries internal resistance, η_{batp} is the number of parallel cells, η_{bats} is the number of series cells, k is the time constant. The total power of the battery will be the sum of *Pcb* and *Pdb* and is given as $P*_{bat}$. The converter efficiency is also linked with the battery power and the losses given as δ_{bat}, thus total battery power is given as-

$$P_{bat}(k) = \delta_{bat}.P \times_{bat}(k) \tag{5}$$

3 Simulation Result and Discussion

The vehicle is powered with two sources ICE and battery. Battery power is treated as auxiliary source. So, whenever the vehicle requires additional power at the time of acceleration, the auxiliary source feeds that additional power. The model of HEV is simulated for the Li-ion battery, which is used as auxiliary energy storage. The battery model used is rin (internal resistance), this Simulink model consists voltage source and internal resistance to model the battery. For wheel power, constant rolling

Fig. 5 Battery resistance:
6 Ah saft lithium ion battery

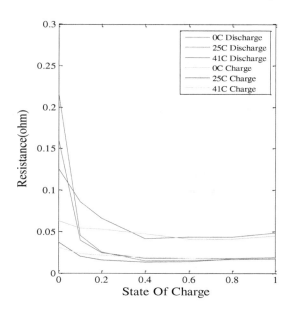

Table 1 Vehicle data for
simulation

Vehicle mass	1350 kg
Max power of ICE	40 KW
Max. battery operating voltage (during charging)	6 V
Min. battery operating voltage (during discharge)	9 V
Modular mass of one ESS	11 kg
Number of modules (strings in series)	5

resistance coefficient is taken and the constant power load accessory model is taken (Fig. 5).

The simulation is performed on ADVISOR with the vehicle data shown in Table 1:

Internal resistance as a function of state of charge-Here the internal resistance of battery varies with the state of charge of battery. It shows the internal resistance of the battery when empty, i.e. at 0 °C, during charging at 25 °C and during maximized charging at 41 °C. The figure shows that the resistance value is higher when *SOC* value is low.

During discharge the value of internal resistance is decreased and reached the lowest point at half charge. There is higher reading of resistance immediately after full charge and discharge. The value of internal resistance of Li-ion battery is mostly flat during empty to full charge value. The battery power decreases when *SOC* is form 0 to 70% and the largest change occur between 0 and 30% of *SOC* (Fig. 6).

Fig. 6 Battery open circuit voltage: 6 Ah saft lithium ion battery

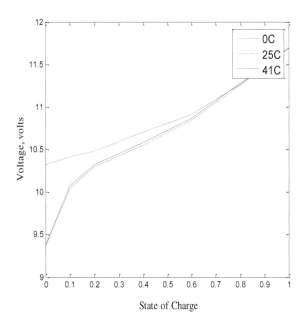

For estimating battery state, the *SOC* and *OCV* relationship determination is necessary. The figure shows the relationship that the *OCV* of the battery is proportional to the *SOC* of the battery for different charging states. Li-ion battery does not have a linear relationship between *OCV* and *SOC* (Fig. 7).

The figure showing instantaneous power and *SOC* curve. *SOF* denotes state of function which gives the relation between state of charge and state of health parameters of the battery, and the output power degradation is expressed as

$$SOF(t) = \frac{P(t) - P_d}{P_{\max} - P_d} \tag{6}$$

$$P(t) = P_{\max}.SOC(t).SOH(t) \tag{7}$$

where

$P(t)$—Instantaneous power supplied by battery P_d-Demanded power $P_{\max}$-The maximum power supplied by the fully charged battery at its maximum capacity.

4 Conclusion

The half-bridge bidirectional dc-dc converter used for controlling power flow in the HEV is simulated and the battery model for the charging/discharging resistance is analyzed. During regenerative braking, the energy is stored and the reverse power

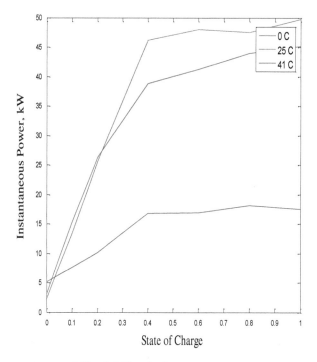

Fig. 7 Instantaneous power 6 Ah saft lithium ion battery

flow is allowed by the converter, thus the motor works as generator and supply the generated power to the battery. This mode of energy transfer needs the converter to operate in buck mode. During the sudden power demand, the propulsion power is supplied by main source and assisted by auxiliary source, thus in this case battery is used to supply the demanded extra power. This mode of transfer is possible when converter operates in boost mode. Further the charging/discharging model for Li-ion battery is simulated and different characteristics are analyzed through the waveforms occurred. The two modes of operation are motoring and generating. When battery is providing power, motoring mode is needed and thus the state of charge decreases. When the energy is supplied to the battery, generating mode is needed and the state of charge increases.

References

1. M. Naresh, S. Member, V.S. Bai, A.K. Pandey, Flow for plug-in hybrid electric vehicle, in *Innovations in Power and Advanced Computing Technologies (i-PACT)* (2017), pp. 1–5
2. O. Kolawole, I. Al-Anbagi, The impact of EV battery cycle life on charge-discharge optimization in a V2G environment, in *2018 IEEE Power and& Energy Society Innovative Smart Grid Technologies Conference (ISGT)* (2018), pp. 1–5. https://doi.org/10.1109/ISGT.2018.8403393

3. J.A. Mane, A.M. Jain, Design, modelling and control of bidirectional DC-DC converter (for EV), in *2015 International Conference on Emerging Research in Electronics, Computer Science and Technology (ICERECT)* (2016), pp. 294–297. https://doi.org/10.1109/ERECT.2015.749 9029
4. S. Barcellona, L. Piegari, Lithium ion battery models and parameter identification techniques. Energies **10** (2017). https://doi.org/10.3390/en10122007
5. P.S. Tomar, M. Srivastava, A.K. Verma, An improved bidirectional dc/dc converter with split battery configuration for electric vehicle battery charging/discharging, in *2018 India International Conference on Power Electronics (IICPE)* (2018), pp. 1–6. https://doi.org/10.1109/IICPE.2018.8709487
6. Z. Chen, H. Lin, B. Li, R. Qi, A novel bidirectional battery energy controller for plug-in hybrid electric vehicle, in *IEEE International Conference on Automation and Logistics (ICAL)* (2011), 173–177. https://doi.org/10.1109/ICAL.2011.6024706
7. K.B. Adam, M. Ashari, Design of bidirectional converter using fuzzy logic controller to optimize battery performance in electric vehicle, in 2015 *International Seminar on Intelligent Technology and Its Applications (ISITIA)* (2015), pp. 201–205. https://doi.org/10.1109/ISITIA.2015.7219979
8. H. Al-Sheikh, O. Bennouna, G. Hoblos, N. Moubayed, Study on power converters used in hybrid vehicles with monitoring and diagnostics techniques, in Proceedings of the Mediterranean Electrotechnical Conference—MELECON (2014), pp. 103–107. https://doi.org/10.1109/MELCON.2014.6820515
9. Y. Zhou, Z. Li, Z. Song, W. Li, The charging and discharging power prediction for electric vehicles, in *IECON 2016-42nd Annual Conference of the IEEE Industrial Electronics Society* (2016), pp. 4014–4019. https://doi.org/10.1109/IECON.2016.7793755
10. S. Stockar, V. Marano, M. Canova, G. Rizzoni, L. Guzzella, Energy-optimal control of plug-in hybrid electric vehicles for real-world driving cycles. IEEE Trans. Veh. Technol. **60**, 2949–2962 (2011). https://doi.org/10.1109/TVT.2011.2158565
11. M. Verasamy, M. Faisal, P.J. Ker, M.A. Hannan, Charging and discharging control of Li-Ion battery energy management for electric vehicle application. Int. J. Eng. Technol. **7**, 482–486 (2018). https://doi.org/10.14419/ijet.v7i4.35.22895
12. F. Pei, K. Zhao, Y. Luo, X. Huang, Battery variable current-discharge resistance characteristics and state of charge estimation of electric vehicle. Proc. World Congr. Intell. Control Autom. **2**, 8314–8318 (2006). https://doi.org/10.1109/WCICA.2006.1713597
13. J. Noppakunkajorn, O. Baroudi, R.W. Cox, L.P. Mandal, Using transient electrical measurements for real-time monitoring of battery state-of-charge and state-of-health, in *IECON Proceedings of the IECON 2012 38th Annual Conference on IEEE Industrial Electronics Society* (2012), pp. 4060–4065. https://doi.org/10.1109/IECON.2012.6389241
14. P. Pathiyil, K.S. Krishnan, R. Sunitha, V.N. Vishal, Battery model for hybrid electric vehicle corrected for self-discharge and internal resistance, in *Proceedings of the 2016 2nd International Conference on Advances in Electrical, Electronics, Information, Communication and Bio-Informatics (AEEICB)*, IEEE (2016)), pp. 214–218. https://doi.org/10.1109/AEEICB.2016.7538276

Impact of Inverter Interfaced DG Control Schemes on Distributed Network Protection

Sudhir Kumar Singh, Rajveer Singh, Sourav Diwania, Ankit Singhal, and Shivam Saway

1 Introduction

Distributed generation (DG), especially inverter interfaced distributed generation are being increasingly installed on roof-top [1, 2], and popularly used by various industries to meet local distribution demand. IIDG can cause various protection issues in case of either series or shunt fault in the network. This is because IIDG do not supply high fault current in the network due to lower rated capacity of semiconductor switches utilized in inverter [3, 4]. Therefore, re-evaluation of conventional protection design parameters is required owing to different fault characteristics of IIDG. Inverter Interfaced DGs has its own protection which is embedded with inverter control to assure the safety concerns of semiconductor switches against overcurrent [5]. Usually, due to internal protection, this current is twice of rated load current [6, 7]. Thus, overall network protection needs to include the effect of internal protection since internal protection allows to feed the fault lower than its rated capacity. There are mainly two types of control schemes that are utilized in IIDG, viz. Voltage control scheme and current control scheme [8]. Accordingly, dual mode inverter is required in IIDG with reliable operation in both islanded and grid connected conditions. Islanded mode enables to work in voltage control mode and when grid supply restores, it switches back to current control mode. In the current control mode, local load voltage is maintained by the grid itself and in voltage control mode, it would be maintained by inverter itself. Islanded condition is more crucial as it is required to maintain a constant voltage across the load by inverter itself. The transition between the above two modes may result in voltage overshoot

S. K. Singh · R. Singh · S. Saway
Jamia Millia Islamia, New Delhi 110025, India

S. Diwania (✉) · A. Singhal
KIET Group of Institutions, Delhi-NCR, Ghaziabad, India
e-mail: souravdiwania123@gmail.com

© The Author(s), under exclusive license to Springer Nature Singapore Pte Ltd. 2021   229
J. Kumar and P. Jena (eds.), *Recent Advances in Power Electronics and Drives*, Lecture Notes in Electrical Engineering 707,
https://doi.org/10.1007/978-981-15-8586-9_21

across load connected, accompanied by supply inrush current into the grid due to voltage frequency mismatch [9]. Accordingly, IIDG should have efficient facility to smooth transition between two modes to avoid voltage overshoot, voltage frequency mismatch. Several design and control topologies have been discussed in the literature for dual mode inverters [10–14]. The control schemes in [15, 16], have separate control loops for both the modes which increase the complexity of the system. Furthermore, most of the control schemes neglect the grid uncertainties [17, 18], and cause voltage overshoot and inrush current into the grid. This paper investigates distribution network protection considering all grid uncertainties. In addition, it also examines the detailed analysis of fault current characteristics of IIDG along with the impact of IIDG on associated feeder protection considering both voltage and current control modes. The effect of IIDG controller on distribution network protection has been already studied in [19–21], in which the authors proposed FCL inclusion on the feeder to limit current feed by IIDG to solve the detection problem in IIDG. Furthermore, several discussed papers did not include the effect of the controller on protection scheme, especially in IIDG, whereas few papers [23, 24], considers the effect of the controller but did not provide an adequate solution.

In this paper, a detailed analysis of both the control schemes has been discussed. This helps to investigate IIDG fault characteristics which are supportive to understand network protection analysis. The result illustrates that limiting fault feeding characteristics of inverters marginally impact primary and back up protection but in case of increased penetration level, IIDG may disturb protection coordination, cause protection issues and also cause fuse recloser coordination issues.

2   Current and Voltage Controlled Mode in IIDG

Many IIDG based commercial units utilize three phase VSI with PWM or SVM to control its switches. The effective controls of IIDG enable them to cope up with wide demand variation of distribution network. There are briefly two types of control schemes/modes (current control and voltage control) to efficient regulation of inverter output. Current control mode, enable the inverter to feed current to grid following a set reference that is locked by PLL. Current control mode in VSI has become an attractive feature to utilize in many applications viz. ac/dc converters, power electronics applications and mainly grid connected inverters [25]. Current control mode also plays a significant role in power quality enhancement. Several current control topologies discussed in various literature [25, 26]. However, this mode suffers from the major drawback of inefficiency to maintain their terminal voltage to the distribution network levels. They mainly depend upon utility grid or any other voltage controlled DGs for maintaining the required voltage, and in the absence of such voltage support, its terminal voltage becomes susceptible, especially in case of shunt fault conditions.

3-ɸ VSI has been most popularly utilized with IIDG with PWM/SVM switching topology to interconnect with utility grid as shown in Fig. 1. 3-ɸ VSI connected to

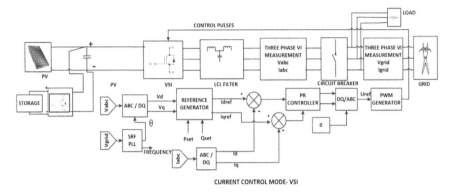

Fig. 1 IIDG connected to grid

the grid with LCL filter. LCL filter before PCC is inserted to compensate the inverter output harmonics. LCL filter, unfortunately, introduces undesirable resonance but proven superior filter as compare to simple L filter. In this mode, the current supplied to the grid varies according to set reference power. MPPT algorithm produces dc current which is utilized to estimate set reference power. Inverter filtered current and 3-φ voltages are converter into d-q coordinates (V_d, V_q, I_d, I_q)

Power formula in d-q coordinates are given below

$$P = \left(V_d I_d + V_q I_q\right) \tag{1}$$

$$Q = \left(V_d I_q + V_q I_d\right) \tag{2}$$

On the basis of the above formulas, I_{dref} and I_{qref} are calculated which is compared with measured I_d and I_q and error is passed to PR controller for regulating the error to zero steady state. Conventional PI controller cannot minimize the resonance problem arise by LCL filter so popular the proportional + resonant (PR) controller has been used for error reduction. The output of this PR controller is converted back to a natural frame of reference and it has been compared with triangular pulses called PWM which would further control six switches of VSI. For grid synchronization, the synchronous reference frame phase locked loop (SRF-PLL) has been used. Active power is set to be 10 KW and Reactive power set to be zero to obtain unity power factor at the grid side. Various system parameters are mentioned in Table 1. Standalone Solar switched to voltage controlled mode due to constant voltage requirement of load in absence of grid as shown in Fig. 2. For any input DC voltage variation, load voltage will remain constant provided the dc bus voltage is much higher than the peak of the ac line to line voltage. The main technical constraint in this mode is to effective control of voltage and frequency due to fluctuating generation or load. In this mode, inverter

Table 1 System parameters

System parameters	Values
Inverter output voltage (Line to line)	500 V
Nominal DC link voltage	700 V
Grid frequency	50 Hz
DC link capacitor	900
Switching frequency	4 kHz
Inverter side inductor	3 mH
Grid side inverter	1.5 mH
Shunt capacitor	30

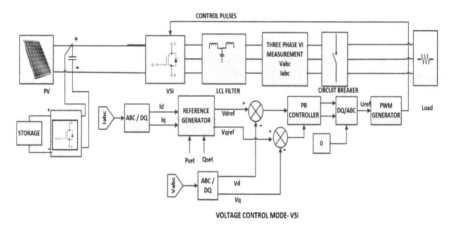

Fig. 2 standalone IIDG

regulates effectively the voltage across the load. The synchronization in this mode during the transition from current control to voltage control is required to generate load voltage reference.

3 Investigation of Voltage and Current Controlled IIDG On Distribution Network Protection

3.1 Current Control Mode

In order to examine the effects of current control scheme in IIDG on the distribution network, a simple model has been constructed in MATLAB whose typical values are given in Table 1. On this model 3-ϕ, a short circuit has been created at PCC. The contribution of the inverter to the fault has been examined. The inverter output current, in case of the current control scheme, is tried to maintain constant. However, the current response solely depends upon fault impedance. In Fig. 3, with a high impedance of 5Ω, the current was maintained constant even after fault occurs at 100 m-s. Figures 4 and 5 illustrate current response for low fault impedance which is assumed to be 1 m Ω and zero fault impedance.

In both the conditions, controller efficiently tries to maintain inverter output current with slight overshoot. Figure 6 illustrates reactive power compensation at PCC due to efficient controller action. Reactive compensation has been utilized by pre-specifying zero reactive power in reference generation.

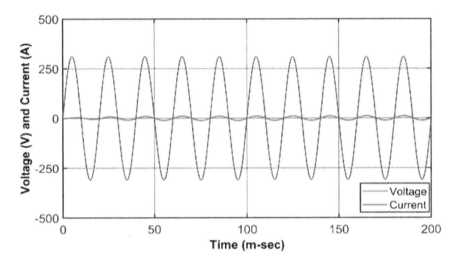

Fig. 3 High impedance fault

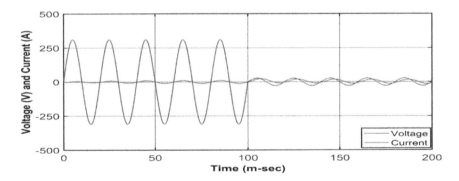

Fig. 4 Low impedance fault

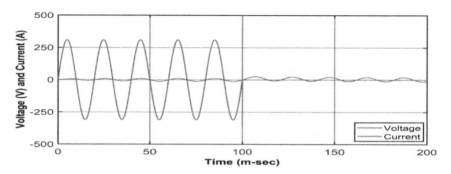

Fig. 5 Zero impedance fault

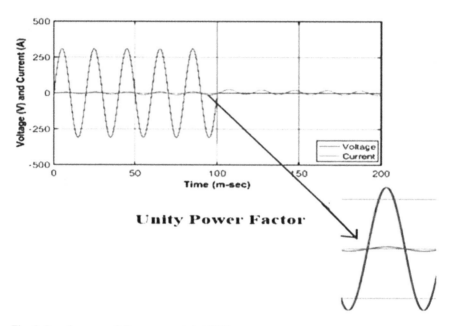

Fig. 6 Reactive power fully compensated at PCC

3.2 Voltage Controlled Mode

To examine the effect of voltage controlled IIDG on distribution system network, it was modeled in MATLAB and three phase fault with 5Ω fault impedance occurred at 100 m-s after LC filter, nearby the three phase load. Figure 7 illustrates the output current of DG increases at fault instant provided the voltage remains constant. Voltage control mode regulates the voltage within the specified range. The voltage controller regulates output voltage within the specified limit only above 5 Ω that is assumed here high impedance fault as shown in Fig. 3. Furthermore, below that value, DG output

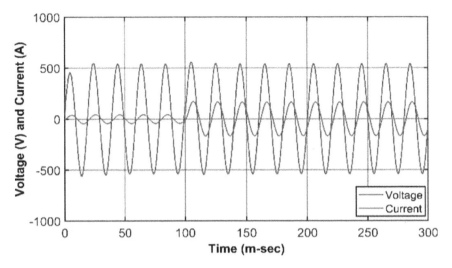

Fig. 7 IIDG output voltage and current at high fault impedance

voltage will no longer be constant and will fall down at fault occurrence as seen in Fig. 8. Hence, voltage control is possible only for high impedance fault and for low impedance and zero impedance fault, and it is not feasible to regulate the output voltage. In order to examine fault characteristics between both modes/schemes, different impedance faults, change of locations and type of faults has been investigated. In all the scenarios, voltage controlled IIDG effectively maintains voltage waveform, however, current would vary depending upon fault location, type and impedance. On the contrary, another mode viz. current controlled IIDG current would

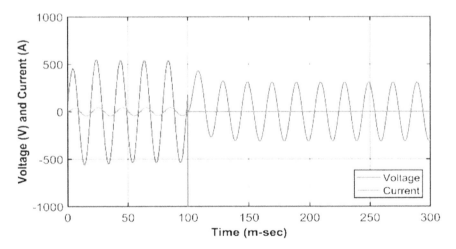

Fig. 8 IIDG output voltage and current at zero fault impedance

maintain constant but voltage varied according to the fault types and location and impedance. Figure 9 illustrates inverter filtered output voltage and current when the fault occurred at two locations 5 and 10 km from inverter terminals. Line impedance is assumed to be 1.25 Ω/km.

Previous discussions regarding current control mode were based on the pure resistive load connected to filter end, therefore, transient overshoot at the time of fault was almost negligible.

Interestingly, the same fault would cause overshoot for the inductive load connected parallel to a preexisting resistive load. Higher the value of VAR in load, higher overshoot has been observed as shown in Fig. 10. Such a transient overshoots have been observed in IIDG during commercial testing [27]. This current overshoot

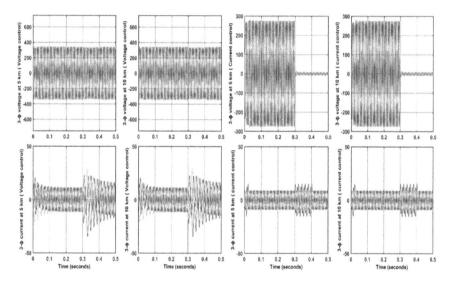

Fig. 9 Inverter filtered voltage/current on different control schemes and at different locations

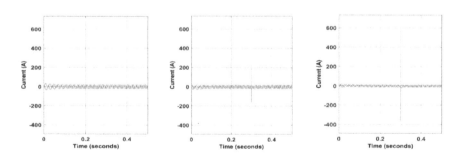

Fig. 10 (IIDG output current under current control with varying load VAR, fault at 0.3 s Left (0 kVAR), middle (5 kVAR), right (10 kVAR)

might trigger inverter internal protection and/or perturb nearby overcurrent relay setting in the distribution network. Transient overshoot, however, decays very fast in few milliseconds depending upon the load VAR.

4 Proposed PR Controller in Inverter Current Loop Over Conventional PI Controller

The current loop of the inverter with PR controller is shown in Fig. 11. PR controller can be obtained by frame transformation [28]. The conventional PI controller implemented in a synchronous frame is denoted here as $G_{DC}(s)$ and can effectively be transformed into stationary reference frame through frequency modulation which is mathematically expressed below

$$G_{AC}(s) = G_{DC}(s - j\omega) + G_{DC}(s + j\omega) \tag{3}$$

where $G_{AC}(s)$ illustrates the equivalent stationary frame transfer function.

$$\text{For ideal PR Controller } G_{AC} = \frac{Y(s)}{E(s)} = \frac{2ks}{s^2 + \omega^2} \tag{4}$$

$$\text{For non-ideal PR controller } G(s) = \frac{Y(s)}{E(s)} = \frac{2ks\omega_f}{s^2 + \omega_f^2 + 2\omega_f s} \tag{5}$$

where K = controller gain and $Y(s)$ and $E(s)$ represents output and error in s-domain. ω = angular frequency.

Equation (4) along with Kp (proportional controller) forms ideal PR controller which is having infinite gain at grid frequency, however, shows zero shift and zero gain at different frequencies, furthermore, it causes stability problems which will be compensated with using Eq. (5), which is non-ideal PR controller having finite (high) gain with very small steady error.

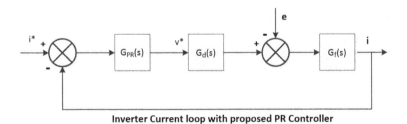

Inverter Current loop with proposed PR Controller

Fig. 11 PR controller embedded in inverter current loop

Fig. 12 Non-ideal-controller
(Kp = 1, K = 20, =
314 rad/s)

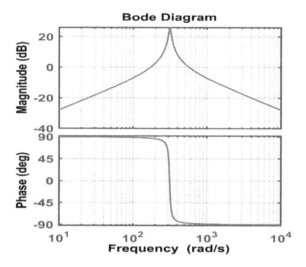

Bode plot for the non-ideal controller is plotted in Fig. 12, with controller gain of 20 and proportional gain of unity. In a comparison of PI, which only deals with either positive sequence or negative sequence, PR controller having zero cross coupling results as a superior tool to deal with grid synchronization [28], and steady state error elimination, so it is embedded in the control mechanism of the converter in further works. PR controller is used to control the grid current i*. When synchronous controller based on PI is transformed back into stationary reference frame through inverse transformation, then PR controller is obtained.

Synchronous controller measurement depends upon the grid phase angle while the PR controller based on grid frequency. In case of unbalanced/fault conditions synchronous controller which is conventional PI required to deal with two control loops viz. positive sequence and negative sequence as there is no correlation among phase angles of positive and negative sequences. However, grid frequency is the same for both positive/negative sequences, so only one resonant controller (PR controller) is required to compensate positive and negative sequence currents simultaneously.

5 Performance Evaluation of Protection Based on Proposed Controller

The proposed PR controller is embedded into an inverter closed loop and the comparison is examined from the conventional PI controller for multiple fault scenarios. Figure 13 illustrates voltage controlled mode and Fig. 14, represents current controlled mode. Figures 13 and 14 suggests that the PI controller failed to detect various fault types because it works on conventional protection topology and do not consider inverter, and thus remained unclear. On contrary, the proposed

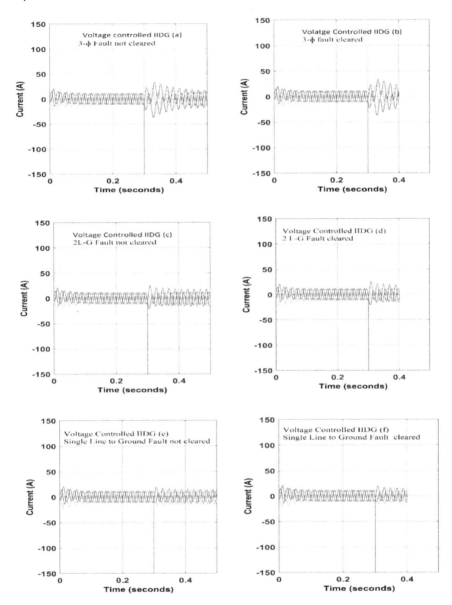

Fig. 13 Impact of voltage controlled mode on different faults (Comparison of the conventional PI controller and PR controller)

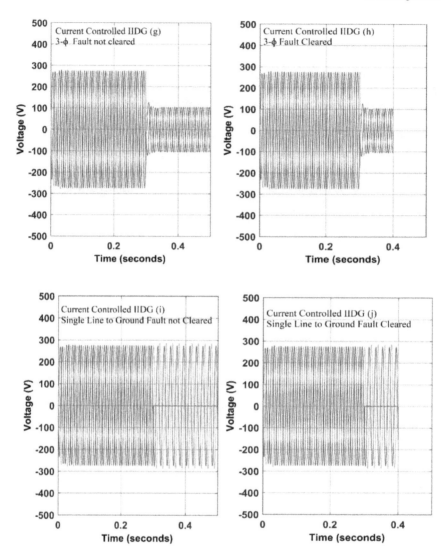

Fig. 14 Impact of current controlled mode on different faults

controller efficiently detects the fault and clears it effectively. Proposed PR controller work in an advanced environment, which took consideration of inverter mode, fault types and more effective damping as compared to PI controller. Figure 14c illustrates 3-ϕ fault under current controlled mode which is detected by under voltage relay, whereas in Fig. 13, voltage controlled mode, the fault is detected by overcurrent relay. So, the proposed controller first identifies the type of fault and the inverter mode then effectively triggers the required relay to clear the fault. All the relay settings were chosen to be within IEEE 1547.

6   Conclusion

In this paper, Fault current from IIDG in distribution network has been examined. In fact, the impact of IIDG control modes voltage controlled/current controlled modes are thoroughly studied and investigated. In addition, the conventional PI controller working on conventional overcurrent protection schemes has been compared with the proposed PR controller. PR controller is a more effective tool to damp out grid uncertainties. Here the proposed controller is embedded in the inverter closed loop and inverter different modes are considered to analyze their effects on various types of fault. This paper deduced the following significant points:

- Proposed controller limits inverter output current effectively on reduced terminal voltage in the case of both three phase symmetrical and L-G, 2L-G unsymmetrical faults.
- A comparative study has been conducted between the conventional PI controller and PR controller and their effects on inverter output current, and the terminal voltage is examined on various types of faults.
- Conventional PI controller is easy to install but its behavior is sluggish and works on conventional protection topology viz. overcurrent protection.

References

1. J. Rocabert, A. Luna, F. Blaabjerg, P. Rodriguez, Control of power converters in AC microgrids. IEEE Trans. on Power Electron. **27**, 4734–4749 (2012)
2. M. Singh, V. Khadkikar, A. Chandra, R.K. Varma, Grid interconnection of renewable energy sources at the distribution level with power-quality improvement features. IEEE Trans. Power Deliv. **26**, 307–315 (2011)
3. A. Abdel-Khalik, A. Elserougi, A. Massoud, S. Ahmed, Fault current contribution of medium voltage inverter and doubly-fed induction machine-based flywheel energy storage system. IEEE Trans. Sustain. Energy **4**(1), 58–67 (2013)
4. Y. Han, X. Hu, D. Zhang, Study of adaptive fault current algorithm for microgrid dominated by inverter based distributed generators, in *Proceedings of 2nd IEEE International Symposium on Power Electronics for Distributed Generation Systems (PEDG)*, 16–18 June 2010, pp. 852–854.
5. C.J. Mozina, Impact of green power generation on distribution systems in a smart grid, in *IEEE/PES Power Systems Conference and Exposition (PSCE)*, Mar 2011, pp. 1–8
6. H. Nikkhajoei, R.H. Lasseter, Microgrid protection, in *IEEE Power Engineering Society General Meeting*, June 2007, pp. 1–6
7. T. Loix, T. Wijnhoven, G. Deconinck, Protection of microgrids with a high penetration of inverter-coupled energy sources, in *2009 CIGRE/IEEE PES Joint Symposium in Integration of Wide-Scale Renewable Resources Into the Power Delivery System* (2009), pp. 1–6
8. M. Baran, L. El-Markaby, Fault analysis on distribution feeders with distributed generators. IEEE Trans. Power Syst. **20**(4), 1757–1764 (2005)
9. M. Rizo, M. Liserre, E. Bueno, F.J. Rodriguez, C. Giron, Voltage control architectures for the universal operation of DPGS. IEEE Trans. Ind. Inform. **11**, 313–321 (2015)
10. L. Qin, Y. Shuitao, F.Z. Peng, Multi-loop control algorithms for seamless transition of grid-connected inverter, in *IEEE Applied Power Electronics Conference and Exposition (APEC)* (2010), pp. 844–848

11. H. Shang-Hung, K. Chun-Yi, L. Tzung-Lin, J.M. Guerrero, Droop-controlled inverters with seamless transition between islanding and grid-connected operations, in *IEEE Energy Conversion Congress and Exposition (ECCE)* (2011), pp. 2196–2201
12. D.S. Ochs, B. Mirafzal, P. Sotoodeh, A method of seamless transitions between grid-tied and stand-alone modes of operation for utility-interactive three-phase inverters. IEEE Trans. Ind. Appl. **50**, 1934–1941 (2014)
13. Z. Qing-Chang, T. Hornik, Cascaded current-voltage control to improve the power quality for a grid-connected inverter with a local load. IEEE Trans. Ind. Electron. **60**, 1344–1355 (2013)
14. Y. Zhilei, X. Lan, Y. Yangguang, Seamless Transfer Of Single Phase Grid-Interactive Inverters Between Grid-Connected And Stand-Alone Modes. IEEE Trans. Power Electron. **25**, 1597–1603 (2010)
15. P. Rodriguez, A. Luna, R.S. Munoz-Aguilar, et al., A stationary reference frame grid synchronization system for three-phase grid-connected power converters under adverse grid conditions. IEEE Trans. Power Electron. **27**, 99–112 (2012)
16. H. Shang-Hung, K. Chun-Yi, L. Tzung-Lin, J.M. Guerrero, Droop controlled inverters with seamless transition between islanding and grid connected operations, in *IEEE Energy Conversion Congress and Exposition (ECCE)* (2011), pp. 2196–2201
17. D. Dong, T. Thacker, I. Cvetkovic, R. Burgos, D. Boroyevich, F. Wang et al., Modes of operation and system-level control of single-phase bidirectional PWM converter for microgrid systems. IEEE Trans. Smart Grid **3**, 93–104 (2012)
18. M. Fatu, F. Blaabjerg, I. Boldea, Grid to standalone transition motion-sensorless dual-inverter control of PMSG with asymmetrical grid voltage sags and harmonics filtering. IEEE Trans. On Power Elec. **29**, 3463–3472 (2014)
19. W. El-Khattam, T. Sidhu, Restoration of directional overcurrent relay coordination in distributed generation systems utilizing fault current limiter. IEEE Trans. Power Del. **23**(2), 576–585 (2008)
20. M.M.R. Ahmed, G. Putrus, L. Ran, R. Penlington, Development of a prototype solid-state fault-current limiting and interrupting device for low-voltage distribution networks. IEEE Trans. Power Del. **21**(4), 1997–2005 (2006)
21. H. Wan, K.K. Li, K.P. Wong, An adaptive multiagent approach to protection relay coordination with distributed generators in industrial power distribution system. IEEE Trans. Industry Applications **46**(5), 2118–2124 (2010)
22. H. Yazdanpanahi, Y.W. Li, W. Xu, A new control strategy to mitigate the impact of inverter-based DGs on protection system. IEEE Trans. Smart Grid **3**(3), 1427–1436 (2012)
23. Y. Pan, W. Ren, S. Ray, R., Walling, M. Reichard, Impact of inverter interfaced generation on overcurrent protection in distribution systems, in *Power Engineering and Automation Conference (PEAM)*, vol. 2, Sept. 2011, pp. 371–376
24. C. Mozina, Impact of green power inverter-based distributed generation on distribution systems, in *Georgia Tech Annual Protective Relaying Conference*, Atlanta, May 2013
25. M. Kazmierkowski, L. Malesani, Current control techniques for three-phase voltage source PWM converters: a survey. IEEE Trans. Ind. Electron. **45**(5), 691–703 (1998)
26. D. Zmood, D. Holmes, Stationary frame current regulation of PWM inverters with zero steady-state error. IEEE Trans. Power Electron. **18**, 814–822 (2003)
27. M.A. Haj-Ahmed, M.S. Illindala, The influence of inverter-based DGs and their controllers on distribution network protection. IEEE Trans. Ind. Appl. **50**(4), 2928–2937 (2014)
28. R.Teodorescu, F. Blaabjerg, M. Liserre, P.C. Loh, Proportional-resonant controllers and filters for grid connected voltage source converters, in *IEEE Proceedings of Electric Power Applications*, vol. 153 (2006), pp. 750–762

Realization of CMOS 0.18 μm Low Noise Amplifier for 2–5 GHz Using Cascode-Cascade Topology

Dheeraj Kalra

1 Introduction

Nowadays there is a demand for highly proficient circuits in the sectors of healthcare, agriculture, industries, IoT, etc. The receiver circuit has to process the received signal in such a manner that the signal quality should not be degraded and the original signal can be recovered from the modulated signal. In the receiver chain, the first block after the antenna is the low noise amplifier (LNA), whose work is to process the signal in such a way that its noise contribution is as low as possible so that the signal to noise ratio can be increased [1]. The next blocks are mixer, filter, etc. and their performance depends on the performance of the LNA as shown in Fig. 1.

As per today's requirement, the circuit should consume less power, i.e. power consumption of the circuit should be as low as possible. Low power is possible due to shrinking the size of CMOS technology day by day [2]. But the undesirable effect of shrinking the size is degradation of mobility and velocity saturation. There are a lot of well known design techniques such as cascode structure which is shown in Fig. 2a. Cascode technique is grouping of Common Source and Common Gate configurations [3].

Common gate MOSFET is mounted on the Common source configuration. The advantage of this topology is to have high bandwidth [4], low noise figure and good linearity. High gain is achieved due to the cascode connections of two configurations. The gain value is the multiplications of transconductance of two MOSFETs. The disadvantages of this technique are to have high overdrive voltage [5]. Another topology is the source degeneration topology which is shown in Fig. 2b. In this topology, an inductor connected at the source helps in canceling out the effect of MOS parasitic capacitances, and hence give good impedance matching at the input

D. Kalra (✉)
GLA University, Mathura, India
e-mail: dheeraj.kalra@gla.ac.in

© The Author(s), under exclusive license to Springer Nature Singapore Pte Ltd. 2021
J. Kumar and P. Jena (eds.), *Recent Advances in Power Electronics and Drives*, Lecture Notes in Electrical Engineering 707,
https://doi.org/10.1007/978-981-15-8586-9_22

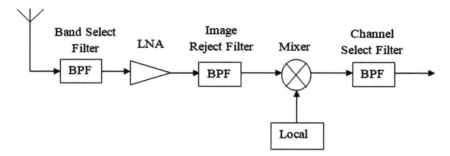

Fig. 1 Receiver architecture

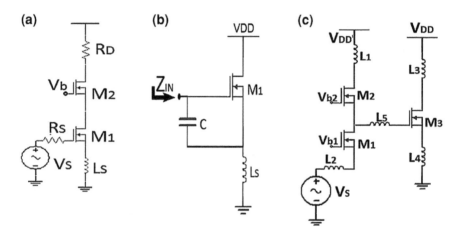

Fig. 2 **a** Cascode topology **b** source degeneration **c** cascade topology

side. Input impedance acts as the resistive component at the resonance frequency [6–7]. Benefits of this topology are to have flat gain and low power consumption. The disadvantages include more circuit area due to the inductor's presence and high noise figure at high frequencies. The new topology, i.e. cascade technique shown in Fig. 2c, in which the first stage is the common gate stage and helps in providing the high gain at a low frequency, while the next stage gives the high gain at a high frequency, so achieves the high bandwidth [8–9]. But due to the presence of more number of MOS, the circuit area and its power consumption has been increased [10]. The design of LNA like that noise contribution should be as low as possible [11].

Noise Factor is given by Noise Factor $= \frac{\text{SNR}_{i/p}}{\text{SNR}_{o/p}}$, Noise Factor $= 1 + \frac{\overline{V_n^2}}{4KTR_S}$ [2].

Where $V_n^2 =$ Noise Voltage, $K =$ Boltzmann Constant, $T =$ Temperature in Kelvin, $R_S =$ Source Resistance. For a simple common source amplifier shown in Fig. 2b, noise factor can be calculated as NF $= 1 + \frac{\gamma}{(R_S||R_P)g_m R_D} + \frac{1}{(R_S||R_P)g_m^2 R_D}$. Matching at the input and output port is very important in designing the LNA as the minimum amount of power must be reflected at the port, this will help be in maximizing the power at the input and output load resistance, respectively [12–14].

2 Realized Circuit

The realized circuit is shown in Fig. 3, which is being simulated in the Advanced Design System (ADS) using 180 nm CMOS technology. At the input side π filter composed of L_1, C_1 and L_2 are connected for the purpose of band pass filter. L_4, C_2, L_3 and C_3 are used to provide the feedback which will give the stability to the circuit, as well as provide the impedance matching. It helps in canceling out the parasitic capacitances of MOS.

Current reuse technology is being used which helps in getting the low power consumption of the circuit and M_1 and M_2 MOS are connected in cascode connection provide the good gain of the circuit. MOSFET M_3 connected in cascade topology using the source degeneration technique which is being used to offer the linearity in the circuit and reducing the value of noise figure at high frequencies. Inductors L_6 and L_7 form the mutual coupling and provide the feedback which gives the stability to the circuit and also grant impedance matching at the input and output port. Proper bias voltages are applied at the R_b resistances for the accurate functioning of the circuits. S-parameters are used to measure the gain, impedance matching and the power reflection at the input and output port. S_{21}, i.e. the ratio of power at the output port to the power at the input port gives the value of power gain to the circuit. There is always a tradeoff between the gain, bandwidth, power dissipation, noise figure and matching for the LNA circuit.

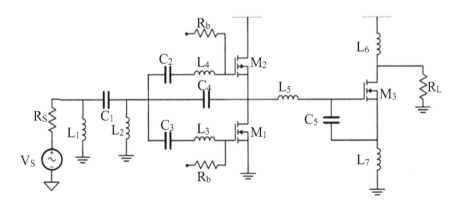

Fig. 3 Realized LNA circuit

3   Simulation Results

The realized circuit is realized using 180 nm CMOS for the frequency range of 2–5 GHz. The results for S-parameters are shown in Fig. 4. The value of S_{21} is positive for the entire frequency range of 2–5 GHz and the maximum value obtained is 12.227 dB. The S_{11} is negative which shows the good impedance matching at the input port. The minimum value of S_{11} is 15.789 dB at 2.520 GHz.

The S_{22}, i.e. the reflection at the output port is also negative and its minimum value is -30.936 dB at 4.1 GHz. S_{12}, i.e. the power reflection from port 2 to port 1, it's minimum value is -98.974 dB at 2 GHz. So the S-parameters results satisfying the requirement of LNA in the receiver circuit. The simulation for Noise Figure versus RF frequency is as shown in Fig. 5. The results show the noise figure minimum value of 2.440 dB at 2.863 GHz and the maximum value is 3.470 dB at 5 GHz (Table 1).

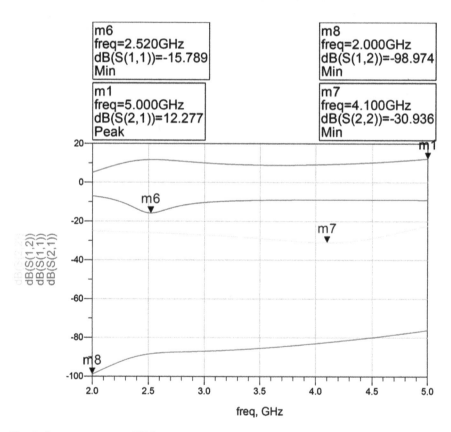

Fig. 4 S parameters versus RF frequency

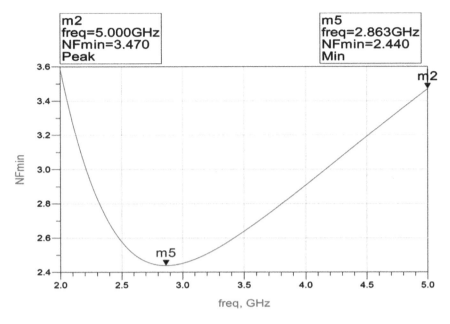

Fig. 5 Noise Figure versus RF frequency

Table 1 Comparison of results with other references

Parameters	[3]	[5]	[6]	This work
Frequency range (GHz)	2–6	0.5–8.2	0.5–7	2–5
S_{21} (dB)	10	25	22	12.277
S_{11} (dB)	<−9	–	−3 to −25	−15.489
NF (dB)	2.3	1.9–2.6	2.3–2.9	2.440
Technology	0.18 μm CMOS	90 nm CMOS	90 nm CMOS	90 nm CMOS

4 Conclusion

The realized circuit is realized in 0.18 μm CMOS technology in ADS software for the frequency range of 2–5 GHz. Cascode topology is used to increase the gain of the circuits, as well as current reuse topology helps in reducing the power consumption of the circuit. Feedback topology helps in providing the stability of the circuit and cascade topology provides the reduction in noise figure at high frequency. The maximum value of the gain, i.e. S_{21} is measured to be 12.277 dB. Input reflection co-efficient, i.e. S_{11} is −15.789 dB at 2.520 GHz whereas the output reflection co-efficient, i.e. S_{22} is −30.936 dB at 4.1 GHz and S_{12}, i.e. the power reflection from port 2 to port 1, it's minimum value is −98.974 dB at 2 GHz. Noise figure minimum value is 2.440 dB at 2.863 GHz and the maximum value is 3.470 dB at 5 GHz.

References

1. M. Kumar, V.K. Deolia, A wideband design analysis of LNA utilizing complimentary common gate stage with mutually coupled common source stage. Analog Integr. Circuits Signal Process. **98**(3), 575–585 (2019)
2. D. Kalra, et al., Design analysis of inductorless active loaded low power UWB LNA using noise cancellation technique. Frequenz **74**(3–4), 137–144 (2020)
3. C.-W. Kim, M.-S. Kang, P.T. Anh, H.-T. Kim, S.-G. Lee, An ultra-wideband CMOS low noise amplifier for 3–5-GHz UWB system. IEEE J. Solid-State Circuits **40**(2), 544–547 (2005)
4. B. Razavi, *RF Microelectronics*, Prentice Hall Communications Engineering and Emerging Technologies Series, 2nd edn. (Prentice Hall Press, USA, 2011)
5. J.-H. Zhan, S. Taylor, A 5 GHz resistive-feedback CMOS LNA for low-cost multi-standard applications, in *IEEE International Solid-State Circuits Conference Digest of Technical Paper (ISSCC)*, Feb (2006), pp. 721–730
6. B. Perumana, J.-H. Zhan, S. Taylor, B. Carlton, J. Laskar, Resistive feedback CMOS low-noise amplifiers for multiband applications. IEEE Trans. Microw. Theory Techn. **56**(5), 1218–1225 (2008)
7. T. Chang, J. Chen, L. Rigge, J. Lin, ESD-protected wideband CMOS LNAs using modified resistive feedback techniques with chipon-board packaging. IEEE Trans. Microw. Theory Techn. **56**(8), 1817–1826 (2008)
8. M. Chen, J. Lin, A 0.1–20 GHz low-power self-biased resistive feedback LNA in 90 nm digital CMOS. IEEE Microw. Wirel. Compon. Lett. **19**(5), 323–325 (2009)
9. F. Silveira, D. Flandre, P. Jespers, A gm/ID based methodology for the design of CMOS analog circuits and its application to the synthesis of a silicon-on-insulator micropower OTA. IEEE J. Solid-State Circuits **31**(9), 1314–1319 (1996)
10. C.C. Enz, E.A. Vittoz, *Charge-Based MOS Transistor Modeling: The EKV Model for Low-Power and RF IC Design* (Wiley, New York, NY, USA, 2006)
11. V. Aparin, G. Brown, L. Larson, Linearization of CMOS LNA's via optimum gate biasing, in *Proceedings of the IEEE International Symposium on Circuits System (ISCAS)*, May (2004), pp. 748–751
12. M. Parvizi, K. Allidina, M.N. El-Gamal, A sub-mw, ultra-low-voltage, wideband low-noise amplifier design technique. IEEE Trans. Very Large Scale Integr. (VLSI) Syst. **23**(6), 1111–1122 (2015)
13. A. Kaveh, M.A. Motie Share, M. Moslehi, Magnetic charged system search: a new metaheuristic algorithm for optimization. Acta Mechanica **224**(1), 85–107 (2013)
14. S.Z. Khong, et al., Unified frameworks for sampled-data extremum seeking control: global optimisation and multi-unit systems. Automatica **49**(9), 2720–2733 (2013)

New Symmetric 9-Level Inverter Topology with Reduced Switch Count and Switching Pulse Generation Using Digital Logic Circuit

V. Thiyagarajan

1 Introduction

Multilevel inverter is one of the most promising and highly reliable voltage source power converters which interconnect the dc system with ac system [1]. Several power electronic switches and dc sources are connected together to create multistep output voltage waveform [2]. The output waveform looks similar to the sinusoidal waveform with minimum harmonic content for a higher number of output levels, and thus eliminates the filter requirements [3]. Multilevel inverters had gained remarkable attractiveness in terms of structure and control techniques due to its better electromagnetic compatibility, less total harmonic distortion, smaller common mode voltage, less dv/dt stress, reduced switching losses, high efficiency and improved power quality [4, 5]. In addition, multilevel inverters have a modular structure, capable of transformerless operation and fault tolerant operation by utilizing multiple redundant switching states with suitable control schemes [5, 6]. Due to these significant capabilities, the multilevel power converters have attracted various industrial applications such as electric traction, electric aircraft power system, uninterrupted power supply system, flexible AC transmission system, photovoltaic power system, electric drives, hybrid electric vehicles and distributed generation [6, 7]. Different classifications of the multilevel inverters used in commercial applications include flying capacitor, neutral point clamped and cascaded H-bridge [6–8]. Among these conventional inverter topologies, cascaded H-bridge inverters have an exact modular structure for a higher number of output levels. There are certain issues in these conventional topologies such as greater number of switches and its associated gate driver circuits, isolated dc sources, voltage balancing problem and so on [8]. Practically, flying capacitor and

V. Thiyagarajan (✉)
Sri Sivasubramaniya Nadar College of Engineering, Kalavakkam 603110, Tamil Nadu, India
e-mail: thiyagarajanv@ssn.edu.in

© The Author(s), under exclusive license to Springer Nature Singapore Pte Ltd. 2021
J. Kumar and P. Jena (eds.), *Recent Advances in Power Electronics and Drives*, Lecture Notes in Electrical Engineering 707,
https://doi.org/10.1007/978-981-15-8586-9_23

neutral point clamped inverters are not suitable for higher output levels, because they have complex inverter structure and higher ratings for power semiconductor devices which make tedious to implement these inverter structures.

Different multilevel inverters using a reduced number of switches were presented in the literature [1–8]. A new symmetric and asymmetric type multilevel inverter with two dc sources and four switches is proposed in [1]. In this, a separate polarity changing unit is used to obtain the negative output levels. This inverter requires a higher number of switches to generate a staircase waveform. Moreover, complex control is implemented even for low rating power switches with a separate driver circuit. A reduced switch count based multilevel inverter to generate all possible combinations of input dc sources is presented in [2]. This inverter has the capability to operate in both lower and higher switching frequencies. However, high spikes may occur in the voltage waveform due to the lack of path for reverse current flow. In [3], a new basic unit with three sources and five unidirectional switches is proposed to generate only positive levels at the output. To obtain the negative levels, H-bridge is used as a polarity changing unit. This would results in higher voltage stress across the switches for higher output results and need higher rate power semiconductor devices. Another multilevel inverter topology with an array of voltage sources in additive nature is proposed in [4]. This inverter produces all levels of the output voltage by switching the appropriate power semiconductor devices. This topology lacks modularity and restrictions on modulation strategy and control method. The basic unit of the inverter proposed in [5], consists of four dc sources and ten switches and create 17-level during asymmetric operation. To create higher output levels, several basic circuits can be connected in series across the load terminals. The main drawback of this topology is not suitable for high voltage applications because of high variety of voltage sources and TSV value. In [6], a new extendable type multilevel inverter with various ratings of both unidirectional and bidirectional switches is presented. This topology reduces the number of switches for higher output levels and extending the use for low voltage and high power application. However, this topology lacks the attribute of combining two dc sources in parallel and does not own the equal load sharing capability. Other inverter topologies using an efficient fundamental switching technique are presented in [7, 8].

This paper aims to develop more reliable 9-level inverter topology with a minimum number of circuit components. Section 2 presented the operation of the proposed symmetrical 9-level inverter topology. A comparison study is done in Sect. 3. The switching strategy using digital logic circuit based multicarrier pulse width modulation (PWM) technique to obtain the switching signals for the proposed 9-level inverter is explained in Sect. 4. Section 5 presents the simulation results of the 9-level inverter and the conclusion is given in Sect. 6.

Fig. 1 Proposed 9-level
inverter

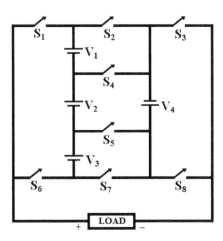

2 Proposed 9-Level Inverter

The proposed 9-level symmetrical multilevel inverter topology with four dc voltage
sources and eight power electronic switches is shown in Fig. 1. Here, the voltage
sources V_1, V_2, V_3 are connected in series and the combination is connected in parallel
with the source V_4. In order to avoid short circuit across the dc sources, the pairs of
switches (S_1, S_6), (S_2, S_4, S_5, S_7), (S_3, S_8) should not be turned ON simultaneously.
To create 9-output levels, the magnitude of all dc sources V_1, V_2 and V_3 should
be equal. The important feature of the proposed symmetric topology is its inherent
ability to create negative levels without additional H-bridge.

Some of the sample output levels obtained at the inverter output terminals are
shown in Fig. 2. The blocking voltage across the switches S_3 and S_8 is V_{dc}. For
switches S_2 and S_7, the blocking voltage is $4V_{dc}$ and for the remaining switches,
this value is equal to $3V_{dc}$. Therefore, the total standing voltage (TSV) across the
switches of the proposed 9-level inverter topology is $22V_{dc}$.

3 Comparison Study

Table 1 presents the comparison between the proposed 9-level symmetric inverter
with other symmetrical topologies presented in the literature. This comparison study
is presented based on the total number switches, ON-state switches and value of
TSV. As shown in Table 1, the proposed 9-level inverter topology inherently creates
negative output levels, however, the topologies presented in [4–6], need an additional
H-bridge circuit as a polarity changing unit to produce negative output levels. The
comparative study shows that the proposed inverter topology creates a 9-level output
with a minimum number of switches. The proposed inverter uses only eight switches,
however, the topology presented in [1], needs 12 switches, the inverter presented in

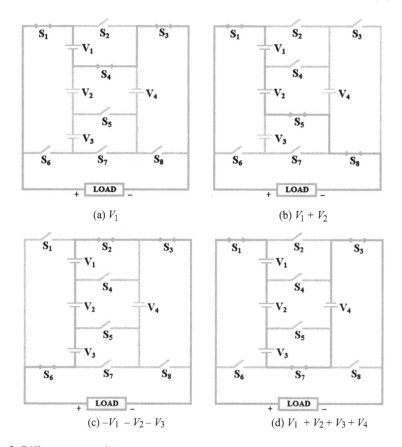

(a) V_1

(b) $V_1 + V_2$

(c) $-V_1 - V_2 - V_3$

(d) $V_1 + V_2 + V_3 + V_4$

Fig. 2 Different output voltages

Table 1 Comparison study

Topology	No. of switches	ON-state switches	TSV	Negative level
[1]	12	4	$28V_{dc}$	H-bridge
[2]	16	8	$29V_{dc}$	H-bridge
[3]	11	4	$27V_{dc}$	H-bridge
[4]	10	5	$28V_{dc}$	Inherent
[5, 6]	10	4	$25V_{dc}$	Inherent
Proposed	8	3	$22V_{dc}$	Inherent

[2], needs 16 switches and the 11 switches are needed for the topology presented in [3]. Additionally, the proposed topology need only three ON-state switches to create 9-output levels. But, the topologies presented in [1, 3, 5, 6], needs four ON-state switches, the topology presented in [4], needs five ON-state switches and the

inverter shown in [2], needs eight ON-state switches. The reduction in the number of conducting switches reduces the inverter losses and increases its efficiency. Also, it is noted that the TSV value for the proposed inverter is $22V_{dc}$. Therefore, the inverter size, cost and complexity is significantly reduced.

4 Switching Pulse Generation

A modified multicarrier based digital logic technique is used to obtain the switching pulses. In this technique, the stage-1 signals C_1, C_2, C_3, C_4 are generated by comparing the sinusoidal reference signal with four different carrier signals. The magnitude of the carrier signals are determined by

$$V_{Ck} = \frac{2k - 1}{2} \text{ where, } k = 1, 2, 3, 4$$

The generation of stage-1 signals during a positive cycle is shown in Fig. 3a.

The following equation helps to obtain the intermediate signals A, B and C from the stage-1 signals:

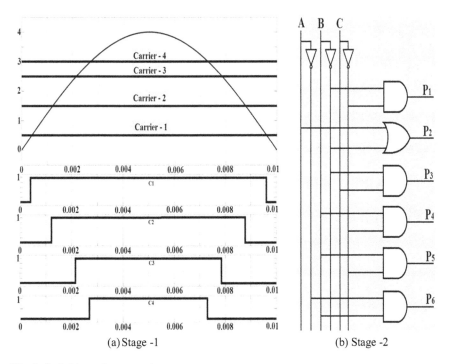

(a) Stage -1   (b) Stage -2

Fig. 3 Switching pulse generation

$$A = C_4$$
$$B = C_2$$
$$C = C_1 - C_3$$

In the next stage, using fundamental digital logic gates such as AND, OR and NOT gates, different stage-2 signals $P_1 - P_6$ are generated as shown in Fig. 3b. Finally, the required switching pulses for the proposed symmetric inverter is obtained by

$$
\begin{aligned}
S_1 &= T_1 & S_5 &= P_4 T_1 + P_3 T_2 \\
S_2 &= P_1 T_1 + P_5 T_2 & S_6 &= T_2 \\
S_3 &= P_2 T_1 + P_6 T_2 & S_7 &= P_5 T_1 + P_1 T_2 \\
S_4 &= P_3 T_1 + P_4 T & S_8 &= P_6 T_1 + P_2 T_2
\end{aligned}
$$

where, T_1 is the signal which is logic-1 during the positive cycle and logic-0 during the negative cycle and T_2 is the signal which is logic-0 during the positive cycle and logic-1 during the negative cycle.

5 Simulation Results

The magnitude of the voltage sources are $V_1 = V_2 = V_3 = V_4 = V_{dc} = 50$ V. The 9-level inverter output voltage waveform and its THD are shown in Fig. 4a and b, respectively.

As expected, the output voltage is of staircase waveform with THD as 9.36%. The output voltage and current waveforms for the proposed 9-level inverter for different series RL load parameters are shown in Fig. 5. For pure resistive load, $R = 35 \ \Omega$, the load voltage THD and load current THD are equal to 9.36%. Also, it is observed that the load current THD is varying between 2.35 and 9.36% as the power factor of the load varies from 0.5 to 1.

6 Conclusion

In this paper, a new symmetrical 9-level inverter topology with reduced number of switches is recommended. The proposed inverter uses four dc voltage sources and eight power switches to generate a 9-level voltage across the series RL load. The proposed inverter structure is compared with several other symmetrical topologies

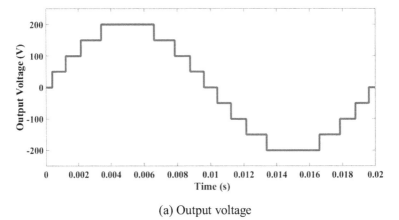

(a) Output voltage

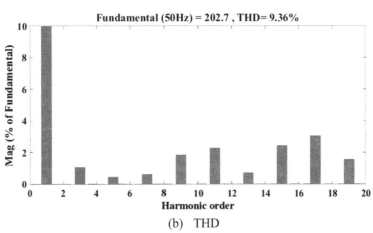

(b) THD

Fig. 4 Simulation results

which indicate the reduction in the number of switches, total number of conducting switches and TSV value, which in turn, achieves higher inverter efficiency and lower switching losses. Finally, the inverter performance is analyzed by generating a 9-level output voltage using Matlab/Simulink.

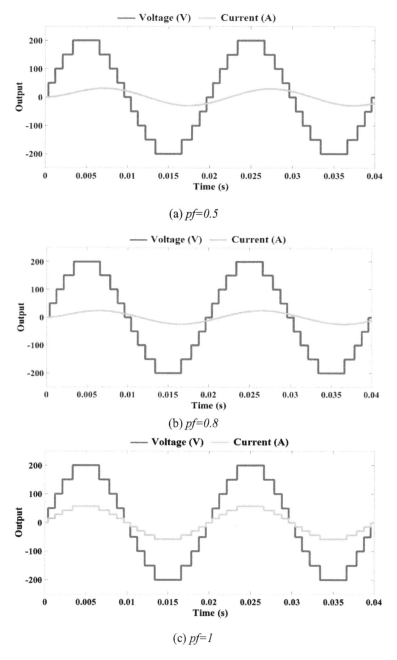

(a) *pf=0.5*

(b) *pf=0.8*

(c) *pf=1*

Fig. 5 Output waveforms

References

1. V. Thiyagarajan, P. Somasundaram, Multilevel inverter topology with modified pulse width modulation and reduced switch count. Acta Polytechnica Hungarica **15**(2), 141–167 (2018)
2. N. Prabaharan, Z. Salam, C. Cecati, K. Palanisamy, Design and implementation of new multilevel inverter topology for trinary sequence using unipolar pulse width modulation. IEEE Trans. Industr. Electron. **67**(5) (2020)
3. E. Babaei, S. Laali, Z. Bayat, A single-phase cascaded multilevel inverter based on a new basic unit with reduced number of power switches. IEEE Trans. Industr. Electron. **62**(2), 922–929 (2015)
4. M. Jayabalan, B. Jeevarathinam, T. Sandirasegarane, Reduced switch count pulse width modulated multilevel inverter. IET Power Electron. **10**(1), 10–17 (2017)
5. V. Thiyagarajan, P. Somasundaram, Modeling and analysis of novel multilevel inverter topology with minimum number of switching components. CMES **113**(4), 461–473 (2017)
6. V. Thiyagarajan, P. Somasundaram, Simulation and analysis of novel extendable multilevel inverter topology. J. Circuits Syst. Comput. **28**(06), 1950089-1-25 (2020)
7. P.K. Kar, P. Anurag, S.B. Karanki, Development of an enhanced multilevel converter using an efficient fundamental switching technique. Int. J. Electr. Power Energy Syst. **119**, 1–13 (2020)
8. D.S. Vanaja, A.A. Stonier, A novel PV fed asymmetric multilevel inverter with reduced THD for a grid-connected system. Int. Trans. Electr. Energy Syst. **30**(4), 1–15 (2020)

Voltage Sag Mitigation Using Distribution Static Compensator

Megha Vyas, Vinod Kumar Yadav, Shripati Vyas, and R. R. Joshi

1 Introduction

Both, commercial and industrial users of electrical power are suffering from regularly or periodically outages due to problem of low quality of power [1]. Two main things in power are quality and reliability and these should be proper. Most severe issues in power system regarding power quality nowdays are voltage sags, harmonics distortion and a less power factor. Amongst various disturbances of power system sags, swells in voltage and harmonics are approximately simple problems for impressionable loads [2]. The compensator stated is also used for reducing harmonic distortion because of the presence of non-linear loads in power system and to reduce the properties of sag and swell in voltage [3]. The author considered DSTATCOM for reducing the voltage sag. This research deals with the combined process of the simulation and modelling to solve power quality issue and also for voltage sag and swell which depends on SPWM with help of DSTATCOM [4]. The main issues a distributed system having the voltage sag & swell and to resolve these problems, many power electronic devices are employed. Distribution STATCOM, that is, the most well-organized and real current conventional power device is employed in the power system and Voltage Source Converter (VSC) switching is done by Sinusoidal Pulse Width Modulation (SPWM) methodology while modelling and simulation of the planned system is done MATLAB environment.

M. Vyas (✉) · V. K. Yadav · S. Vyas · R. R. Joshi
Department of Electrical Engineering, College of Technology and Engineering, Udaipur,
Rajasthan 313001, India
e-mail: megha.vyas14@gmail.com

© The Author(s), under exclusive license to Springer Nature Singapore Pte Ltd. 2021   259
J. Kumar and P. Jena (eds.), *Recent Advances in Power Electronics
and Drives*, Lecture Notes in Electrical Engineering 707,
https://doi.org/10.1007/978-981-15-8586-9_24

2 Power Quality

2.1 Survey of Issues in the Power Quality

Disturbances in power affect all the electronic devices, even some sensitive & costly devices, like relays and changeable speed drive contractors further the whole electrical system [4]. Disturbances in power quality results in many types of problems like sags and swells in system voltage, flickers, distortion in harmonics, impulse transients, short and long interruptions, spikes in voltage and noise [5]. As per IEEE "electronic equipment's powering and grounding should be favourable to the equipment" [6] (Figs. 1, 2 and 3).

At the present time, issue of power quality is a very burning issue because most of the devices are semiconductor dependent and controlled by power electronic devices, further suffering by disturbances with poor power quality. In the current time, industrial and commercial users are using electrical power as important raw material and disturbances like sags and swells in voltage are never acceptable for proper functioning of their costly equipment's [7]. In power system, almost 85% problems of power quality are because of sag in voltage and the very common problem in industrial equipment is due to voltage dip. Hence, to recover these power quality problems we have used FACTS devices. With the use of FACTS controllers', power can flow along such line during both the normal and fault condition [8]. Injection of reactive power to a load point of common coupling can be done to reduce the sag in voltage. A greater speed and reactive power can be achieved with the help of

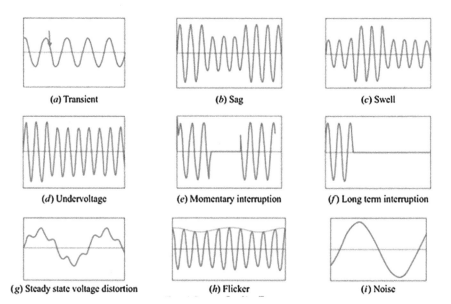

(a) Transient   (b) Sag   (c) Swell

(d) Undervoltage   (e) Momentary interruption   (f) Long term interruption

(g) Steady state voltage distortion   (h) Flicker   (i) Noise

Fig. 1 Power quality events

Fig. 2 Classification of
power quality problems

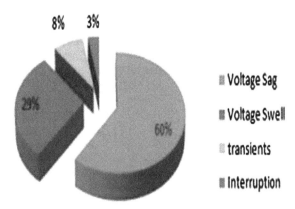

Fig. 3 Power quality
problem evolution process

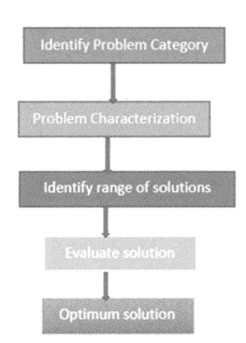

DSTATCOM to mitigate the problem of sag and swell in voltage. Reliable power
quality and harmonic reduction can be also achieved with the help of DSTATCOM
[9, 10] (Table 1).

Table 1 Power quality terms [11–15]

Terms	Description
Fluctuation in voltage	Variation in voltage in a regular way or frequently changes in series
Harmonic	It is a voltage or current which is proportional to system's fundamental frequency, induced due to the effect of non-linear loads
Inter harmonic	Usually a type of waveform distortion, a forced signal output on supply voltage by equipments
Electrical noise	Normally a high frequency intrusion, induced by a due to some factors like, by the electrical motor or arc welding which is exceptionable noise with current waveform or the voltage of system
Transients	"Surges" which originate in the re-striking or pre-striking of the switch during either opening or closing operations
Voltage sag	Collapse in AC voltage at a desired frequency from 0.5 cycles to 1 min's period. These are induced by frequently increment in load like short circuits or faults, turning on of electric heaters, starting of motors, or induced by undesired surges in impedance sources
Voltage swell	These are induced by a sudden decrease circuit load by a weak voltage regulator, yet also be induced due to movable spoiled connection of neutral
Noise—random transients	Impulses that are transported along with specified AC current

2.2 Power Quality Issues

At present time, power quality is a very big issue, because a very small disturbance can make system totally crash, can break communication of system and some other problems like harmonics, flicker, transients and these things result into more costing to power operators. The system should maintain sinusoidal waveform of bus voltage at specified frequency and voltage [16]. These problems of quality of power can be put into categories like short, long and continuous periods. Following are the issues in power quality:

1. Flicker
2. Less efficiency
3. Disturbances created inside the system.
4. Grounding and Bonding
5. Lower quality process
6. Interruption
7. Due to overheating of equipment, losses increase and health of equipment reduces.
8. Process shutdown
9. Equipment loss.

Due to issues in grounding and powering, data and processing difficulties can occur that can affect manufacturing and service qualities.

2.3 Approaches to Mitigate Power Quality Issues

1. **Grounding & bonding**: In grounding, a circuit is connected to ground and in bonding, interconnections of conductive path ensures basic electrical potential amongst the bonding parts.
2. **Correct wiring**: It is used for assuring correct wiring within a limit and should be inspected for insecure, misplaced or faulty joints at board.
3. **Redeveloping the technologies**: Basically, for bettering the power quality by joining electrical loads, like Voltage Sag, Transients, variation in frequency [17].

3 Methodology

DSTATCOM is the greatest significant controller for distributed networks. This is extensively employed for controlling network voltage, decrease harmonics, lesser number of surges and fewer requirements for load compensating [15]. It uses PWM technique, a power electronics converter and contains voltage source converter with 2-step self commutation, a dc storing device, a shunt connected coupling transformer equivalent with the distributed network over a coupled transformer [16]. Formation alike agrees the system to engage/produce controlled active and reactive power. The DSTATCOM has been applied mostly for the directive of voltage, power factor improvement and removal of harmonics in current [17]. Devices like these are activated to offer the steady regulated voltage by a collateral controlled converter. DSTATCOM is employed for limiting the voltage on assembly opinion and controlling is done by SPWM technique and needs rms voltage value dimensions at the point of load. It contains a VSC i.e. a grounded compensator employed for the alteration sags in bus voltage [18]. The efficiency of DSTATCOM to mitigating the voltage can be dependent on the following:

- Line impedance value
- The degree of fault at load line.

Various DSTATCOM's parts are presented in Fig. 4, which has a DC link capacitor, a transformer with inverter output feeding to line voltage, some extra inverter components and an ac filter [18].

3.1 Mathematical Modelling of DSTATCOM

The I_{sh}, injected shunt current adjusts the voltage sag with correcting drop in voltage around the network impedance and I_{sh} may be maintained by correcting output converter voltage in the network. The I_{sh} may be presented like

$$I_{sh} = I_L - I_S \tag{1}$$

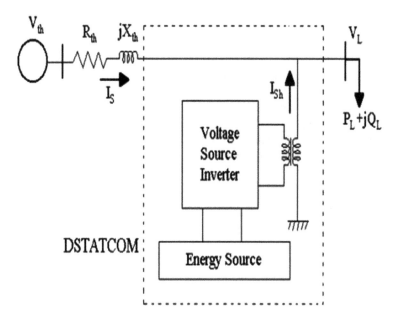

Fig. 4 Structure of DSTATCOM

The source current is

$$I_s = \frac{V_{th} - V_L}{Z_{th}}$$

The shunt current injected is

$$I_{sh} = I_L = \frac{V_{th} - V_L}{Z_{th}} \tag{2}$$

Further we can write

$$I_{sh} < \eta = I_L < (-\theta) - \frac{V_{th}}{Z_{th}} < (\alpha - \beta) - \frac{V_L}{Z_{th}} < (-\beta) \tag{3}$$

The injecting power to the DSTATCOM may be presented

$$S_{sh} = V_L I_{sh}^* \tag{4}$$

where, I_{out} = Output current
  I_L = Load current
  I_s = Source current
  V_L = Load voltage
  V_{th} = Thevenin voltage

Z_{th} = Impedance (Z_{th} = R + jX).

These expressions present DSTATCOM efficiency with modified voltage sag and based on Z_{th} or degree of fault of load bus. The current I_{sh} has been reserved in quadrature along the V_L, the desired voltage improvement may be proficient in the absence of any active power to power system. At another side, if I_{sh} is decreased, the equal voltage improvement can also be attained to the lesser injection of power into the power system.

4 DSTATCOM's Simulink Model

DSTATCOM is linked with distributed system for improving its operation. Distribution system's working according to power quality difficulties is analyzed with the help of DSTATCOM and studied with Simulink software in MATLAB environment.

In Fig. 5, Simulation of DSTATCOM is presented for the proposed system including a transmission system 230 kV, 50 Hz and showed with a Thevenin equivalent, connected to 3-phase X-mer's primary feeding in star/star/star, 230/11/11 kV. Secondary side is linked with a fluctuating load at 11 kV. A 2-step DSTATCOM is connected to tertiary of 11 kV to transfer rapid voltage at the load point in the system. At dc side a capacitor of 75 μF gives DSTATCOM energy storing abilities. For controlling the operation interval of DSTATCOM a Circuit Breaker is employed.

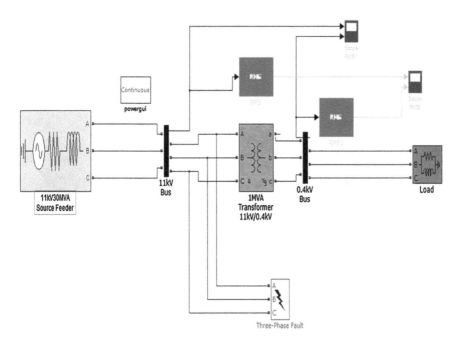

Fig. 5 simulation with or without DSTATCOM

4.1 Simulation Result DSTATCOM Performance During 3-φ Faults

In this case 3-Φ fault is created for voltage sag since the duration of the sag depends on the fault resistance, therefore, faults of different resistance are created and faults for different time interval are also created. Under 8 different cases the DSTATCOM has been analyzed and each case consists of line voltage magnitude in p.u. line voltage waveform and reactive current compensation provided to the distribution network. The load parameters for the system is considered as $R = 0.3\Omega$ and $L = 0.48$H. DSTATCOM of 19 kV energy storage device is considered explained as under:

Condition 1. If $R_f = 0.70 \ \Omega$, If $T_f = 0.55$ s $- 0.65$ s

By applying a resistance of 0.70 Ω, a 3-Φ fault is induced for a period of 0.55 s-0.65 s time and now, the system is operating in absence of DSTATCOM. Because of this, it results a drop in voltage from 1p.u. to 0.684p.u. in magnitude and 31% sag in voltage as presented in Fig. 6a. In second condition, when the DSTATCOM is present, then the voltage increases 0.684p.u. to 0.948 p.u. in magnitude because of giving a reactive current to the system and also, the voltage sag reduced 31% to only 6% which is presented in Fig. 6b.

The voltage magnitude in p.u.3-Φ waveform of the line and reactive current injection in p.u. without DSTATCOM compensation is shown in Fig. 7a. The voltage magnitude in p.u.3-Φ waveform of the line and reactive current injection in p.u. with DSTATCOM compensation is shown in Fig. 7b).

The Reactive current without DSTATCOM compensation, respectively, is shown in Fig. 8a. The Reactive current with DSTATCOM compensation, respectively, is shown in Fig. 8b.

In condition (1), of result shown in the faults are created for the same depth as the duration for fault (width of faults) is varied. As concluded from the above conditions that such drastic results can occur during the events like motor starting, and transformer energizing.

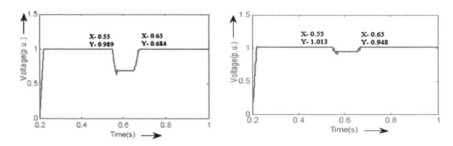

Fig. 6 **a** Line voltage magnitude without DSTATCOM compensation **b** Line voltage magnitude with DSTATCOM compensation

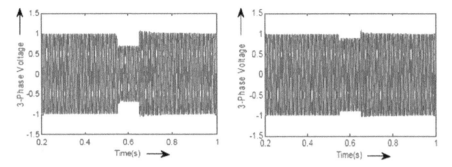

Fig. 7 **a** 3-Φ line voltage without DSTATCOM compensation **b** 3-Φ line voltage in presence of DSTATCOM

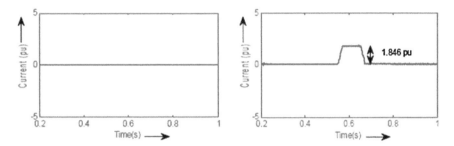

Fig. 8 **a** Reactive current without DSTATCOM compensation **b** reactive current with DSTATCOM compensation

5 Conclusion

In this research, the voltage sag issue and its effect on crucial and non-linear loads is studied. The method for designing of DSTATCOM's different parts is stated. By employing the SPWM technique, the controlling of voltage source converter is done. Modelling of DSTATCOM is implemented using the MATLAB/Simulink software. The results truly state the role of DSTATCOM for completely reducing the voltage sags and for a rapid dynamic response. The future work focuses on the use of multilevel inverter rather than DSTATCOM.

References

1. H. Hingorani, Introducing custom power. IEEE Spectrum **32**(6), 41–48 (1995)
2. M. F. Faisal, Power quality management program: TNB'sexperience. Distrib. Eng. Depart. TNB (2005), IEEE Std. 1159–1995. Recommended Practice for Monitoring Electric Power Quality

3. P. Boonchiam, N. Mithulananthan, Understanding of dynamic voltage restorers through MATLAB simulation. Thammasat Int. J. Sci. Tech **11**(3) (2006)
4. A. Ghosh, G. Ledwich, *Power Quality Enhancement Power Devices* (Kluwer Academic Publishers, 2002)
5. O. Anaya-Lara, E. Acha, Modeling and analysis of custom power systems by PSCAD/EMTDC. IEEE Trans. Power Delivery **17**(1), 266–272 (2002)
6. A. Visser, J. Enslin, H. Mouton, Transformer-less series sag compensation with a cascaded multilevel inverter. IEEE Trans. Ind. Electron. **49**(4), 824–831 (2002)
7. P. Loh, M. Vilathgamuwa, S. Tang, H. Long, Multilevel dynamic voltage restorer. IEEE Power Electron. Lett. **2**(4), 125–130 (2004)
8. E. Acha, V. G. Agelidis, O. Anaya-Lara, T.J.E. Miller, *Power Electronic Control in Electrical Systems* (Newnes Power Engineering Series, 2002)
9. P. Kumar, N. Kumar, A.K. Akella, Review of D-STATCOM for stability analysis. IOSR J. Electr. Electron. Eng. **1**(2), 01–09 (2012), ISSN: 2278-1676
10. D. Cholleti, Implementation of SPWM technique in DSTATCOM for voltage sag and swell. Int. Electr. Eng. J. **5**(12), 1649–1654 (2014)
11. H.A. Mohammad, M. Narender Reddy, Mitigation of voltage SAG and voltage swells by controlling the DSTATCOM. Int. J. Ethics Eng. Manage. Educ. **2**(1), 20–24 (2015)
12. M. Nair, *Power Quality Text Book*, 11th edn is bn no. 978-81-910618-6-4, Balaji learning published
13. A. Singh, B. Singh, *An Application of Iterative Learning Control Applied to Distribution Static Compensator System* (ITM Gurgaon, IIT Delhi)
14. S. Elango, E. Chandra Sekaran, Mitigation of voltage sag using distribution static compensator (D-Starcom). 978-1-61284-764-1/11/$26.00 ©2011, IEEE
15. S.K. G. Nawaz, S. Hameed, Mitigation of power quality problems using DSTATCOM **1**, 2347–2812, ISSN
16. P. Jyotishi, p. Deeparamchandani, Mitigate voltage sag/swell condition and power quality improvement in distribution line using D-STATCOM. Int. J. Eng. Res. Appl.
17. P. Hariyani Mehul, Voltage stability with the help of D-STATCOM, in *National Conference on Recent Trends in Engineering and Technology* (B. V. M. Engineering College, V. V. Nagar, Gujarat, India, May 2011), pp. 13–14
18. Y. Wang, J. Tang, X. Qiu, Analysis and control of D-STATCOM under unbalanced voltage condition, in *2011 International Conference on Mechatronic Science, Electric Engineering and Computer*, (Jilin, China, August 19–22, 2011)

Design and Practical Implementation of 750 W Solar Integrated Electric Vehicle

M. Indeevar Reddy, Meda Sai Rahul, Sai Kumar, M. Tejaswini, T. Dinesh, and K. S. Pratap

1 Introduction

Nowadays most of the companies are manufacturing Electric Vehicles. As of now, this number is very little. These electric vehicles are charged from the grid and need charge stations to travel for longer distances. A model of this type of electric vehicle is available on the Delhi roads which are popular and cheap. It can carry a load of 6 passengers weight, which is about 600 kilograms along with the body of the vehicle [1], Passengers feel comfortable traveling in these Electric Vehicles as they don't produce noise and smoke which cause pollution. As a result, these are widely used on the Delhi roads. Almost 4 lakhs of electric vehicles are registered in India. A recent forecast predicted that EVs will overtake traditional vehicle sales by 2038, in the world [2]. It also increases share in power demand of 2–3.5% of which at

M. Indeevar Reddy (✉)
Department of Electrical Engineering, Delhi Technological University, Delhi 110042, India
e-mail: indeevarreddymeegada1998@gmail.com

M. Sai Rahul · S. Kumar · M. Tejaswini · T. Dinesh · K. S. Pratap
Department of Electrical and Electronics Engineering, JNTUA College of Engineering, Kalikiri, India
e-mail: sairahulmeda15ka1a0228@gmail.com

S. Kumar
e-mail: dandasaikumar@gmail.com

M. Tejaswini
e-mail: mekkalatejeswani4@gmail.com

T. Dinesh
e-mail: dinesht43@gmail.com

K. S. Pratap
e-mail: pratapmnit@gmail.com

© The Author(s), under exclusive license to Springer Nature Singapore Pte Ltd. 2021   269
J. Kumar and P. Jena (eds.), *Recent Advances in Power Electronics and Drives*, Lecture Notes in Electrical Engineering 707,
https://doi.org/10.1007/978-981-15-8586-9_25

least half of that is being supplied by solar. This type of EVs makes use of batteries that are charged from conventional energy sources. This not only keeps a burden on the grid but also reduces power quality. Instead, solar power is implemented in these stations to charge these vehicles. But considering the storage complexity, it is not economical. Hence, we go for an electric vehicle that is integrated with the PV panel. It avoids the disadvantage of using grid energy, and to an extent, it reduces the demand on the grid. The vehicle is designed with a PV panel on its top [3], it is continuously connected through the MPPT charge controller to the battery bank [4]. It charges the battery as long as the sun is available in the sky. A major portion of the charging is taken from solar energy. It is also provided with charging ports used to charge the vehicle from the grid [5]. The design and pratical implementation are done considering the efficiency, cost of the vehicle, and robustness.

This paper is arranged in five sections commencing with an introduction followed by Sect. 2, which presents the information about the PV panel. In Sect. 3, the design of the vehicle, both electrical and mechanical components are discussed. Section 4, deals with real-time implementation and graphical analysis of vehicle efficiency on different roads, followed by a conclusion in Sect. 5.

2   Photovoltaic Module

The photovoltaic module acts as an energy source. It absorbs solar radiation and converts it into electric energy. This energy produced is utilized for charging batteries or a battery bank. This stored charge is used to drive electric appliances. Each photovoltaic module consists of numerous PV cells. Each cells has specific voltage rating (0.5–0.6 volts) and are connected in parallel and series to produce a PV module of specific rating. Such modules are again interconnected to produce large voltage output and forms into a solar array or PV system.

A single monocrystalline PV module of 328.6 Watts power rated is used as an energy source for the vehicle (Fig. 1 and Table 1).

3   Design Configuration of the Vehicle

Various equipment is used to design the vehicle includes BLDC motor, MPPT charge controller, battery bank, and body of the vehicle. The sequence Connection diagram of solar-powered EV is presented in the Fig. 2.

Fig. 1 Testing of solar panel

Table 1 Parameters and values of PV module [6]

Specification	Rating
Maximum power (P_{max}) in watts	328.6 W
Open circuit voltage (V_{oc}) in volts	45.5 V
Short circuit current (I_{sc}) in Amp	9.39A
Current at maximum power(I_{mp}) in Amp	8.79A
voltage at maximum power (V_{mp}) in volts	37.4 V
Module efficiency	17.0%
Cell efficiency	18.6%
Shunt resistance (R_{sh}) in ohms	81.9Ω
Series resistance (R_s) in ohms	0.052Ω

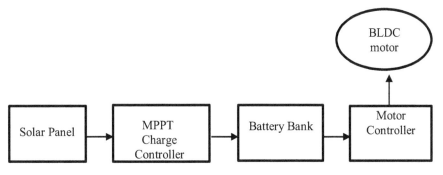

Fig. 2 Connection diagram of solar powered EV

Fig. 3 BLDC motor
coupled with sprocket wheel

3.1 Brush Less DC Motor

A BLDC motor of 48 volts, 19 Amp, 750 watts rated is used as driving force for
the vehicle. This motor is controlled by the motor controller. BLDC motor has wide
range of speed control charcterstics compared to other type of AC motors. It has
a sprocket wheel of 16 teethes attached to its shaft. The motor is coupled with the
long shaft of rear wheel through a gearing wheel arrangement. The motor works
only when it is supplied with the required voltage of 48 Volts connected through the
controller. Motor speed can be varied by rising the throttle. As we rise the throttle, a
signal is sent to the controller by which the speed of the motor is controlled (Fig. 3).

3.2 MPPT Charge Controller

A Maximum Power Point Tracker (MPPT) is designed to track the maximum power
from the solar panel at any instant of time. It is a power electronic device. It works in
constant current, constant voltage, and in constant power modes to charge batteries.
It makes the use of BUCK-BOOST converter and works with the help of an inbuilt
algorithm. It has four terminals. It is fed from the solar panel and connected to the
battery bank [4, 5, 1, 7, 8, 9] (Fig. 4 and Table 2).

This device can be used for the solar panels having voltage rating of 12–60 V and
charges lead acid battery packs. This device is also used for PV power generation
systems, distributed households, turbines, solar street lamps, and solar wind car
[11, 12].

Fig. 4 MPPT device (model name: MPT-7210A)

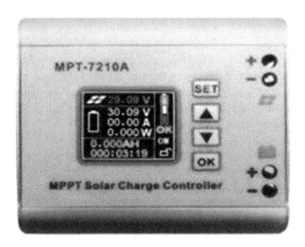

Table 2 Specifications of MPPT [10]

Parameters	Values
Input voltage range	12–60 V
Output voltage key range	24 V/36 V/48 V/72 V
Output current range	0–10Amp
Maximum output power range	0–100Watts

3.3 Battery Bank

Four lead acid batteries, each battery is of 12 Volts and 26 Ah rated are connected as are connected in series to produce 48 Volts output to run 48 V BLDC motors. These batteries are charged from both solar panel and the grid. Each battery has to be charged to 10% of the rated current. An external charger is provided to charge the vehicle from the grid. MPPT is connected to this battery bank (Table 3).

Table 3 Specifications of Battery [13]

Parameter	Rating
Nominal voltage	12 V
Charging voltage	13.75 V
Rated capacity (20-hour rate)	26 Ah
Charging current	10% of the rated current
Dimensions	
Total height	175 mm
(With terminals) Length	167 mm
Width	126 mm
Weight approx	9 kg

3.4 Body of the Vehicle

An auto chassis (body of the auto without a mechanical engine is chosen) which weighs about 150 kg with three wheels, one wheel at the front and the other two are rear wheels. It has an aerodynamic structure that eliminates air friction. It has a seating capacity of 5 persons. It is provided with a rare drum brake mechanism along with a front brake near the handle.

Weight to power ratio calculations are carried out to know the power rating of the motor to carry the required load. These calculations are carried out considering the rolling resistance force and aerodynamic force (Fig. 5).

4   Weight to Power Ratio Calculations

Consider the mass of the vehicle 375 kg and assumed speed is 8.33 Kmph.

The rolling resistance is defined as the force acting against the rolling motion whenever rolling objects rolls over the road surface. Equation governing rolling resistance force is given by the Eq. (1).

Fig. 5 Solar-powered electric vehicle that has been designed and implemented

$$F_{acc} = \mu r \times W \tag{1}$$

where
W—is vehicle Weight = Mass × Gravity,
μr—rolling resistance coefficient (0.0004–0.003)
Frolling = 375 × 0.01 × 9.81 (for car tires $\mu r = 0.01$)
= 36.78 N.

The Aerodynamic or air resistive force is simply a force exerted by the air against the moving objects, i.e., vehicles on the road. This opposing force is given by the Eq. (2).

$$F_{drag} = \frac{1 \times Cd \times Across \times \rho \times (v)^2}{2} \tag{2}$$

Cd—aerodynamic drag coefficient,
Across—the frontal area (sqft),
ρ—Density of Air mass = 1.2 Kgm3 (as per ISA),
v—speed of the vehicle in kmph.

$$F_{drag} = \frac{1 \times 0.32 \times 1 \times 1.2 \times (8.33)2}{2} = 13.32 \text{ N}$$

The drag force is detectable at speed of 40 km/h.

Force due to acceleration (*Facc*) of the vehicle is calculated from the below formulae.

According to newton's second law of motion.

$$F_{acc} = M \times a \tag{3}$$

$$= 375 \times 0.0925 = 34.6 \text{ N}$$

The net motive force required by the vehicle to prevail over these conflicting forces is given by

Ftotal = *Frolling* + *Fdrag* + *Facc* (adding equations: 1 + 2+3)
Ftotal = 34.6 + 13.2 + 36.78
Ftotal = 84.78 N

Total power required is P = 84.78 × 8.33 = 706 Watts.

Since 706 watts is required to carry a load of 375 kg. we choose the next highest value of the motor available in the market, i.e., 750 Watts.

4.1 Performance Analysis

To test the performance of each device and the vehicles, the following test and trails were performed. The values of the PV module are taken for every hour in a day along with the readings of the MPPT shown on the MPPT display. The variations in the MPPT set values and the output values are also tabulated in Table 4.

The above table shows changes in the output of the PV module according to the timings in a day. It also shows the MPPT output values which are maintained at a constant level for the variations in the PV output (Fig. 6).

Table 4 PV output voltage parameters at different temperature and timings along with MPPT values

S. No	Time and Temp(°C)	PV output voltage (V)	MPPT set value (V)	MPPT voltage (V)	Current set value (AMP)	Current (AMP)	Power (watts)
1	09:45 AM 28 °C	37.77	54	50.39	2.6	2.53	127.4
2	10:58 AM 29 °C	39.80	54	51.8	2.6	2.6	133
3	11:25 AM 30 °C	40.28	54	52.36	2.6	2.52	125.25
4	12.54 PM 32 °C	41.20	54	53.03	2.6	2.6	138
5	02:00 PM 37 °C	41.09	54	54.01	2.6	2.01	113.7

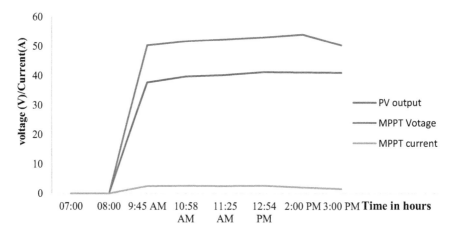

Fig. 6 Graph showing variations in PV output voltage and time along with MPPT output voltage and current values

Table 5 Speed parameters of the vehicle on different road surfaces

| Time (Sec) | High | Medium | Low | Backward speed on plane road | Speed on the irregular road surface | Mud rough surface road |
	Plane road speed (kmph)	Plane road speed (kmph)	Plane road speed (kmph)			
0	0	0	0	0	0	0
4	12	10	9	9	4	12
6	15.5	15.4	11	7.7	5.5	14.9
8	17	15.4	11	7.4	8.4	16.3
12	19.5	15.6	11.2	7.5	11.5	17.4
14	18.4	–	–	–	–	–

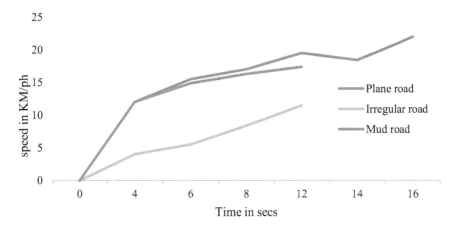

Fig. 7 Graph showing variations in the speed of the vehicle on the different road surfaces concerning time (sec)

The vehicle is driven on different roads with different loads in the vehicle. A test drive is conducted and the vehicle covered a distance of 15 km with a speed of 20 Kmph until the battery indicator is low. The test performed for a week and the speed readings on different road surfaces are recorded in Table 5 and (Fig. 7).

5 Conclusion and Future Work

Designing solar integrated EVs is still an evolutionary work. The vehicle presented in this paper is robust in construction and economical. The vehicle performance is evaluated from both mechanical and electrical aspects and the results indicate that the vehicle has a better prospect for a short distance in both urban and rural areas than traditional grid-powered electric bikes. They can also be used for longer distance by increasing PV panel efficiency and using large capacity smaller in size batteries.

As solar energy is the main source to charge the battery bank, the vehicle is completely independent of the traditional grid, this will help in reducing the burden on the grid and at the same time, this is one of the best solutions for green transportation for the country like India and Bangladesh. These vehicles eliminate fuel costs and electricity charges, thus making them economically viable compared to other solutions. Solar integrated Electric Vehicles are mainly useful to the people who live in tribal areas where there is no availability of vehicles that run with fuel.

The efficiency of the vehicle can be increased to a great extent by employing solar panels having good efficiency. With the recent advancement the efficiency of the solar panel (IBC) has been increased to 22.8% and from the energy storage prospect graphene batteries are the most suited type, but it is important to keep an eye on the economical prospects in designing a vehicle.

References

1. S. Sadagopan, S. Banerji, P. Vedula, M. Shabin and C. Bharatiraja, A solar power system for electric vehicles with Maximum power point tracking for novel energy sharing, in *2014 Texas Instruments India Educators' Conference (TIIEC)* (Bangalore, 2014), pp. 124–130. https://doi.org/10.1109/TIIEC.2014.029
2. T. Pachal, A.K. Dewangan, Electric solar vehicle—rayracer. Thesis work. https://doi.org/10.13140/rg.2.1.4276.5201,april-2016
3. P. Padmagirisan, V. Sankaranarayanan, Powertrain control of a solar photovoltaic-battery powered hybrid electric vehicle. Front. Energy **13**, 296–306 (2019). https://doi.org/10.1007/s11708-018-0605-8
4. B. Revathi, A. Ramesh, S. Sivanandhan, T.B. Isha, V. Prakash, G. Saisuriyaa, Solar charger for electric vehicles, in *2018 International Conference on Emerging Trends and Innovations In Engineering And Technological Research (ICETIETR)* (Ernakulam, 2018), pp. 1–4. https://doi.org/10.1109/icetietr.2018.8529129
5. S. Manivannan, E. Kaleeswaran, Solar powered electric vehicle, in *2016 First International Conference on Sustainable Green Buildings and Communities (SGBC)*, (Chennai, 2016), pp. 1–4. https://doi.org/10.1109/sgbc.2016.7936074
6. Solar panel information is available online https://www.loomsolar.com/collections/solar-panels
7. A. Goel, V. Rajput, S. Bajaj, R. Mittal, A. Dube, Ruchika, Solar hybrid electric vehicle— A green vehicle for future impulse, in *2016 3rd International Conference on Computing for Sustainable Global Development (INDIACom)* (New Delhi, 2016), pp. 2794–2800
8. Tejas H. Panchal, Vedang Y. Baxi, Simulation and Analysis of Various Techniques Used for MPPT Solar Charge Controller. Int. J. Res. Sci. Innovation-IJRSI **3**(1), 90–93 (2015)
9. G.R. Chandra Mouli, P. Bauer, M. Zeman, System Design for a Solar Powered Electric Vehicle Charging Station for Workplaces, vol. 168 (2016), pp. 434–44315. https://doi.org/10.1016/j.apenergy.2016.01.110
10. MPPT information is available online, PDF https://www.yampe.com/images/pdf/2907.pdf
11. J.V. Merin et al., Hybrid MPPT solar-wind electric vehicle with automatic battery switching, in *2018 IEEE 10th International Conference on Humanoid, Nanotechnology, Information Technology, Communication and Control, Environment and Management (HNICEM)* (Baguio City, Philippines, 2018), pp. 1–6

12. M. Nirmala, K. Malarvizhi, G. Thenmozhi, Solar PV based electric vehicle. Int. J. Innovative Technol. Exploring Eng. (IJITEE) **8**(22) (2018)
13. Battery information is available online https://www.quanta.in/images/pdf/Quantamanual-new.pdf

Controlling Mode Transition Noise Occurred at Ground Rail in Data Preserving MTCMOS Shift Register

Anjan Kumar

1 Introduction

The most commonly used sub-threshold leakage power reduction strategy in cutting edge integrated circuit is a MTCMOS technique also called as gated power technique [8]. In this technique, an additional high threshold transistors are connected between the source terminal of all NMOS of logical block and ground line. To meet timing constraint, logical block is designed using a low threshold transistor and a high threshold transistor is applied to limit the current flow in standby mode. Since transistors with both high and low threshold voltage are used to design the overall circuit, hence this technique is named as a multi-threshold CMOS (MTCMOS) technique. In combinational circuit data retention in sleep mode is not important as output at any time does not depend on previous output, but in sequential circuit data preservation is very much needed as output at any time depends both on the present, as well as the previous output. Very few techniques exist to preserve the data while reducing both sub-threshold leakage current and fluctuations on ground line [4, 9].

To apply the gated power technique in sequential circuits, some modifications are needed either in logical block or in footer high threshold transistor. Modification in a logical block is not feasible as an area of overall design increases [12]. In sequential MTCMOS design, fluctuations occur at the ground rail when making transitions from non-operatory mode to operatory active mode. This is because of the large current flowing through high threshold sleep transistors [5]. The fluctuations on the ground line is an essential reliability challenge for VLSI integrated circuits. In 2007, Zhiyu Liu et al. proposed SRAM flip flop with extra data retention cells connected in the slave part of the master-slave D FF. SRAM flip flop is effective in reducing

A. Kumar (✉)
GLA University, Mathura, India
e-mail: anjan.kumar@gla.ac.in

© The Author(s), under exclusive license to Springer Nature Singapore Pte Ltd. 2021   281
J. Kumar and P. Jena (eds.), *Recent Advances in Power Electronics*
and Drives, Lecture Notes in Electrical Engineering 707,
https://doi.org/10.1007/978-981-15-8586-9_26

sub-threshold leakage current and preserve the data in non-operatory mode but area overhead and fluctuations on the ground line during mode transitions are the major limitations. These limitations allow us to discover new low leakage MTCMOS design which can hold data in sleep mode and decrease fluctuations in mode transitions.

In this paper, data preserving MTCMOS is proposed using an additional high V_{th} P type MOS transistor connected in parallel with two high V_{th} stacked NMOS transistor. The proposed technique is applied to a 16-bit shift register. The simulation result is very much encouraging as fluctuations on the ground line during mode transitions are reduced significantly.

The paper is written in the following sequence. In the second section, prevalent sequential MTCMOS techniques are discussed. In the third section, proposed design is discussed. Analysis of the proposed design is discussed in section IV. Section V and VI show Simulation results and conclusion, respectively.

2 Exhisting Data Preserving MTCMOS Shift Register Circuit

In this section, various prevalent MTCMOS techniques used in the sequential circuit are briefly discussed.

2.1 Mutoh Technique

In Mutoh MTCMOS technique leakage power is reduced by eliminating sneak leakage path. Any path which bypasses off state sleep transistor is called sneak leakage path. To reduce sneak leakage path distributed sleep transistors are used in both footer and header. In sleep mode, data retention is made possible by connecting two inverters in cross-coupled fashion as shown in Fig. 1 [12].

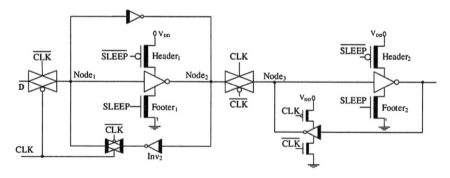

Fig. 1 Mutoh FF with facility of data retention [12]

Fig. 2 Tri-mode technique [5]

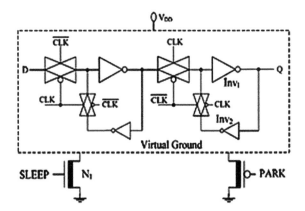

Although this technique is able to preserve the data while decreasing sub-threshold leakage power in the non-operatory sleep mode, during the non-operatory to operatory mode transition high fluctuations occur at the ground rail. Due to high fluctuations, wrong data is latched which decreases the reliability of overall circuitry [13]. Since several high threshold transistors are used, hence the overall area becomes high.

2.2 Tri-Mode Design

In this technique, an additional PMOS (high V_{th}) is attached in parallel with the NMOS(High V_{th}) transistor [5, 9]. An intermediate mode called park mode is inserted to preserve the data. Maximum leakage saving occurs in non-operatory mode when both park transistors and sleep transistors are off. Before complete reactivation, first park transistor is turned on, which provide sufficient voltage to retain the previous data and then complete reactivation occurs by turning on sleep transistor N1

This technique uses two approaches to decrease fluctuations at the ground rail. first by using small-sized high threshold NMOS and PMOS and second stepwise discharge of virtual ground rail potential (Fig. 2).

3 Proposed MTCMOS Shift Register

A modified stacked transistor-based MTCMOS shift register is proposed. The proposed design consists of two high threshold transistors connected in series as shown in Fig. 3. Transistor directly connected to ground behaves like a diode as gate and drain are connected. To control the node potential of VGND2, an additional capac-

Fig. 3 Proposed design

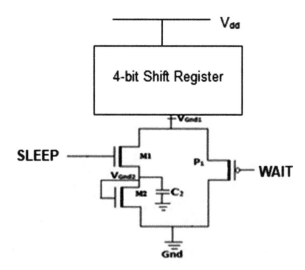

itor is connected between two high threshold transistors. This extra PMOS (high V_{th}) transistor is used to insert an additional mode between non-operatory mode to operatory active mode. In this mode, data retention is achieved as sufficient voltage is provided at a virtual node by PMOS. This intermediate mode is called as data preserving wait mode.

4 Analysis of Proposed Mtcmos Shift Register

4.1 Method to Reduce Fluctuations at Ground Rail During Non-Operatory Mode to Operatory Mode

Proposed design follow two important steps

Step 1:-
In non-operatory sleep mode, the voltage across VGND1 is approximately equal to VDD as all the high threshold transistors are turned off. In non-operatory sleep to active mode transition, an additional wait mode is inserted by turning on a high threshold PMOS transistor. In this mode, data is also preserved as sufficient voltage is there at VGND1. The addition of this intermediate wait mode also reduces the sudden fluctuation as a parasitic capacitor connected at VGND1 node is discharged up to the threshold voltage of the PMOS transistor.

Step 2:-
Fluctuation on ground line is again decreased by controlling the transition current flowing through high threshold stacked transistors during the complete activation

process. This is done by connecting the second sleep transistor (M2) and additional capacitor C_2 connected at VGND2.

During the wait to operatory mode change, first NMOS (M1) is made in on state and PMOS is made in the off state. In non-operatory to operatory mode transition voltage greater than the threshold voltage of M1 is applied on SLEEP input. Now parasitic capacitor at virtual ground starts discharging and an external capacitor starts charging. When an external capacitor reaches up to the threshold voltage of the second NMOS it will also turn on.

The first transistor turns on in saturation region while the second transistor turns on in the active region as $V_{DS} > V_{GS}-V_{th}$ and drain current across second NMOS transistor can be written as

$$I_d = \mu_n C_{ox} \frac{W}{L} \left[\frac{V_{DS}^2}{2} - V_{th} V_{DS} \right] \tag{1}$$

From Eq. (1), it is concluded that, the drain current is controlled by controlling drain to source voltage. Since virtual ground voltage decreases V_{DS} also decreases. Decreases in the drain to source voltages reduce drain current. Which in turn, reduces fluctuation at ground rail.

4.2 Method to Decrease the Sub-Threshold Leakage Power in Non-Operatory Mode

In non-operatory sleep mode, all high threshold transistors are turned off. From sub-threshold current equations, it is seen that there are three voltage components (V_{GS}, V_{DS} and V_{BS}) which affect the leakage current

$$I_{sub} = I_0 \cdot e^{\frac{q}{nk_BT}(V_{GS} - V_{THO} + \gamma V_{BS} + m V_{DS})} \left(1 - e^{\frac{-qV_{DS}}{k_BT}} \right) \tag{2}$$

$$I_0 = \mu \cdot \frac{W}{L} \sqrt{\frac{q\varepsilon_{Si}NDEP}{2\Phi_S}} \left(\frac{k_BT}{q} \right)^2 \tag{3}$$

V_{THO} threshold voltage at zero body bias, γ and m is body effect coefficient and DIBL coefficient, respectively. From Eqs. (2) and (3), it is concluded that as two voltage component (V_{DS} and V_{BS}) affect the leakage current. Sub-threshold current decreases exponentially if V_{DS} and V_{BS} increases. In proposed circuit drain to source voltage of First NMOS(M1) decreases as virtual ground node potential(VGND2) increases due to drain current of the first NMOS transistor(M1). So the net V_{DS} of first NMOS transistor decreases and V_{BS} becomes negative. Due to these two effects, the overall sub-threshold current gets reduced.

5 Simulations Results

The simulation is performed on Tanner EDA V13.0 using 90-nm MTCMOS technology [4]. Threshold voltage of different transistors is given in Table 1.

First of all, 4-bit shift register is designed using master-slave D FF and the proposed techniques are applied. To check the effectiveness of the proposed MTCMOS technique result is compared using two existing techniques (Mutoh, Tri-mode) using the same 8-bit shift register. The size of the transistor shift register is accomplished by keeping the rising and falling times equal to each other and likewise high to low and low to high propagation delay. The sizing of additional high V_{th} transistor is performed to get an equal sum propagation delays (clock-to-q delay) and setup time with individual FF and a register. The sizes of sleep transistors are tabulated in Table 2.

5.1 Fluctuations on Ground Line Comparison

Here the fluctuation on ground line by different MTCMOS circuits is compared. In order to perform fluctuation on ground line analysis and additional 40-pin DIP parasitic impedance model is connected on both header and footer.

The peak value of fluctuation on the ground line using various gated power circuit configurations are tabulated in Table 3. Now supply voltage is varied from 1.2 v to 0.6 V and the corresponding fluctuation on the ground line is listed in Table 4. The simulation result clearly shows that the proposed circuit configuration is most

Table 1 Parameters for different transistor

Parameter	Value (mV)
Low V_{th}(NMOS) (Used in shift register)	65
High V_{th}(NMOS) (M1 and M2 transistors)	290
Low V_{th}(PMOS) (Used in shift register)	−55
High V_{th}(PMOS) (P1 Transistors)	−255
Capicitor (C_2)	4.5pf

Table 2 The width of header and footer in different circuit configurations

Circuit configurations	Header (PMOS) (μm)	Footer (NMOS) (μm)
Mutoh	700	500
Tri-mode	–	12.0
Proposed	–	12.0

Table 3 Value of the ground rail fluctuations in (mv)

High threshold PMOS size (μm)	0.12	28.0
Mutoh	201.1	
Tri-mode	22.40	19.48
Proposed	5.75	4.18

Table 4 Variation in fluctuations on the ground rail in mV at different supply voltage (Content of shift register is 0 in all FF)

Voltage (V)	Mutoh	Tri-mode	Proposed
1.2	222.6	30.46	8.14
1.0	201.1	22.40	5.75
0.8	190.3	17.08	2.79
0.6	182.1	12.87	1.88

Fig. 4 Variation in fluctuations on ground rail at different supply voltage

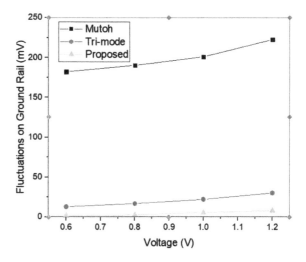

effective in reducing fluctuation on the ground line. There is an approximately 84.65% reduction in fluctuation on the ground line in comparison to commonly used Tri-mode technique as depicted in Fig. 4.

The sizes of PMOS transistor connected in parallel is used to calculate the data preserving voltage on ground rail. As the width of a high threshold transistor increases, resistance gets reduced, thereby decreasing the virtual ground voltage. Thus, the voltage shift on the virtual ground rails of the proposed and the Tri-Mode techniques are reduced as transistor widths increases, and hence the fluctuation on ground rail get reduced as shown in Table 4.

In Mutoh shift register, the total size of high threshold transistor is larger as compared to other technique, and hence very high instantaneous current is produced, and therefore, fluctuations on the ground line becomes high.

The proposed data preserving register gives minimum fluctuation on the ground line as compared to other data preserving shift register discussed in this paper. This is due to the extra circuitry connected in addition to transistor M1. The voltage fluctuation at the ground rail is decreased by a significant amount as shown in Fig. 4.

5.2 Leakage Current Comparison

A comparison of leakage current is shown in Tables 5 and 6. Result clearly shows that maximum leakage shaving is achieved in non-operatory sleep mode. The Leakage current in non-operatory sleep mode is reduced by 98.67% over the Mutoh technique. Figure 5 indicates leakage current relation with supply voltage variation. As supply voltage decreases, leakage current in standby mode also decreases.

In non-operatory sleep mode, the proposed technique generates minimum leakage current because of the stacking effect. The proposed 16-bit shift register reduces the leakage current by 4.97x as compared to the tri-mode technique in non-operatory sleep mode. On the other hand, in data preserving mode, both proposed and tri-Mode shift registers generate almost equal leakage current.

Table 5 Comparison of leakage current for different techniques (Content of shift register is 0 in all FF and supply voltage of 1 V)

Circuit configurations	Non-operatory mode	Data retention wait mode
	Leakage (nA)	Leakage (nA)
Mutoh	643.394	
Tri-mode	8.743	1018.871
Proposed	2.545	1016.165

Table 6 Change in Leakage current (nA) when Vdd is varying (Non-operatory mode)

Supply voltage (V)	Tri-mode	Proposed
1.2	10.435	4.132
1	8.743	2.545
0.8	7.960	1.919
0.6	7.155	0.818

Fig. 5 Variation in fluctuations on ground rail at different supply voltage

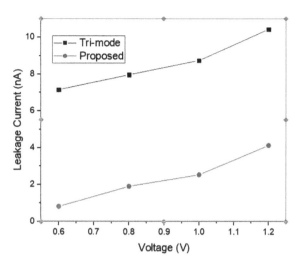

6 Conclusion

In this article, modified stacked transistor-based data retention shift register is presented. The proposed design decrease the fluctuations at the ground rail while preserving the data. Maximum leakage current minimization is recorded in non-operating sleep mode. To preserve the data additional wait mode is inserted by connecting extra PMOS in parallel with a stacked NMOS transistor. Using the proposed design subthreshold leakage current is minimized up to more than 80% as compared to the most commonly used tri-mode design. However, in data preserving mode both tri-mode and proposed technique generate almost equal leakage current. Fluctuations at the ground rail during mode transition is also affected by the proposed design. Result showed 70% decrement in fluctuations at the ground rail as compared to tri-mode design.

References

1. A. Abdollahi, F. Fallah, M. Pedram, A robust power gating structure and power mode transition strategy for mtcmos design. IEEE Trans. Very Large Scale Integr. (VLSI) Syst. **15**(1), 80–89 (2007)
2. S. Bandyopadhyay, S. Saha, U. Maulik, K. Deb, A simulated annealing-based multiobjective optimization algorithm: Amosa. IEEE Trans. Evol. Comput. **12**(3), 269–283 (2008)
3. M.H. Chowdhury, J. Gjanci, P. Khaled, Controlling ground bounce noise in power gating scheme for system-on-a-chip, in *2008 IEEE Computer Society Annual Symposium on VLSI* (IEEE, 2008), pp. 437–440
4. H. Jiao, V. Kursun, Ground bouncing noise suppression techniques for data preserving sequential mtcmos circuits. IEEE Trans. Very Large Scale Integr. (VLSI) Syst. **19**(5), 763–773 (2010)

5. Jiao, H., Kursun, V.: Tri-mode operation for noise reduction and data preservation in low-leakage multi-threshold CMOS circuits, in *IFIP/IEEE International Conference on Very Large Scale Integration-System on a Chip* (Springer, Beriln, 2010), pp. 258–290

6. H. Jiao, V. Kursun, Multi-phase sleep signal modulation for mode transition noise mitigation in MTCMOS circuits, in *2012 International SoC Design Conference (ISOCC)* (IEEE, 2012), pp. 466–469

7. J. Kao, A. Chandrakasan, MTCMOS sequential circuits, in *Proceedings of the 27th European Solid-State Circuits Conference* (IEEE, 2001), pp. 317–320

8. S. Kim, S.V. Kosonocky, D.R. Knebel, Understanding and minimizing ground bounce during mode transition of power gating structures, in *Proceedings of the 2003 International Symposium on Low Power Electronics and Design* (2003). pp. 22–25

9. S. Kim, S.V. Kosonocky, D.R. Knebel, K. Stawiasz, M.C. Papaefthymiou, A multi-mode power gating structure for low-voltage deep-submicron CMOS ICS. IEEE Trans. Circuits Syst. II: Express Briefs **54**(7), 586–590 (2007)

10. Z. Liu, V. Kursun, New mtcmos flip-flops with simple control circuitry and low leakage data retention capability, in *2007 14th IEEE International Conference on Electronics, Circuits and Systems* (IEEE, 2007), pp. 1276–1279

11. Z. Liu, V. Kursun, Characterization of wake-up delay versus sleep mode power consumption and sleep/active mode transition energy overhead tradeoffs in MTCMOS circuits, in *2008 51st Midwest Symposium on Circuits and Systems* (IEEE, 2008), pp. 362–365

12. S. Mutoh, T. Douseki, Y. Matsuya, T. Aoki, S. Shigematsu, J. Yamada, 1-v power supply high-speed digital circuit technology with multithreshold-voltage CMOS. IEEE J. Solid-State Circuits **30**(8), 847–854 (1995)

13. M. Pattanaik, B. Raj, S.,Sharma, A. Kumar, Diode based trimode multi-threshold cmos technique for ground bounce noise reduction in static CMOS adders, in *Advanced Materials Research*. vol. 548 (2012), pp. 885–889. Trans Tech Publications

14. S. Solanki, A. Kumar, R. Dubey, Stacked transistor based multimode power efficient MTCMOS full adder design in 90 nm CMOS technology, in *2016 International Conference on Communication and Signal Processing (ICCSP)* (IEEE, 2016), pp. 0663–0667

15. S.P. Thomas, A. Jose, Transistor full adder: a comparative performance analysis. Recent. Trends Electron. Commun. Syst. **5**(3), 22–31 (2019)

Modelling of Lithium-Ion Battery Using MATLAB/Simulink for Electric Vehicle Applications

Raja Kumar Sakile, Pawan Kumar, and Umesh Kumar Sinha

1   Introduction

Due to the lack of fossil fuels, crude oil price is increased day by day; in order to reserve the fossil fuels, we should go for Electric Vehicles (EVs). EVs are more popular due to less pollution and less running coast, but EVs are facing a major problem that is the Energy storage problem. For energy storage purposes, batteries are used. Therefore, the modelling of batteries is very important. EV application purpose considers any battery, most probably Lithium-ion battery as it has many advantages. Lithium-ion battery should convert into electrical equivalent circuit model and electrochemical model. Electrochemical models are not preferred for modelling due to more non-linear equations as they increase the complexity [1]. Accurate battery information is required such as SOC and remaining useful life of the battery. SOC means the charge available in the battery. Measurement of accurate Voltage and Current are important to manage the energy consumption of the system [2]. Accurate information is required to guide the circuit designers to prevent the overcharging and discharging of the battery [3].

Rechargeable batteries play an important role in EV applications. They can convert chemical energy into electrical energy. Rechargeable batteries are used in many electrical and electronic applications, such as Toys, Television, Smartphones and Laptops. Electric vehicles operation purpose higher rating batteries are used and the developed modelling is used to estimate the different types of parameters [4]. Therefore, the

R. K. Sakile · P. Kumar (✉) · U. K. Sinha
Electrical Engineering Department, NIT Jamshedpur, Jharkhand, India
e-mail: Pawank064@gmail.com

R. K. Sakile
e-mail: 2018rsee006@nitjsr.ac.in

U. K. Sinha
e-mail: uksinha.ee@nitjsr.ac.in

© The Author(s), under exclusive license to Springer Nature Singapore Pte Ltd. 2021   291
J. Kumar and P. Jena (eds.), *Recent Advances in Power Electronics
and Drives*, Lecture Notes in Electrical Engineering 707,
https://doi.org/10.1007/978-981-15-8586-9_27

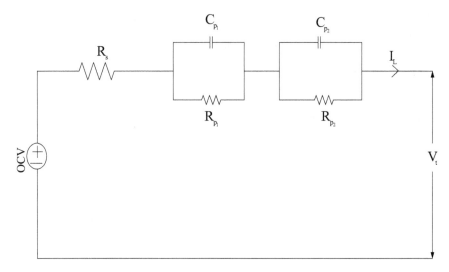

Fig. 1 Electrical circuit model of the battery model

life span of the battery is reduced. When the battery is in the stationary position, the calculation of parameters are easy. If the battery is in the moving position, then the calculation of parameters should be converted into a more complex one and some factors affect the system like climate change and temperature. Therefore, the lifespan of the battery is reduced for that proper design [5].

This paper is arranged as follows. Section 2 Introduces the battery modelling and presents the scenario of a rechargeable battery. Section 3 consists of the development of a battery model and Sects. 4 and 5 are the proposed model of a battery and conclusion, respectively (Fig. 1) [6].

2 Battery Modelling

A battery is a chemical device where many chemical factors are involved to design the modelling. Therefore, a battery can be replaced with an electrical model. The model can be developed in a MATLAB/Simulink environment. The battery parameters mostly depend on SOC and Current. Battery current and initial SOC, both the inputs are given to the OCV for the calculation of valid and accurate SOC, where V_{p_1} and V_{p_2} are the voltage across the RC parallel circuits, respectively. V_{int} is the voltage across the Shunt resistance and C_{p_1}, C_{p_2} are the polarization capacitances (Fig. 2) [7].

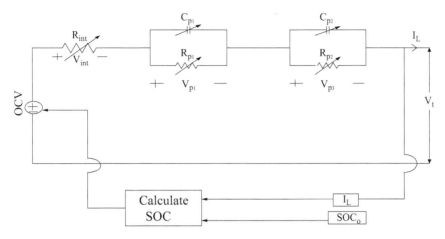

Fig. 2 Variable electrical model

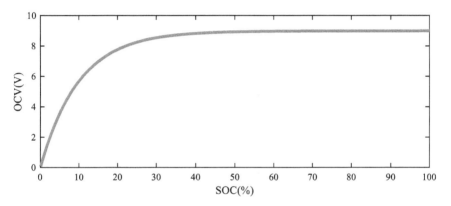

Fig. 3 SOC-OCV relationship curve

The circuit parameters (R_{int}, R_{p_1}, R_{p_2}, C_{p_2} and C_{p_2}) are identified from the Continuous Discharge curve [8, 9]. The SOC and OCV relationship is shown in Fig. 3. Due to some nonlinear parameters, the linear curve should be distributed (Fig. 4).

3 Proposed Model

The variable electrical circuit is proposed for the simulation. To control the battery voltage, five subsystems are required. Those are Calculation of OCV, Combination of RC values individually, Calculation of SOC, Voltage across the shunt resistance and voltage across the RC parallel networks. The proposed block diagram is as shown in Fig. 5. The subsystems are explained in the following [5].

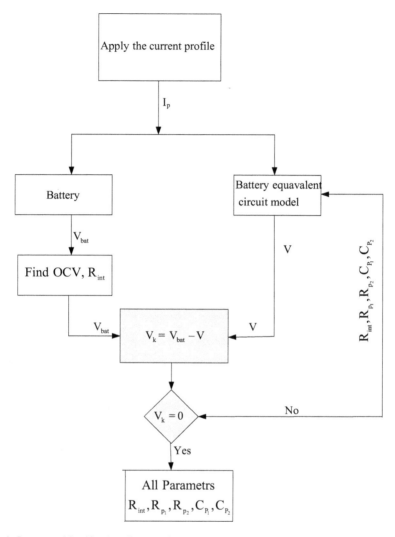

Fig. 4 Parameter identification diagram [3]

(1) **SOC Calculation**:

For the calculation of SOC, the subsystem contains two inputs which are Initial SOC and current. The output is considered as the SOC. The subsystem for SOC calculation is as shown in Fig. 5 [4].

$$x(t) = x(t)_0 - \int \frac{I * 100}{Q * 3600} \, dt \qquad (1)$$

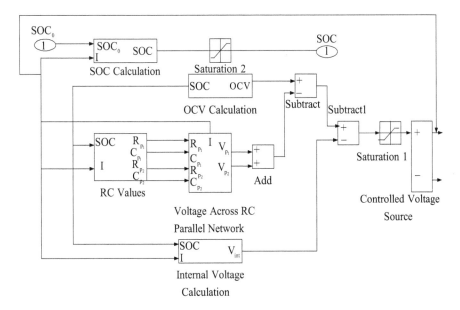

Fig. 5 Proposed model for MATLAB/Simulink

The above Eq. (1) is used to calculate real-time SOC, where $x(t)$, $x(t)_0$ and I are real-time SOC, initial SOC and the current, respectively. Q represents the capacity of the battery [5] (Fig. 6).

(2) **Calculation of OCV:**

The calculation of OCV is the important parameter in the subsystem. OCV is directly proportional to the SOC. The relationship can be established between OCV and SOC is a mathematical polynomial equation as shown in Fig. 7 [10, 11].

Fig. 6 Calculation of OCV

Fig. 7 Simple RC network

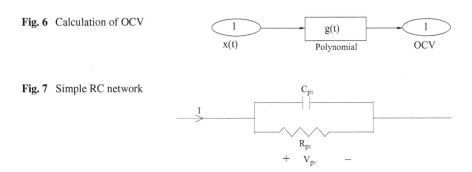

(3) **RC Values**:

RC values are more dependent on Current and Real SOC. RC parallel network values ($R_{p_1}, C_{p_1}, R_{p_2}, C_{P_2}$) are determined by using 2-D lookup tables, otherwise, Fig. 4 is used to determine all the parameters.

(4) **Voltage across RC Parallel Network**:

To determine V_{p_1}, consider a single RC network and write the related equations. Similarly, write the related equations to the second RC network. Finally, combine the equation blocks according to the written equations [12].

$$I = \frac{V_{p_1}}{R_{p_1}} + SC_{p_1} V_{p_1}$$

$$I = SC_{p_1}(V_{p_1} + \frac{V_{p_1}}{SC_{p_1} R_{p_1}})$$

$$\frac{I}{SC_{p_1}} = V_{p_1}(1 + \frac{1}{SC_{p_1} R_{p_1}})$$

$$V_{p_1} = (\frac{1}{S}) * \left[\frac{I}{C_{p_1}} - \frac{V_{p_1}}{R_{p_1} C_{p_1}} \right] \tag{2}$$

(5) V_{int} **Calculation**:

Internal Voltage or Voltage across shunt resistance can be determined by using Eq. (3). The internal resistance depends on SOC and Current.

$$V_{\text{int}} = I * R_{\text{int}} \tag{3}$$

After connecting five subsystem blocks, the terminal voltage can be determined by using Eq. (4) (Fig. 8).

$$V_t = OCV - V_{p_1} - V_{P_2} - V_{\text{int}} \tag{4}$$

4   Results and Discussion

The Battery parameters depend on current profile and SOC. The related waveforms of the proposed model are as shown in Fig. 8. The proposed model contains five subsystems, and each subsystem gets more results but the accuracy should be very less. The current profile is applied to the Circuit to get the accurate SOC. When the SOC is higher, then the current is discharging in nature and the SOC reduces towards zero and the current is charged vice versa. The related waveforms are presented below.

5 Conclusion

The proposed model has been developed in MATLAB/Simulink. The model was designed and explained, with each subsystem in detail. The proposed model gives accurate results such as Remaining useful life of battery and Accurate SOC. The Developed model is applicable to all type of Chemical related batteries. The designers develop battery models without any complex computation. This simple and easy model gives accurate results in the field of Electrical Vehicles and also Green technology application purposes. The developed model is suitable as an energy storage system. Lithium ferro phosphate battery is used for the calculation of accurate SOC in order to know the performance of the battery and battery model parameters.

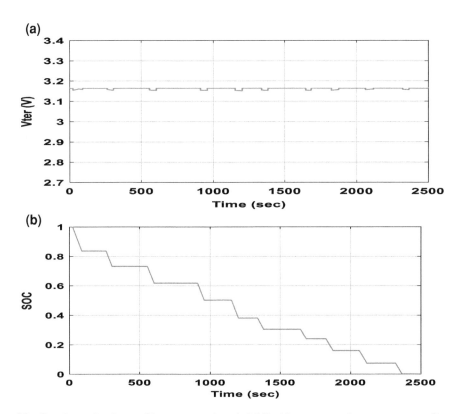

Fig. 8 **a** Internal voltage with respect to time, **b** SOC with respect to time, **c** current profile, **d** OCV-SOC curve

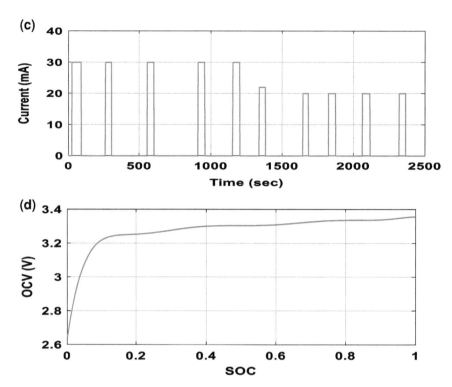

Fig. 8 (continued)

References

1. D. Liu, L. Li, Y. Song, L. Wu, Y. Peng "Hybrid state of charge estimation for lithium-ion battery under dynamic operating conditions. Electr. Power Energy Syst. 0142–0615 (2019)
2. A. Vasebi, S.M.T. Bathaee, M. Partovibakhsh, Predicting state of charge of lead-acid batteries for hybrid electric vehicles by extended Kalman filter. Energy Convers. Manag. **49**, 75–82 (2008)
3. R. Sakile, U.K. Sinha, Estimation of SOC (state- of-charge) and SOH (state-of- health) of lithium-ion battery by using adaptive techniques. IJITEE J. **9**(3) (2020)
4. L. Yang, Y. Cai, Y. Yang, Z. Deng, Supervisory long-term prediction of state of available power for lithium-ion batteries in electric vehicles. Appl. Energy 0306–2619 (2019)
5. L.W. Yao, J.A. Aziz, P.Y. Kong, N.R.N. Idris, Modelling of lithium-ion battery using MATLAB/Simulink. IEEE Conf. 978-1-4799-0224-8
6. Y. Shen, Adaptive online state-of-charge determination based on neuro-controller and neural network. Energy Convers. Manage. 0196–8904 (2009)
7. M.U. Cuma, T. Koroglu, A comprehensive review on estimation strategies used in hybrid and battery electric vehicles Renew. Sustain. Energy Rev. **42**, 517–531 (2015)
8. Z. Ma, N. Murgovski, S. Cui, Predictive energy management for electric variable transmission HEV. IFAC PapersOnline **52**(5), 417–422 (2019)
9. I. Baccouc, S. Jemmali, A. Mlayah, B. Manai, N.E.B. Amara, Implementation of an improved coulomb-counting algorithm based on a piecewise soc-ocv relationship for soc estimation of li-ion battery. Int. J. Renew. Energy Res. (2017)

10. G.L. Plett, Review and some perspectives on different methods to estimate state of charge of lithium-ion batteries. J. Automot. Saf. Energy **10**(3) (2019)
11. A. Vasebi, M. Partovibakhsh, S.M.T. Bathaee, A novel combined battery model for state-of-charge estimation in lead-acid batteries based on extended Kalman filter for hybrid electric vehicle applications. J. Power Sources **174,** 30–40 (2007)
12. T. Weigert, Q. Tian, K. Lian, State-of-charge prediction of batteries and battery–supercapacitor hybrids using artificial neural networks. J. Power Sources 0378–7753 (2010)
13. H.B. Sassi, F. Errahimi, N. Es-Sbai, C. Alaoui, Comparative study of ANN/KF for on-board SOC estimation for vehicular applications. J. Energy Storage 2352–152X (2019)
14. I.H. Li, W.Y. Wang, Merged fuzzy neural network and its applications in battery state-of-charge estimation. IEEE Trans. Energy Convers. **22**(3) (2007)
15. H.E. Hongwen, R. Xiong, X. Zhang, F. Sun, J. Fan, State-of-charge estimation of the lithium-ion battery using an adaptive extended kalman filter based on an improved the venin model. IEEE Trans. Veh. Technol. **60**(4) (2011)

Quality Factor Based Analysis of Radial Distribution System for Active Compensation

Kamala Kant Mishra and Rajesh Gupta

1 Introduction

Exponentially rising nonlinear loads in low voltage distribution system is a challenge to maintain power quality within the standard as per IEEE 519-1992. The harmonic loads present in the system may be voltage source harmonic load (VSHL), current source harmonic load (CSHL), and may be varying from VSHL dominant to CSHL dominant and vice versa. Both types of loads prevail in the low voltage distribution system (LVDS) [1–5]. To compensate the system, various active and passive filters have been reported by authors depending upon the system configuration and suitability of filters [6–14]. The compensation characteristic depends upon various system parameters and operating conditions. The use of active filters has application issues and was addressed [13]. Further various control strategies have been reported for filters considering multiple factors. The performance of active filters for load voltage control is also compared [15–22]. The factors which decide the performance of the system and need to be analyzed are source, feeder, and load. The feeder may be stiff or non stiff depending upon the impedance. This has an impact on the quality of power available at the PCC. Similarly, source impedance also has an impact on the quality of voltage available at the PCC. The source may be ideal or distorted. Source impedance has a role to play in the quality of power. The analysis of the radial distribution system needs analysis of source, feeder, and load.

K. K. Mishra (✉)
Kashi Institute of Technology Varanasi, Varanasi, India
e-mail: kkmishramnnit@gmail.com

R. Gupta
Motilal Nehru National Institute of Technology Allahabad, Prayagraj, India
e-mail: rajeshgupta@mnnit.ac.in

© The Author(s), under exclusive license to Springer Nature Singapore Pte Ltd. 2021
J. Kumar and P. Jena (eds.), *Recent Advances in Power Electronics and Drives*, Lecture Notes in Electrical Engineering 707,
https://doi.org/10.1007/978-981-15-8586-9_28

In this paper, a radial distribution system is considered for analysis with two PCC. One near the source itself, where the impedance of the transformer is considered as source impedance. Another PCC is considered at a point where feeder impedance has considerable value. The above mentioned concerned have been taken into account for the analysis. A radial distribution system is modeled and frequency domain analysis is done for performance evaluation. The analysis is further justified through the simulation results. The simulation results are experimentally validated.

2 Modeling of the Proposed Distribution System

Single line diagram for a radial distribution system considered for the analysis is shown in Fig. 1. The system is showing two load buses with the point of common coupling (PCC) denoted by PCC1 and PCC2 at these two buses. The impedance between the source and the load comprises of fixed and variable impedance.

2.1 Linear Load Connected System

The layout in Fig. 1, is reduced to an equivalent circuit for analysis in Fig. 2.

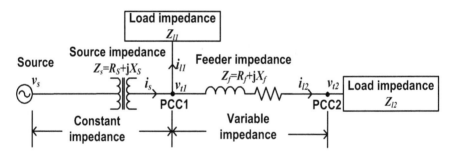

Fig. 1 Single line diagram of a radial distribution system considered for analysis

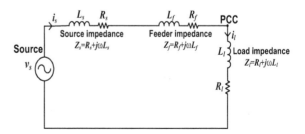

Fig. 2 Equivalent circuit of a radial distribution system under analysis

Following differential equation can be written by applying KVL equation in the circuit shown in Fig. 2.

$$\left(L_f + L_s\right)\frac{di_l(t)}{dt} + \left(R_f + R_s\right)i_l(t) + L_l\frac{di_l(t)}{dt} + R_l i_l(t) = v_s(t) \tag{1}$$

Taking Laplace transform of (1) with zero initial condition, we get

$$sL_f i_l(s) + sL_s i_l(s) + R_f i_l(s) + R_s i_l(s) + sL_l i_l(s) + R_l i_l(s) = v_s(s) \tag{2}$$

or (2) can be further written as

$$i_l(s)\left[\left(R_f + R_s + R_l\right) + s\left(L_f + L_s + L_l\right)\right] = v_s(s) \tag{3}$$

Considering the output $v_o(t)$ across the RL load, we get

$$i_l R_l + L_l\frac{di_l(t)}{dt} = v_o(t) \tag{4}$$

Following is obtained after taking the Laplace transform

$$i_l(s)[R_l + sL_l] = v_o(s) \tag{5}$$

Output to input voltage transfer function can be written as

$$\frac{v_o(s)}{v_s(s)} = \frac{R_l + sL_l}{\left(R_f + R_s + R_l\right) + s\left(L_f + L_s + L_l\right)} \tag{6}$$

From (6), the voltage transfer function (with output as terminal voltage v_{t2}) is written as

$$\frac{v_{t2}(s)}{v_s(s)} = \frac{R_l + sL_l}{\left(R_f + R_s + R_l\right) + s\left(L_f + L_s + L_l\right)} \tag{7}$$

The voltage transfer function (with output as terminal voltage v_{t1}) may be written by considering $Z_f = 0$ and linear load as

$$\frac{v_{t1}(s)}{v_s(s)} = \frac{R_l + sL_l}{(R_s + R_l) + s(L_s + L_l)} \tag{8}$$

2.2 Nonlinear Load Connected System

There are two types of nonlinear loads—voltage source and current source harmonic loads. A simple bridge rectifier circuit feeding dc load is shown in Fig. 3. The supply voltage v_l is fed to the bridge rectifier consisting of four diodes D_1, D_2, D_3, D_4. The dc output voltage V_d drives the dc current I_d. The load is a parallel combination of capacitor C and resistor R. The ac current consumed by the load is i_l [23]. This circuit is one of the important load considered under VSHL.

During the rectifier conduction period, $i_l(t)$ can be expressed as

$$i_l(t) \approx c\frac{d|v_l(t)|}{dt} + \frac{|v_l(t)|}{R} \tag{9}$$

Taking the Laplace transform of dc side current and voltage with zero initial condition, we get

$$I_d(s) = Csv_d(s) + \frac{v_d(s)}{R}$$

$$= v_d(s)\left[Cs + \frac{1}{R}\right]$$

$$= v_d(s)\left[\frac{1 + sCR}{R}\right] \tag{10}$$

or it can be further written as

$$v_d(s) = I_d(s)\left[\frac{R}{1 + sCR}\right] \tag{11}$$

Figure 4 shows a nonlinear load connected distribution system. Where v_l is equivalent distorted load voltage. Following can be written after writing KVL and taking Laplace transformation for the system shown in Fig. 4. Assuming nonlinear load to act like a parallel RC circuit, then following is analyzed in s-domain.

Fig. 3 Circuit diagram of a voltage source harmonic load for modeling

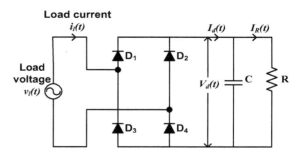

Fig. 4 A nonlinear load connected distribution system

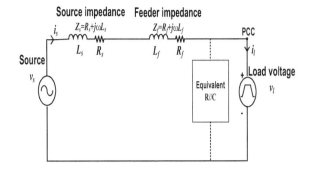

$$i_l(s)\left[(R_f + R_s) + s(L_f + L_s)\right] + v_l(s) = v_s(s) \tag{12}$$

From (11) and (12), the voltage transfer function at PCC2 is

$$\frac{v_{t2}(s)}{v_s(s)} = \frac{R}{(R_f + R_s + R) + s(L_f + L_s + CRR_s + CRR_f) + s^2(CRL_s + CRL_f)} \tag{13}$$

The voltage transfer function at PCC1 is written by considering $Z_f = 0$. From Eq. (13), the voltage transfer function at PCC1(v_{t1}) is written as

$$\frac{v_{t1}(s)}{v_s(s)} = \frac{R}{(R_s + R) + s(L_s + CRR_s) + s^2(CRL_s)} \tag{14}$$

3 Frequency Domain Analysis

3.1 Effect of Ideal and Distorted Source

The source impedance cannot be neglected irrespective of the load. The source current depends upon the nature of the load. The transformer impedance is considered as the part of the source and its impedance represents source impedance. Depending upon the nature of the load, the voltage at source terminal (PCC1) may be sinusoidal or distorted.

To analyze further, the feeder impedance is assumed zero, resulting in the proposed layout of the distribution system with the load connected at PCC1 only. The circuit for the system with loads at PCC1 is shown in Fig. 5.

For the analysis purpose, the model of the distribution system with zero feeder impedance is considered. The source may be considered to be ideal or distorted depending upon the quality of source/terminal voltage V_{t1} (V_s). The X/R ratio of the

Fig. 5 Circuit diagram for analysis of source with nonlinear connected load at PCC1

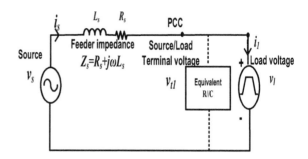

source as well as for the load is considered variable for the purpose of the study. Assuming constant VSHL (with constant $(X/R)_{load}$ ratio), the $(X/R)_s$ ratio is varied. The quality of voltage (sinusoidal or distorted) V_{t1} changes as $(X/R)_s$ ratio is varied. Where $(X/R)_s$ is the X/R ratio of the source. At some value of this ratio, the source behaves as an ideal source and at other value, it behaves as a distorted source.

The voltage at the PCC1 depends upon source/transformer impedance, feeder impedance, and load impedance. The load is connected at PCC1. The transformer impedance is assumed to be the part of the source impedance. To correlate source, feeder, and load impedance together, factor "a" and factor "b" are introduced. The factor "a" and factor "b" are defined below

$$a = \left[\left(\frac{X}{R} \right)_{load} \div \left(\frac{X}{R} \right)_s \right] \tag{15}$$

and

$$b = \left[\left(\frac{X}{R} \right)_{load} \div \left(\frac{X}{R} \right)_f \right] \tag{16}$$

For performing frequency domain analysis, the Eq. (13), is considered. The transfer function is the source impedance of the system. The analysis of the system is carried out using MATLAB software. The data as mentioned in Table 1, is used in (13), and Bode plot is drawn with varying values of source resistance and inductance. The load is linear with $(X/R)_{load} = 6.37$. The Bode plot for voltage transfer function at PCC1 with varying X_s/R_s is plotted as shown in Fig. 6.

From the plot shown in Fig. 6, it is clear that as the factor "a" decreases, it acts like a non stiff source and the source appears distorted. For higher values of "a", it acts like a stiff source and the source appears ideal.

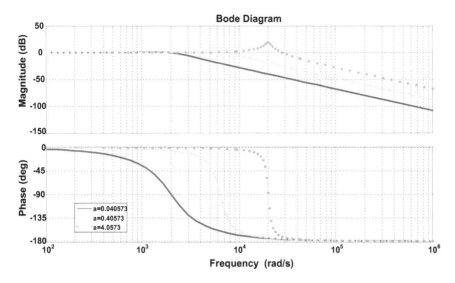

Fig. 6 Bode plot for the voltage transfer function at PCC1 with varying factor 'a' (X_s/R_s)

3.2 Effect of Weak and Strong Feeder

In this section, the proposed layout is analyzed at the PCC2. The load is considered at PCC2 with source and feeder impedance. The source impedance is constant with $(X/R) = 1.57$, making it an ideal source. The feeder impedance is varied depending upon the distance referring to Fig. 1. The distribution system has feeder impedance consisting of two parts, one as variable part and another one is a constant part. The constant part of the feeder impedance is the impedance of transformer/source, whereas the variable part is the impedance between the source terminal (PCC1) and load terminal (PCC2), i.e., the source impedance is the constant part, whereas the feeder impedance is the variable part.

For frequency domain analysis, the Eq. (7), is considered for study. The transfer function is the voltage transfer function at PCC2. The analysis of the system is carried out using MATLAB software. The data mentioned in Table 2, is used and Bode plot is drawn with varying values of feeder resistance and inductance.

The Bode plot for voltage transfer function at V_{t2} with varying "b" is shown in Fig. 7. From this plot, it is clear that as the value of "b" increases, the feeder acts like a stiff feeder (strong feeder). In other words, with an increase in the value of X_f/R_f, it becomes non stiff feeder (weak feeder).

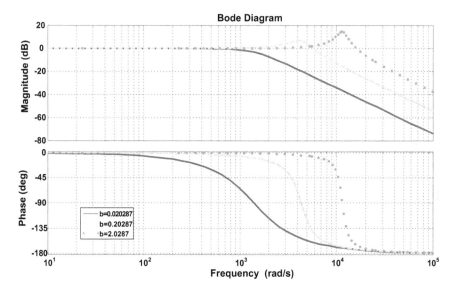

Fig. 7 Bode plot of voltage transfer function at PCC2 with varying "b" (X_f/R_f)

4   Simulation Results

4.1   Effect of Ideal and Distorted Source

The model as explained in Fig. 5, is considered for study using PSCAD simulation. The details of the source and load parameters are given in Table 1.

The simulation results with varying $(X/R)_s$ are shown in Fig. 8.

Where $(X/R)_f$ is the X/R ratio of the feeder. The X_l and R_l are reactance and resistance of the load. The X_s and R_s are reactance and resistance of source/transformer and X_f and R_f are reactance and resistance of the feeder impedance. Keeping load impedance constant, as the source impedance increases, the factor "a" decreases. From Fig. 8a, it is clear that with higher values of X_s/R_s (=157), the factor "a" decreases and the voltage (V_{t1}) THD increases to 5.17% resulting into crossing the

Table 1 Parameters of source and load

Voltage source (ideal), V_s	230 V single phase, 50 Hz
Source impedance, Z_s	$R_s = 0.001$ Ω, $0.000005 \leq L_s \leq 0.005$ H and $1.57 \leq X_s/R_s \leq 1570$
Load impedance, Z_l (Constant) [Voltage source type harmonic load]	A bridge rectifier with resistance R and capacitor C connected in parallel in the output circuit $R = 1$ Ω and $C = 500$ μF $(X/R)_{load} = 6.37$

(a)

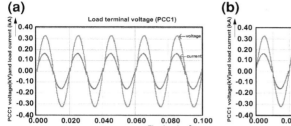

(b)

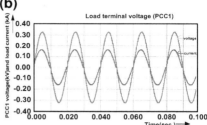

Fig. 8 **a** PCC1 bus voltage at $a = 0.040573$ ($X_s/R_s = 157$). The Voltage THD = 5.17% at PCC1 resulting in a distorted voltage source. The load current scale factor is 0.5. **b** PCC1 voltage with $a = 4.057325$ ($X_s/R_s = 1.570$). The Voltage THD = 0.08% at PCC1 resulting in an ideal voltage source. The load current scale factor is 0.5

limit imposed by IEEE 519-1992 of 5% for low voltage distribution system. In this case, the source voltage is assumed to be distorted at PCC1.

Similarly, in Fig. 8b, the value of "a" increases with a decrease in value of X_s/R_s (=1.57). In this case, the voltage (v_{t1}) THD decreases to 0.08% resulting in an ideal voltage source at the PCC1.

Considering the model of the distribution system with transformer and load connected at PCC1, the transformer impedance may be considered as part of the source impedance. Based upon this, the PCC1 voltage represents source terminal voltage for analysis of the PCC2 voltage.

The variation of source voltage THD with a change in X/R ratio of source/transformer in comparison to X/R ratio of load is shown in Fig. 9. The PCC1 voltage THD is plotted against the factor "a". The simulation using PSCAD is carried out for varying "a". The results are plotted as shown in Fig. 9. From Fig. 9, it is clear that as the factor "a" increases, the source shifts its quality from distorted to sinusoidal source.

When "a" is 0.4, the THD falls below 1% and considered to be an ideal voltage source at PCC1. When $a = 4$, the PCC1 voltage THD becomes almost zero. At lower values of "a" the source voltage at PCC1 behaves as a distorted source.

Fig. 9 PCC1 bus voltage (V_{t1}) THD with variation in factor "a"

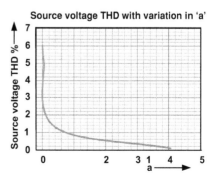

4.2 Effect of Weak and Strong Feeder

The model as explained in Fig. 3, is considered for study using PSCAD simulations. The detailed parameter of the source, feeder, and load considered in the simulation study is listed in Table 2.

Based upon the details as shown in Table 2, the simulation is carried out with variable feeder impedance. The source impedance is considered constant with $X_s/R_s = 1.570$. The feeder impedance is varied with variation of line inductance causing variation in X/R ratio of the feeder. The X_f/R_f is varied from 3.14 to 314 with constant $(X/R)_{load}$ (=6.37). The simulation result of load voltage THD at PCC2 is shown in Fig. 10a, b. From these figures, it is clear that the increase in the value of "b" shifts non stiff feeder into stiff feeder (strong). It is an important outcome of the analysis of the compensation. There is a need for compensation only at the lower values of "b".

The load voltage THD variation with change in the value of factor "b" obtained through simulation is shown in Fig. 11.

The PCC2 voltage THD is plotted against the factor "b". From Fig. 11, it is clear that as the factor "b" increases, the feeder shifts its quality from non stiff (weak) to stiff (strong) feeder. When "b" is 1.3, the THD falls below 1% and becomes a stiff feeder at "b" $= 2$ with THD becoming 0.28%. At lower values of "b" the feeder behaves as a weak feeder.

Table 2 Parameters of source, feeder and load

Voltage source (ideal), V_s	230 V single phase, 50 Hz
Source impedance, Z_s	$R_s = 1$ mΩ, $L_s = 5$ μH
Feeder impedance, Z_f	$R_f = 1$ mΩ, 10 μH $\leq L_f \leq 1$ mH A 10 km long feeder. $3.14 \leq (X_f/R_f) \leq 314$
Load impedance, Z_l (Constant) [Voltage source type harmonic load]	A bridge rectifier with resistance R_l and capacitor C_l in parallel with $R_l = 1$ Ω and $C_l = 500$ μF $(X/R)_{load} = 6.37$

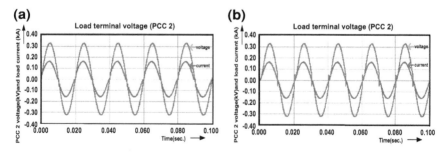

Fig. 10 **a** Load bus voltage (PCC2) at $b = 2.028662$ ($X_f/R_f = 3.14$). The voltage THD $= 0.28\%$ at PCC2 (stiff feeder). Load current scale factor 0.5. **b** Load bus voltage (PCC2) at $b = 0.020287$($X_f/R_f = 314$). The voltage THD $= 6.35\%$ at PCC2 (non stiff feeder). Load current scale factor 0.5

Fig. 11 Load bus voltage
THD (PCC2 voltage THD)
with variation in factor "*b*"

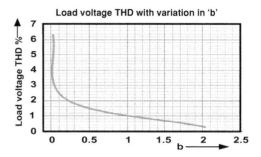

5   Experimental Results and Analysis

The simulation results under Sects. 4.1 and 4.2, are verified with the help of laboratory experimental setup. The simulation results of Sects. 4.1 and 4.2, are validated through experimental results. A nonlinear load is connected in the LVDS and results were obtained at various values of factor "*a*" and "*b*". The LVDS model with transformer is considered for the study. The transformer impedance represents source impedance.

5.1   Ideal and Distorted Source

The LVDS is analyzed for ideal and distorted source. The PCC1 voltage and current at varying values of "*a*" are shown in Fig. 12a, b. The load is connected at PCC2 and the source impedance is varied keeping load impedance constant. Figure 12a shows that the source voltage is almost ideal with voltage THD of 1.23% at $a = 2.59$. As the value of "*a*" is increased, the source voltage becomes distorted. Figure 12b shows that the source voltage becomes distorted with voltage THD $= 13.52\%$ at $a = 0.13$.

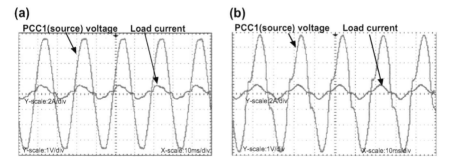

Fig. 12   a PCC1 (source) voltage at $a = 2.59$ representing ideal voltage source, **b** PCC1 (source) voltage at $a = 0.13$ representing the distorted source

(a)   **(b)**

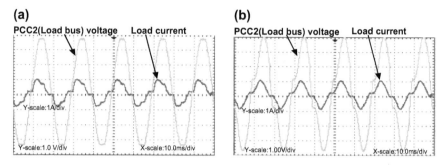

Fig. 13 **a** PCC2 (load bus) voltage at $b = 3.22$, **b** PCC2 (load bus) voltage at $b = 0.323$

From Fig. 12a, b, it is clear that the source voltage becomes more distorted as the value of "a" decreases resulting in a distorted voltage source. The experimental result validates the simulation results of Sect. 4.1.

5.2 Weak and Strong Feeder

The LVDS is analyzed for the weak and strong feeder. The PCC2 voltage and current at varying values of "b" are shown in Fig. 13a, b.

From Fig. 13a, b, it is clear that the PCC2 (load bus) voltage becomes more distorted as the value of "b" decreases resulting into distorted load bus voltage. The non stiffness (weakness) of the feeder increases with increase in the value of "b". Higher the value of "b" results into stronger feeder. The experimental result validates the simulation results of Sect. 4.2.

6 Conclusions

In this paper, a new layout of the distribution system is proposed for the LVDS, separating out feeder impedance into two parts, the constant part represents transformer impedance (source impedance) and the variable part represents the feeder impedance. Two models of the radial distribution system are presented. The classification of source and feeder has been shown with the help of simulation results. When the source is considered at the input of the transformer, the transformer impedance is considered as source impedance. To analyze the proposed system, factors of $(X/R)_{\text{load}}/(X/R)_{\text{source}}$, i.e., X/R of load to X/R of source at PCC1 and $(X/R)_{\text{load}}/(X/R)_{\text{feeder}}$, i.e., X/R of load to X/R of feeder, for PCC2 are considered. From the analysis, it is clear that the voltage at PCC1 will be ideal when X/R of load is higher than the X/R of source for PCC1 connected load. In this case, the loads at PCC2 are subjected to being fed by the ideal voltage source. Conversely, when X/R of load is lower than the X/R of the source for

PCC1 connected load, the loads at PCC2 are subjected to distorted source voltage. In case of loads at PCC2 being subjected to an ideal voltage source, a higher value of *X/R* of load in comparison to *X/R* of feeder results into stiff feeder causing non distorted voltage at PCC2. On the other hand, a lower value results into weaker (non stiff) feeder causing the distorted voltage at PCC2. Transfer function approach is adopted to verify the classification and proposed layout of the distribution system. Finally, experimental results are obtained to validate the simulation and analytical results for the distribution system.

References

1. S. Bhattacharyya, J.M.A. Myrzik, W.L. Kling, Consequences of poor power quality—an overview, in *UPEC 2007* (2007)
2. S. Kawasaki, G. Ogasawara, Influence analyses of harmonics on distribution system in consideration of non-linear loads and estimation of harmonic source. J. Int. Counc. Electr. Eng. **7**(1), 76–82 (2017). https://doi.org/10.1080/22348972.2017.1324267
3. F.E. Postigo Marcos, C.M. Domingo, T.G.S. Román, B. Palmintier, B.-M. Hodge, V. Krishnan, F. de Cuadra García, B. Mather, A review of power distribution test feeders in the United States and the need for synthetic representative networks. Energies **10**, 1896 (2017). https://doi.org/10.3390/en10111896
4. Y. Du, L. Du, B. Lu, R. Harley, T. Habetler, A review of identification and monitoring methods for electric loads in commercial and residential buildings, in *IEEE Proceedings on Energy Conversion Congress and Exposition*, 12–16 Sept. 2010 (2010), pp. 4527–4533
5. D. Venkatesh, S. Kumar, D.V.S.S. Siva Sarma, M. Sydulu, Modeling of nonlinear loads and estimation of harmonics in industrial distribution system, in *15th National Power Systems Conference (NPSC)*, IIT Bombay (2008)
6. M. Tarnini, New series active power filters for computers, loads and small non-linear loads, in *Proceedings of WCECS2009*, USA (2009)
7. H. Akagi, Active harmonic filters. Proc. IEEE **93**(12), 2128–2143 (2005)
8. P. Salmerón, S.P. Litrán, Improvement of the electric power quality using series active and shunt passive filters. IEEE Trans. Power Deliv. **25**(2), 1058–1067 (2010)
9. J.A. Pomilio, S.M. Deckmann, Characterization and compensation of harmonics and reactive power of residential and commercial loads. IEEE Trans. Power Deliv. **22**(2), 1049–1055 (2007)
10. A. Marini, L. Piegari, S.S. Mortazavi, M.-S. Ghazizadeh, Active power filter commitment for harmonic compensation in microgrids, in *Proceedings of the 45th IEEE IECON (IECON 2018)*, Lisbon Portugal (2019), pp. 7038–7044
11. A. Ghosh, G. Ledwich, *Power Quality Enhancement Using Custom Power Devices* (Kluwer Academic Publishers, USA, 2002)
12. V.N. Tul'skii, A.S. Vanin, M.A. Tolba, A.A. Zaki Diab, Arrangement of reactive power compensation units in the radial distribution network of Moscow oblast. Russ. Electr. Eng. **89**(6), 402–408 (2018). ISSN 1068-3712
13. O. Fatih Kececioglu, H. Acikgoz, M. Sekkeli, Advanced configuration of hybrid passive filter for reactive power and harmonic compensation. Springer Plus **5**, 1228 (2016). https://doi.org/10.1186/s40064-016-2917-7
14. J.K. Phipps, A transfer function approach to harmonic filter design. IEEE Ind. Appl. Mag. **3**(2), 68–82 (1997)
15. F.Z. Peng, Application issues of Active power filters. IEEE Trans. Ind. Appl. Mag. **4**, 21–30 (1998)

16. R. Gupta, A. Ghosh, A. Joshi, Multi-band hysteresis modulation and switching characterization for sliding mode controlled cascaded multilevel Inverter. IEEE Trans. Industr. Electron. **57**(7), 2344–2353 (2010)
17. R. Gupta, A. Ghosh, A. Joshi, Characteristic analysis for multi sampled digital implementation of fixed-switching-frequency closed-loop modulation of voltage-source inverter. IEEE Trans. Industr. Electron. **56**(7), 2382–2392 (2009)
18. R. Gupta, Generalized frequency domain formulation of the switching frequency for hysteresis current controlled VSI used for load compensation. IEEE Trans. Power Electron. **27**(5), 2526–2535 (2012)
19. M. Abuzied, A. Hamadi, A. Ndtoungou, S. Rahmani, K. Al-Haddad, Sliding mode control of three-phase series hybrid power filter with reduced cost and rating, in *Proceedings of the 44th IEEE IECON (IECON 2018)*, Washington DC, USA (2018), pp. 1495–1500
20. R. Gupta, A. Ghosh, Reduced order LQG controller for distribution static compensator used for load voltage control of distribution system. Lect. Model. Simul. AMSE Ser. A **8**(3), 33–43 (2007)
21. R. Gupta, A. Ghosh, A. Joshi, Cascaded multilevel control of DSTATCOM using multiband hysteresis modulation, in *IEEE Power Engineering Society General Meeting*, Jun. 18–22, 2006 (2006), pp. 1–7
22. R. Gupta, A. Ghosh, A. Joshi, Performance comparison of VSC-Based shunt and series compensators used for load voltage control in distribution systems. IEEE Trans. Power Deliv. **26**(1), 268–278 (2011)
23. J. Young, L. Chen, A.B. Nassif, W. Xu, A frequency domain harmonic model for compact fluorescent lamps. IEEE Trans. Power Deliv. **25**(2), 1182–1189 (2010)

Analyzing the Effect of Control Modes Operation of Multiple Facts Controllers on System Performance

Ramanaiah Upputuri, Shuvam Sahay, Shaik Riyaz, and Niranjan Kumar

1 Introduction

In order to enhance efficiency, reliability, and versatility of transmission systems, FACTS controllers are integrated into the power system. FACTS solutions allow existing transmission network capacity to be increased by the power grid while remaining or increasing the operating margins required for grid stability.

In [1],a novel Holomorphic Embedding Load-Flow Method (HELM) FACTS controllers modeling for overcoming the difficulties in traditional numerical methods by white germ an power series solutions was proposed. The authors [2, 3] discussed power injection and impedance insertion model of different FACTS controllers. The authors [4–7] investigated different population-based evolutionary algorithms for SVC, TCSC, TCSC-SVC, and UPFC optimal position and rating and then minimized the active power generation cost, active power losses, deviation in load voltage, loading of line, and installation cost by obtained optimal positions. A novel sparsity-constrained OPF problem for the optimal number of positions, types, and setting values of controllers like SVC, TCSC, and TCPS. This type of problems can be solved by an ADMM-IPM-STO algorithm [8].

R. Upputuri · S. Sahay (✉) · S. Riyaz · N. Kumar
Department of Electrical Engineering, NIT Jamshedpur, Jamshedpur, Jharkhand, India
e-mail: shuvam.sahay90@gmail.com

R. Upputuri
e-mail: 2018rsee005@nitjsr.ac.in

S. Riyaz
e-mail: nitjsrriyaz@gmail.com

N. Kumar
e-mail: nkumar.ee@nitjsr.ac.in

The optimal location of FACTS controllers reduces the cost of generation, bus voltage deviation, and active/reactive power losses as single weighted objective. The controllers enhance the voltage security in the system by Adaptive Multi-Objective Differential Evolution algorithm by preventive and corrective control with multi-objective optimization like minimization of voltage stability index, real power control variable, load shedding, and investment of controllers, and finally, a Fuzzy decision is implemented for best compromised solution [9, 10].

Further, the Whale Optimization Algorithm (WOA) for the reactive power planning in the presence of series and shunt controllers is presented in [11]. The optimal reactive power dispatch planning is enhanced by most favorable position of FACTS controllers in power system and also minimized single & multi-objectives by an efficient Quasi-Oppositional Chemical Reaction Optimization (QOCRO). The capacity and most favorable position of TCSC, the reactive power planning of power system, is enhanced with multi-objective optimization by an efficient Adaptive Differential Evolution Algorithm [12].

Based on the vigilant literature review, it is found that most of the literature confined their research to model controllers using power injections, current injections, voltage based injections, and impedance equivalent circuits. And also, the effect of these devices is analyzed on different aspects of power system operation and control. Hence, in this paper, a methodology to optimal locates multiple multi-type controllers in a given system so as to strengthen the system performance. This is accomplished by using an effective optimization algorithm, and this problem is resolved by keeping the system equality and inequality constraints within the limits. The considered objective function in a given system is total active power loss. The location of controllers is identified in such a way that the security of system is increased by minimizing system severity in terms of total power losses. Using the proposed methodology, the optimization problem can be solved with no difficulty. This method of approach is conducted on IEEE-30 standard test system with carrying numerical results.

The remaining paper is arranged as Sect. 2 present modeling of controllers and optimal location of these controllers. Sections 3 and 4 carry power system optimal operation along with optimization problem formulation with constraints and implementation methodologies. Section 5 deals result analysis followed by the conclusion in Sect. 6.

2   Modeling of FACTS Controllers

In this section, various series, shunt, and shunt-series controllers modeling are presented with its power flow incorporation procedure. These devices are modeled in such a way that the magnitude of voltage at the bus nodes, flow of active and reactive power through transmission lines can be controlled. For this, the load flow problem is solved by modifying the respective Jacobian elements in the Newton Raphson (NR) power flow algorithm. The complete SVC, TCSC, and UPFC modeling and its controlled mode operations are presented as follows

$$\begin{bmatrix} \Delta P_i^{new} \\ \Delta Q_i^{new} \end{bmatrix} = \begin{bmatrix} J_1^{new} & J_2^{new} \\ J_3^{new} & J_4^{new} \end{bmatrix} \begin{bmatrix} \Delta \delta_i \\ \frac{\Delta V_i}{V_i} \end{bmatrix} \qquad (1)$$

where $\Delta P_i^{new} = \Delta P_i^{old} + \Delta P_i^{FACTs}$, $\Delta Q_i^{new} = \Delta Q_i^{old} + \Delta Q_i^{FACTs}$
$J_1^{new} = J_1^{old} + J_1^{FACTs}$, $J_2^{new} = J_2^{old} + J_2^{FACTs}$
$J_3^{new} = J_3^{old} + J_3^{FACTs}$, $J_4^{new} = J_4^{old} + J_4^{FACTs}$.

In deregulated environment, the lateral and bilateral contractual agreements require to control the required magnitude of voltage as well as the flow of reactive and active power at particular bus and transmission line, respectively, by using FACTS controls in control mode operation.

2.1 Power Injection Model of SVC

Generally, SVCs are installed at load buses to enhance voltage at bus to a specified value by controlling the control variable [13]. The voltage angle and susceptance of SVC at installed bus are unknown parameters. The SVC is installed at bus-i as shown in Fig. 1.

The current and reactive power drawn by SVC is

$$I^{svc} = jB^{svc}V_i; \ Q^{svc} = Q_i = -V_i^2 B^{svc} \qquad (2)$$

The control modes in SVC are V_{state} ON & OFF. If V_{state} OFF, the installed bus acts as PQ bus (i.e., state variable V_i). If V_{state} ON, the installed bus acts as PV bus (i.e., state variable B^{svc}). Generally, SVC operates in V_{state} ON because for maintaining specified voltage at that bus.

It is observed that $J_1, J_2,$ and J_3 will not affect, because there is no change in bus voltage angle and active power, but J_4 is affected because reactive power changes due to variable susceptance model of SVC.

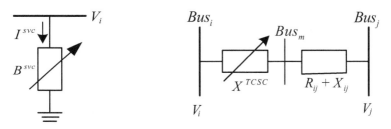

Fig. 1 Variable susceptance SVC model

$$J_4^{FACTs} = \frac{\partial Q_i}{\partial B^{svc}}; \quad B^{svc} = -V_i^2 B^{svc} = Q_i \qquad (3)$$

2.2 Power Injection Model of TCSC

Generally, TCSCs are installed to regulate transmission line power flow [13]. These tend to improve power system stability, voltage profile, etc. Here, TCSC is installed in a line in between the buses i & j as shown in Fig. 2. But in load flow problem a fictitious bus (m) is added for TCSC.

The modified power equations at bus-i are

$$P_i^{TCSC} = |V_i||V_m||Y_{im}|\cos(\theta_{im} + \delta_m - \delta_i);$$
$$Q_m^{TCSC} = -|V_m||V_i||Y_{mi}|\sin(\theta_{mi} + \delta_i - \delta_m) \qquad (4)$$

The modified power equations at bus-m are

$$P_m^{TCSC} = |V_m||V_i||Y_{mi}|\cos(\theta_{mi} + \delta_i - \delta_m);$$
$$Q_m^{TCSC} = -|V_m||V_i||Y_{mi}|\sin(\theta_{mi} + \delta_i - \delta_m) \qquad (5)$$

The control modes in TCSC are P_{state} ON & OFF. If P_{state} OFF, the flow of power in the installed transmission line is based on whatever the initial value of TCSC within the limits and if P_{state} ON, the transmission line power flow is regulated to specified value by controlling the X^{TCSC}, i.e., the linearized equation includes additional row and column.

$$J_1^{FACTs} = \frac{\partial P_i^{TCSC}}{\partial \delta_m}; \quad J_2^{FACTs} = \frac{\partial P_i^{TCSC}}{\partial V_m}V_m; i \neq m; \quad J_3^{FACTs} = \frac{\partial Q_i^{TCSC}}{\partial \delta_m}; i \neq m;$$
$$J_4^{FACTs} = \frac{\partial P_i^{TCSC}}{\partial V_m}V_m \qquad (6)$$

$$\left[\Delta P_{im}^{X^{t csc}}\right] = \left[\frac{\partial P_{im}^{X^{TCSC}}}{\partial \delta_i} \quad \frac{\partial P_{im}^{X^{TCSC}}}{\partial \delta_m} \quad \frac{\partial P_{im}^{X^{TCSC}}}{\partial V_i}V_i \quad \frac{\partial P_{im}^{X^{TCSC}}}{\partial V_m}V_m \quad \frac{\partial P_{im}^{X^{TCSC}}}{\partial X^{TCSC}}X^{TCSC}\right]\left[\begin{array}{c}\Delta X^{TCSC}\\X^{TCSC}\end{array}\right] \qquad (7)$$

Fig. 2 Variable series impedance power flow model of TCSC

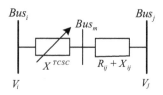

Fig. 3 Schematic diagram
of UPFC

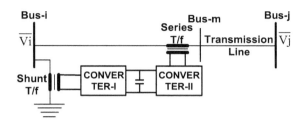

If P$_{state}$ OFF, then the diagonal element is equal to one and remaining elements of
that row and column are zeros.

2.3 Power Injection Model of UPFC

Generally, UPFCs are installed to regulate the flow of both active & reactive in
transmission lines as well as magnitude of voltage at the installed bus nodes. Here,
Fig. 3 shows UPFC is installed in a line between buses i & j. But in load flow problem
a fictitious bus (m) is considered for UPFC.

The shunt and series converters of UPFC are represented as voltage sources and
these voltage sources can be stated as

$$E_{vr} = V_{vr}\angle\delta_{vr}; E_{cr} = V_{cr}\angle\delta_{cr}.$$

The power injection model is derived from these voltage sources. For this, first
convert shunt and series converters into current injections and then converted into
power injections models [14].

2.4 Power Flow Solution Procedure

These devices should be positioned in an appropriate position to evaluate the impact
of these devices on the given power system and Newton Raphson power flow problem
has to be conducted. To this end, the steady-state system load flow equation must
be solved by changing the corresponding power mismatch equations and Jacobian
elements [15].

2.5 Optimal Location Methodology

These should be optimally placed aiming to optimize the operational benefits of SVC, TCSC, and UPFC. In this article, the approach [14] is based on the fuzzy composition of power system voltage deviation index and the severity index.

The position that yields minimum FVDI, FLLI, and FSSI is considered the best position for installing SVC, TCSC, and UPFC, respectively, after measuring this situation.

3 Formulation of Optimization Problem

The motive of the Optimal Power Flow (OPF) problem is to optimize (system control parameters like continuous parameters, e.g., real power generation (P^g) and magnitude of voltage (V^g) of the generators and discontinuous parameters, e.g., tap-changing settings (Tap^T) and shunt capacitor rating (Q^{Cap})) the objective function Total Power Loss (TPL) while fulfilling a set of system physical limitations (Thermal, Equipment, etc.) and constraints (equality and inequality). Finally, this OPF problem that gets non-linear optimization problem is described as

$$Min\ f(x, u).$$

Subjected to $g(x, u) = 0$ & $h(x, u) \leq 0$.

where f is the minimized objective function, g is the equality constraints set, h is the inequality constraints set; x stands for state vector variable and u stands for control vector variable.

3.1 Objective Formulation

The minimization of TPL objective function is the most realistic problem in planning and operation of power system. The mathematical representation as

$$P_{loss} = \sum_{n=1}^{nline} G_n(V_i^2 + V_j^2 - 2V_i V_j \cos(\delta_i - \delta_j)) \tag{8}$$

where G_n is the transmission line conductance, V_i, V_j are magnitudes of bus voltage δ_i, δ_j are angles at buses i and j.

4 Implementation Methodology

4.1 Optimization Algorithm

A technique based on a Unifomly Distributed Two-Stage Particle Swarm Optimization (UDTPSO) option in [13] is taken into consideration for solving OPF problem. This approach independently works on type of problem, number of equality and inequality constraints and number of buses in the system.

4.2 Procedure: Step by Step

For solving the proposed problem, follow the following steps.

1. Generate randomly initial population between the maximum and minimum limits of the particles.
2. Update bus and line data.
3. Perform methodology of NR load flow solution with control modes SVC, TCSC and UPFC.
4. Compute initial solutions and the fitness values, respectively.
5. Determine the best local solutions and select the best global solution.
6. Apply two steps of initialization and execute algorithmic operations.
7. Update the population and repeat the process from step 2–5 until the maximum iterations have been completed.
8. Store output values for the control variables considered.

5 Result Analysis

To study the proposed approach effectively, we have taken IEEE 30 test system [13]. In this work, the authors considered the controlled reactive and active power flow is 110% of base load flow of connected transmission line and magnitude of voltage at the connected bus is 1.05 p.u.

Modul-1: Identification of an optimum position for installing FACTS controllers.

Modul-2: Effect of multiple SVCs, TCSCs, and UPFCs control mode operation on performance of power system using OPF solution.

Modul-1: First, the optimal positions of the proposed FACTS controllers are identified based on the method shown in Sect. 2.4. For the system, the feasible locations of SVC, TCSC, and UPFC are 16, 20, and 20, respectively. In this procedure, the FVDI, FLLI, and FSSI indexes are diminished by maintaining equality and inequality limits using UDTPSO algorithm with SVC, TCSC, and UPFC, respectively. The FACTS controllers SVC, TCSC, and UPFC are located at minimum values of FVDI, FLLI,

Table 1 Optimized FVDI values for IEEE-30 test system with SVC in possible positions

S. No	SVC bus	FVDI value	TPL value
1	22	6.9041	9.6851
2	7	7.1023	16.4815
3	21	7.1134	14.0609
4	26	7.1656	13.0139
5	3	7.6611	15.3661

Table 2 Optimized FLLI values for IEEE-30 test system with TCSC in possible positions

S. No	Sending end of TCSC	Receiving end of TCSC	FLLI value	TPL value
1	27	30	315.9122	4.1078
2	27	29	318.4418	4.2827
3	18	19	319.6088	4.1976
4	14	15	321.1393	3.8794
5	15	18	325.2647	4.3185

Table 3 Optimized FSSI values for IEEE-30 test system with UPFC in possible positions

S. No	Sending end of UPFC	Receiving end of UPFC	FSSI value	TPL value
1	3	4	332.9646	4.3667
2	4	6	335.0143	4.5289
3	12	16	336.256	4.1698
4	21	22	344.0532	4.7496
5	12	15	345.7165	4.3003

and FSSI, respectively. Due to page limit, the most five minimum FVDI, FLLI, and FSSI values for the possible positions are shown in Tables 1, 2, and 3.

Modul-2: The optimal power flow results of multiple SVCs, TCSCs, and UPFCs with different control modes as shown in Tables 4, 5, and 6.

From Table 4, it is noticed that when the device operates in control mode, both total active power loss and generation fuel cost are increased when compared to the device that does not operate in control mode. As the number of devices is included in the system, both total active power loss and generation fuel cost are decreased when all the devices do not operate in control mode and increase when they are in control mode.

From Table 5, it is noticed that when the device operates in control mode, total active power loss decreases but generation fuel cost increases, when compared to the device that does not operate in control mode. As the number of devices is included in the system, total active power loss and generation fuel cost are decreased when all the devices do not operate in control mode and total active power loss decreases but generation fuel cost increases when they are in control mode.

Table 4 OPF results with multiple SVCs in control modes of IEEE-30 test bus system

S. No	Control parameters		1 SVC		2 SVC		3 SVC	
			Vstate = 0	Vstate = 1	Vstate = 0	Vstate = 1	Vstate = 0	Vstate = 1
1	Power generation (MW)	$P_1{}^g$	51.59576	51.83102	51.57218	51.98921	51.53404	52.10019
		$P_2{}^g$	79.99784	79.99014	80	80	79.99589	79.95231
		$P_5{}^g$	49.99673	50	49.98412	49.99794	50	50
		$P_8{}^g$	35	35	35	35	35	35
		$P_{11}{}^g$	30	30	30	29.9946	30	30
		$P_{13}{}^g$	40	40	40	39.9915	39.9927	40
2	Generator voltages (p.u.)	$V_1{}^g$	1.1	1.07079	1.09973	1.03248	1.09999	1.03287
		$V_2{}^g$	1.09785	1.06663	1.09667	1.02749	1.09844	1.02731
		$V_5{}^g$	1.08072	1.04760	1.08179	1.00165	1.08261	1.00150
		$V_8{}^g$	1.08757	1.05273	1.08668	1.01291	1.08948	1.01412
		$V_{11}{}^g$	1.1	1.09804	1.1	1.1	1.1	1.09801
		$V_{13}{}^g$	1.09994	1.03631	1.09170	1.04827	1.1	1.04769
3	SVC susceptance (p.u.)	$B_1{}^{svc}$	0.25	0.12058	0.25	0.24866	0.11210	0.10683
		$B_2{}^{svc}$	–	–	0.11178	0.00981	0.12197	0.00369
		$B_3{}^{svc}$	–	–	–	–	0.16427	0.15542
4	Total generation (MW)		286.5903	286.8212	286.5563	286.9733	286.5226	287.0525
5	Generation cost ($/h)		967.8574	968.4079	967.7196	968.749	967.6885	968.879
6	TPL (MW)		3.1903	3.4212	3.1563	3.5733	3.1226	3.6525

From Table 6, it is noticed that both total active power loss and generation fuel cost are increased, when devices operate in control mode, as well as the number of devices, include in the system.

6 Conclusions

From the analysis presented in this paper, it has been concluded that the power system performance has been enhanced with multiple multi-type controllers when compared to single device. For this, the controllers have been placed in different locations obtained after optimizing the developed severity index in terms of total power losses objective while maintaining system limits. Due to this, the performance of the system has been enhanced not only in terms of security but also in terms of system severity minimization. The presented methodology has proven its effectiveness in solving complex optimal power flow problem with single/multiple multi-type controllers while satisfying both device and system limits. From the results, it is identified that

Table 5 OPF results with multiple TCSCs in control modes of IEEE-30 test bus system

S. No	Control parameters		1 TCSC		2 TCSC		3 TCSC	
			Pstate = 0	Pstate = 1	Pstate = 0	Pstate = 1	Pstate = 0	Pstate = 1
1	Power generation (MW)	$P_1{}^g$	51.87407	51.85456	51.89076	51.8513	51.89076	51.86009
		$P_2{}^g$	80	80	80	80	80	80
		$P_5{}^g$	50	50	50	50	50	49.991011
		$P_8{}^g$	35	35	34.99294	35	34.99194	35
		$P_{11}{}^g$	30	30	30	30	30	30
		$P_{13}{}^g$	39.98492	40	39.97478	40	39.97478	40
2	Generator voltages (p.u.)	$V_1{}^g$	1.1	1.1	1.1	1.1	1.1	1.1
		$V_2{}^g$	1.09661	1.09703	1.09454	1.09711	1.09831	1.09878
		$V_5{}^g$	1.07738	1.07876	1.07352	1.08076	1.07968	1.07990
		$V_8{}^g$	1.08716	1.08662	1.08407	1.08767	1.08856	1.08707
		$V_{11}{}^g$	1.1	1.1	1.1	1.1	1.1	1.1
		$V_{13}{}^g$	1.07218	1.07289	1.07200	1.07103	1.07746	1.07318
3	TCSC reactance (p.u.)	$X_1{}^{tcsc}$	−0.16935	0.00773	−0.30427	−0.5	−0.01392	−0.5
		$X_2{}^{tcsc}$	–	–	−0.26818	−0.35	−0.13090	−0.35
		$X_3{}^{tcsc}$	–	–	–	–	0.01360	0.03
4	Total generation (MW)		286.859	286.8546	286.8585	286.8513	286.8562	286.8511
5	Generation cost ($/h)		968.4802	968.509	968.4423	968.5102	968.4385	968.5110
6	TPL (MW)		3.4590	3.4546	3.45848	3.4513	3.4575	3.4511

this implemented approach operates without dependent on the nature of the objectives and it can be applied to any type of system. The obtained results have been analyzed with supporting numerical results.

The total active power loss and generation fuels cost change depending on control mode of the device, type of device, and number of devices incorporated in the system.

Table 6 OPF results with multiple UPFCs in control modes of IEEE-30 test system

S. No	Control parameters		1 UPFC		2 UPFC		3 UPFC	
			Pstate = 0 Qstate = 0 Vstate = 0	Pstate = 1 Qstate = 1 Vstate = 1	Pstate = 0 Qstate = 0 Vstate = 0	Pstate = 1 Qstate = 1 Vstate = 1	Pstate = 0 Qstate = 0 Vstate = 0	Pstate = 1 Qstate = 1 Vstate = 1
1	Power generation (MW)	P_1^g	52.1929	52.21269	52.33130	52.19739	52.33205	52.23642
		P_2^g	80	79.99899	79.79782	80	79.96301	80
		P_5^g	50	50	50	50	49.99699	50
		P_8^g	34.99379	35	34.98084	35	34.91996	35
		P_{11}^g	30	29.99685	30	30	30	30
		P_{13}^g	40	39.99869	39.99344	40	39.89438	40
2	Generator voltages (p.u.)	V_1^g	1.03582	1.03008	1.06241	1.02044	1.06153	1.02166
		V_2^g	1.03222	1.02898	1.06001	1.01787	1.05907	1.01964
		V_5^g	1.01594	1.01286	1.03855	1.00030	1.03919	1.00490
		V_8^g	1.02215	1.01881	1.04754	1.01216	1.04625	1.01773
		V_{11}^g	1.1	1.09995	1.1	1.1	1.1	1.1
		V_{13}^g	1.03408	1.03265	1.02803	1.00759	1.04853	0.99443
3	Series source voltage magnitude (p.u.)	Vcr1	0.04	0.02566	0.04	0.01832	0.04	0.01734
		Vcr2	–	–	0.00544	−0.03	0.00465	0.03212
		Vcr3	–	–	–	–	0.00554	0.03246
4	Series source voltage angle (deg.)	Tcr1	−87.1236	−141.683	−87.1236	−106.877	−87.1236	−107.753
		Tcr2	–	–	−87.1236	68.9662	−87.1236	−118.585
		Tcr3	–	–	–	–	−87.1236	−37.5966
5	Shunt source voltage magnitude (p.u.)	Vvr1	1	0.99141	1	0.99668	1	0.99603
		Vvr2	–	–	1.09104	1.00281	1.1	0.99588
		Vvr3	–	–	–	–	0.98791	1.02615
6	Shunt source voltage angle (deg.)	Tvr1	−2.6809	−2.2123	−2.4073	−2.3486	−2.42178	−2.3288
		Tvr2	–	–	−2.0817	−3.0079	−2.09424	−3.0527
		Tvr3	–	–	–	–	−3.64910	−4.5009
7	Total generation (MW)		286.8511	287.2072	287.1034	287.1974	287.1064	287.2364
8	Generation cost ($/h)		968.457	969.3397	968.6233	969.3284	968.6258	969.4218
9	TPL (MW)		3.4511	3.8072	3.7034	3.79739	3.70639	3.83642

References

1. B. Mohsen, G. Eskandar, Holomorphic embedding load-flow modeling of thyristor based FACTS controllers. IEEE Trans. Power Syst. **32**(6), 4871–4879 (2017)
2. S. Ravindra, Ch.V. Suresh, S. Sivanagaraju, V.C.V. Reddy, Power system security enhancement with unified power flow controller under multi-event contingency conditions. Ain Shams Eng. J. **8**(1), 9–28 (2015)
3. A. Elmitwally, A. Eladl, Planning of multi-type FACTS devices in restructured power systems with wind generation. Electr. Power Energy Syst. **77**, 33–42 (2016)
4. A. Mohamad, A. Ibrahim, Voltage quality in delta Egypt network and its impact in oil industry. Energy Rep. **5**, 29–36 (2019)
5. A. Rahul, S.K. Bharadwaj, D.P. Kothari, Population based evolutionary optimization techniques for optimal allocation and sizing of thyristor-controlled series compensator. J. Electr. Syst. Inf. Technol. **5**(3), 484–501 (2018)
6. G. Noradin, A.P. Ali, E. Ali, A PSO-based fuzzy long-term multi-objective optimization approach for placement and parameter setting of UPFC. Arab. J. Sci. Eng. **39**(4), 2953–2963 (2014)
7. K. Kavitha, R. Neela, Optimal allocation of multi-type FACTS devices and its effect in enhancing system security using BBO, WIPSO & PSO. J. Electr. Syst. Inf. Technol. **5**(3), 777–793 (2018)
8. D. Chao, F. Wanliang, J. Lin, N. Shuanbao, FACTS devices allocation via sparse optimization. IEEE Trans. Power Syst. **31**(2), 1308–1319 (2016)
9. J. Preetha, D. Devaraj, Adaptive multi objective differential evolution with fuzzy decision making in preventive and corrective control approaches for voltage security enhancement. J. Rank. Inst. **355**(11), 4553–4582 (2018)
10. N. Ehsan, P. Mahdi, A. Hamdi, An efficient particle swarm optimization algorithm to solve optimal power flow problem integrated with FACTS devices. Appl. Soft Comput. J. **80**, 243–262 (2019)
11. S. Raj, B. Bhattacharyya, Optimal placement of TCSC and SVC for reactive power planning using Whale optimization algorithm. Swarm Evol. Comput. **40**, 131–143 (2018)
12. W.S. Sakr, A.E. Ragab, M.A. Ahmed, Optimal allocation of thyristor-controlled series compensators by adaptive differential evolution algorithm. IET Gener. Transm. Distrib. **10**(15), 3844–3854 (2016)
13. E. Acha., C.R.F. Esquivel, H.A. Perez, C.A. Camacho, in *FACTS Modeling and Simulation in Power Networks* (Wiley). ISBN 0–47085271-2
14. U. Ramanaiah, N.K. Anamika, Optimal operation of unified power flow controller under controlled conditions, in *IEEE International Conference on Power Electronics, Control and Automation (ICPECA)*, New Delhi, India (2019)
15. Ch.V. Suresh, S. Sivanagaraju, Analysis and effect of multi-fuel and practical constraints on economic load dispatch in the presence of unified power flow controller using UDTPSO. Ain Shams Eng. J. **6**, 803–817 (2015)

Consistency of Extended Kalman Filtering and Particle Filtering Techniques for the State Estimation of Brushless DC Motor

Appalabathula Venkatesh, Shankar Nalinakshan, and M. Tony Aby Varkey

1 Introduction

Brushless DC motors gained very much attention because of its special features like less maintenance and absence of brushes, revolutionary commutators, when compared to the conventional brushed motors. The first BLDC Motor was designed in 1960s to meet the reliability and non-sparking motor. Integration of electronic regulators further increases the efficiency and maximum power output of BLDC Motors (BLDCM). During 1980s as the permanent magnetic materials usage is increased, there is a development of permanent magnet brushless DC motors along with highly reliable electronic regulators which develop more power output than brushed motors. The main problem during armature reaction is the "commutation process" which mainly reduces the efficiency of the motors [1, 2]. The present day BLDCM has special features like better efficiency, extended operating cycles, reduced noise levels because of absence of revolutionary rectifier, better mechanical characteristics, quick dynamic responses which draws special interest on more research topics on BLDCM.

Reference adaptive systems method is available for solving tracking problems [3, 4]. The main problem in tracking problems is Gaussian noise signals which are

---

A. Venkatesh · S. Nalinakshan (✉)
Electrical and Electronics Engineering, The National Institute of Engineering, Mysuru, Karnataka, India
e-mail: shankar.nalinakshan@nie.ac.in

A. Venkatesh
e-mail: venkatesh15793@gmail.com

M. Tony Aby Varkey
Electronics and Communication Engineering, Presidency University, Bengaluru, Karnataka, India
e-mail: tonyaby.varkey@presidencyuniversity.in

© The Author(s), under exclusive license to Springer Nature Singapore Pte Ltd. 2021   327
J. Kumar and P. Jena (eds.), *Recent Advances in Power Electronics and Drives*, Lecture Notes in Electrical Engineering 707,
https://doi.org/10.1007/978-981-15-8586-9_30

introduced into the system due to various internal/external disturbances which make the system as highly non-linear.

EKF fails in such cases where system is highly non-linear, PF has attains much priority in selecting such conditions where Gaussian noise signals, which is a probability based state estimator.

2 Special Applications of BLDCM

BLDC Motors are used for Electronics Power Steering Systems (EPSSs) applications which detects the position of rotor and applies the optimum torque to drive the steering system. HVAC's applications like heating, ventilation and air conditioning systems are controlled by PWM technique.

BLDCM is integrated with hybrid vehicles to drive the drive/power-trains. Convenient in attaining peak point efficiency and simple for rotor cooling. BLDCMs are mostly used as a Motor Control Systems (MCSs) because of those special characteristics like power to weight ratio, efficiency, cost of controllers, cost of motor, heat content developed in the motor, max or peak point energy tracking, etc. BLDCM draws the special attention among different motors [5].

3 BLDCM with Adaptive Filtering Techniques

The problem in testing any system is its non-linearity nature either due to external/internal disturbances which made the systems as unstable systems. So many of the cases, we will linearize the systems by making them as a reduced order system [6].

Then, we will analyze the linear systems but in practical concern most of the systems are non-linear only, and mainly the problem with non-linear systems is a Gaussian input doesn't necessarily produce a Gaussian output [7]. In order to directly adapt non-linear model, one can go for filtering techniques like Kalman Filter, Particle Filtering, etc. [8]. In practical, robust control of plants is designed in a manner to meet the desired specifications even in the presence of disturbances, non-linearities and with the parametric uncertainties. These AFAs are helpful in these conditions.

3.1 Kalman Filtering Techniques

In general, a non-linear system mostly generates a non-linear outputs. The Riccati equations which is developed by Kalman and Bucy can develop a steady-state stable output even if a system is unstable. Weiner–Kolmogorov filter is also called as Kalman

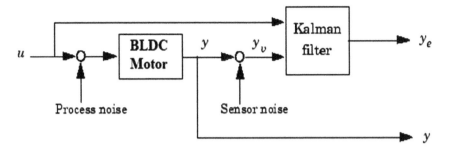

Fig. 1 Block diagram representation of BLDCM with KF

Filter. Kalman Filter will provide the optimal estimation algorithm to CTS/DTS based upon the estimated states [9].

KF mainly consists of two differential equations:

1. State Estimator
2. Covariance of Estimation error.

The block diagram representation of BLDCM with KF along with the estimated and actual states representation is as shown in Fig. 1.

Benefits of Kalman Filters:

- Kalman Filter provides running measure of accuracy of predicted position.
- Though the dynamic system is unstable, it provides stable state estimates.
- It can handle discrete time intervals between the measurement states.
- Uses point predictions instead of using probability in measurements.
- It uses output dynamics for the parameter optimization.

3.2 Extended Kalman Filtering Technique

EKF is the extension work of Kalman Filter. KF faces issues when the system is non-linear in nature inorder to nullify problems in KF. Extended Kalman Filter (EKF) is developed, and it is applicable to both linear as well as non-linear systems also for getting the optimal estimation and optimal solution. The EKF method is applicable to different types of non-linear systems in different formats based upon the nature of state estimators. Those are Continuous-Continuous EKF and Discrete-Continuous EKF, but Discrete-Continuous EKF is the most advantageous and frequently used algorithm because of the continuous measurement data availability nature of dynamical systems and the discrete data from the microprocessors in the modern robust control system applications. Complexity in estimating due to localized uncertainties adaptive sampling rate is used in EKF so that one can call EKF as Adaptive sampling rate filtering technique (ASRFT).

In general for the state estimator with error convariance matrices along with the dynamic system modelling makes the overall estimation and optimal techniques as unrealizable in computational process. The developed ASRFT makes the overall Sensorless BLDC state and output measurements as computational better algorithm.

State-Space Modelling of Continuous-Continuous EKF
Plant Equations

$$\frac{d}{dt}(x(t)) = f(x(t), u(t), t) + G(t) * W(t) \tag{1}$$

Output/Measurement equation $y(t) = h(x(t), t) + V(t)$ (2)

Initialization $\hat{x}(t_0) = \hat{x}_0$; $P_0 = E\left[\hat{x}(t_0)\left(\hat{x}(t_0)\right)^T\right]$ (3)

Kalman gain, Updated and Riccatti equations are as follows

$$K_{ekf}(t) = P(t)C^T R^{-1}(t)$$

$$\frac{d}{dt}(\hat{x}(t)) = f(\hat{x}(t), u(t), t) + K_{ekf}(t)\left[y(t) - h(\hat{x}(t), t)\right] \tag{4}$$

$$\frac{dP}{dt} = AP + PA^T - PC^T R^{-1}CP + GQG^T \tag{5}$$

Similarly, state space modelling of EKF in Continuous-Discrete form is also expressed, where Plant dynamics are expressed in Continuous form and measurements are expressed in discrete interval of time [9].

State-Space Modelling of BLDCM
State-Space representation of BLDCM with state Eq. (6) and output measurement Eqs. (7) is as expressed as follows [10].
State equation of BLDCM

$$x(k + 1) = A_d x(k) + B_d u(k) \tag{6}$$

which internally expressed as

$$f_1 = x_1(k + 1) = A_{11}x_1(k) + A_{12}x_2(k) + B_{11}u(k)$$
$$f_2 = x_2(k + 1) = A_{21}x_1(k) + A_{22}x_2(k) + B_{21}u(k) \tag{7}$$

Output/Measurement equation of BLDCM $y(k) = C_d x(k)$
In above state space modelling equations

$$x(k) = \begin{bmatrix} x_1(k) \\ x_2(k) \end{bmatrix} \rightarrow \text{Present state estimator } x_1(k) \rightarrow \text{Speed } x_2(k) \rightarrow \text{Mechanical}$$

position

where sampling time $T = 1$ s while converting CTS form of BLDCM into DTS form of BLDCM. The simulation results which are shown in Figs. 2, 3, and 4 show the BLDCM with EKF Block Diagram, Output and State estimations with EKF.

The simulation results which are shown in Figs. 3 and 4 are the state estimators and the output measurements under Gaussian white noise signals of a non-linear Sensorless BLDC motor. The estimated values are approximately reaching the actual values without much disturbances. To improve the performance of EKF, there are some modifications done to the EKF [11].

Based on that Kalman Filter techniques are further divided as

(a) Iterated EKF (IKF)
(b) Linearized KF (LKF)
(c) Unscented KF (UKF).

In IKF, method calculation of measured states, Covariance Matrix and Kalman Gain are done repeatedly every time by the updated estimation in each stage. In

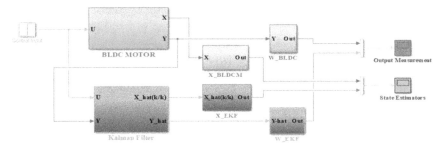

Fig. 2 Block diagram representation of BLDCM with KF

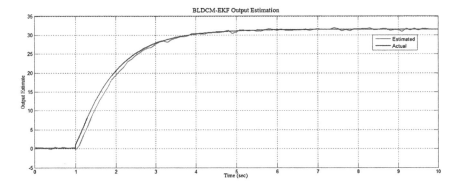

Fig. 3 BLDCM-EKF output measurement estimations

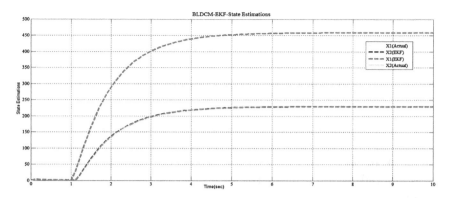

Fig. 4 BLDCM-EKF state estimations

LKF, assumed state trajectory $\bar{x}(t)$ is utilized for the present state measurement. The assumed value should be selected in such a way that it should be very close to best available state. EKF is little bit accurate than LKF. To avoid chattering with EKF at the starting stage, one can choose LKF at staring and then better to switch to EKF like starting methods adopted to the motors which act during the starting stages inorder to avoid large inrush of currents, after attaining pickup speeds switched to the removal of starting/auxiliary devices.

In UKF,as earlier discussed point EKF that produces chattering those tuning difficulties in EKF is overcome by change of characteristics of noise pdf through non-linear transformation. Gaussian distribution properties can be propagated easily rather than covariance matrix. Among all UKF has better value which is closer to the True value. The comparison is shown in Fig. 5.

Recommendations while tuning with EKF:

- Choose R depends upon the sensor types.
- Inorder to make convergence of simulation results at faster rate, assign P_0 with sufficiently higher value.
- The values of matrix Q is to be considered randomly (mostly as a positive definite matrix).
- Data rejection methods are applied if the measured value is out of bounds of actual output.

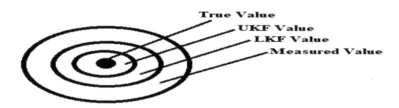

Fig. 5 Comparison of UKF, LKF with EKF

Other than EKF, IKF, LKF and UKF, some filtering techniques are also famous in some special tracking applications like h-Infinite Filter, Particle Filter and Gaussian Filter methods as tracking techniques.

3.3 Particle Filtering Technique

Particle Filter technique is applicable where the Kalman Filter faces issues due to highly non-linear influencing disturbances which causes the state estimations as beyond certain unacceptable limits. In Particle Filtering technique, we can choose the particles as a distributed ones in state space based upon the Probability Distribution Function (PDF) of the states. PF is completely a non-linear state estimator which uses Bayesian Probability Approach to calculate the state estimator. Particle Filtering Technique procedural steps to apply for any system along with the Gaussian white noise signals is explained [12, 13].

Steps involved in solving Particle Filter Technique
Step 1: The state and output measurement equations for the BLDCM are as follows

$$x(k+1) = Gx(k) + Hu(k) + x\_N$$
$$y(k) = Cx(k) + Du(k) + x\_R \tag{8}$$

Step 2: Assume the initial pdf of the state estimations $p(x(0))$ as known values.

Step 3: Generate the N initial particles randomly. The random particles are denoted by $x_{0,p}^+$ where $p = 1, 2, 3, 4, 5, ..., N$

Step 4: Set the Iteration count $k = 1$.

(a) For the known initial known values of the state estimations update the state equation and measurement equations
(b) Calculate the likelihood of the each particle by using the following formula

$$q_i \approx \frac{1}{(2\pi)^{\frac{m}{2}}|R|^{\frac{1}{2}}} * \exp\left(\frac{-[y - h(x(k))]^T R^{-1}[y - h(x(k))]}{2}\right) \tag{9}$$

(c) Based upon the obtained likelihoods which are calculated on the above step

$$q_i = \frac{q_i}{\sum_{m=1}^{N} q_m} \tag{10}$$

(d) Perform resampling process by generating set of posteriori particles.

Step 5: Now, We can calculate the any required state and measurement of the probability Distribution Function (PDF) based upon the set of updated particles by using Eq. (11)

$$p(x(k), y(k)) \approx \sum_{j=1}^{N} w_k^j \delta\left(x_k - x_k^j\right) \qquad (11)$$

where w_k^j, x_k^m → Weight and State of jth particle δ → Delta Dirac Function

The block diagram representation of BLDC Motor with PF algorithm is as shown in Fig. 6. The simulation results which are attached in Fig. 7 are the Particle Filter implementation to the BLDCM for its state estimations along with Actual state estimates of BLDCM. For the simulation of BLDCM with Particle Filtering, the following initial assumptions are made for tuning the Sensorless BLDCM. Consider state-space modelling equations of BLDCM with the state and output measurement vectors G, H, C_d, D_d PF is invented to implement the Bayesian Estimator [14]. The tuning parameters of PF technique are listed in Table 1.

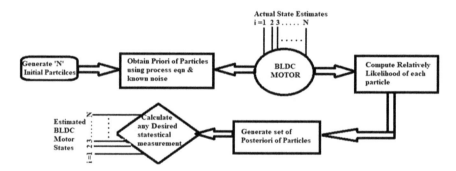

Fig. 6 Block diagram for the implementation of PF technique to BLDC motor

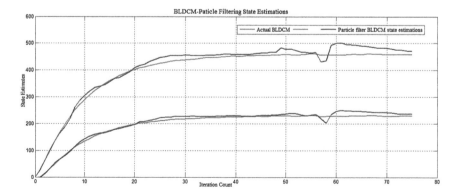

Fig. 7 BLDCM-particle filtering state estimations

Table 1 PF tuning parameters

Parameter	Assumed value
Initial actual state measurement	0.1
System-noise covariance	1
Output-noise covariance	1
No of iterations and particles	75 and 50
Initial State Variance	2
Truncated Limits	76

4 Consistency Tests

The measurement factor for any methods which are adopted for the optimization of performance parameters of any device/machine is concluded based upon its "consistency in reaching the True Value even if we perform the simulation multiple times".

The consistency of Filtering techniques are determined by

- Root Mean Square Error (rmse) value Test,
- Mean Square Error (mse) values,
- Signal Bound Test,
- Normalized Error Square (NES) test,
- Auto-Correlation Test,
- Crammer Rao Inequality Test. Consistency Tests are applied to the obtained simulation results of Sensorless BLDCM and the comparison is done to give the justification on adoption of Filtering Techniques.

Mean Square Error Test For EKF and PF Method

The consistency test by using mse test is performed [15] and resulting error values are calculated for the output measurement and state estimation values of the Sensorless BLDCM with EKF and PF techniques. The mean square error for the state and output measurements are states then PF has less than EKF. The obtained simulation results are tabulated in Table 2, which shows that PF has better performance than EKF technique.

Auto-Correlation Test To Compare EKF-PF Methods

The following simulation results in Fig. 8 marked over a region suggest that BLDCM with Particle Filtering Technique has better consistency when compared to BLDCM

Table 2 Comparison of EKF-PF consistency by MSE Test

Consistency test (mse)	EKF	PF
Measurement	0.5312	0.461992
State estimation-1	0.7266	0.494547
State estimation-2	0.1946	0.125133

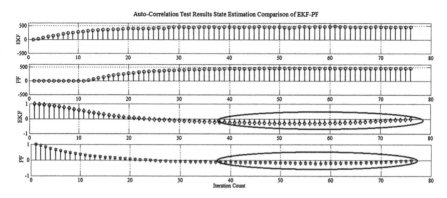

Fig. 8 Comparison of EKF-PF consistency by auto-correlation test

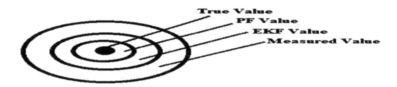

Fig. 9 Comparison based upon consistency tests

with EKF. Axes 1 and 2 represent the state estimates and Axes 3 and 4 represent auto-correlation results for state estimates of BLDCM with no of iterations in X-Axis.

Similarly, we can implement the remaining Consistency Tests based upon the simulation results which are obtained in EKF and PF techniques. Based upon the Consistency tests, we can conclude that PF has more consistency in reaching the True value when compared to EKF method. The following Fig. 9 represents the above conclusion by comparing the True value, Measured value, EKF value and PF values in a pictorial way.

Conclusions and Future Scope

In this paper, adaptive filtering techniques EKF and PF are adopted to BLDCM state and output measurements. The simulation results are attached and based upon the obtained results. The consistency tests are applied to determine the errors to obtain the suitability of desired filtering technique. PF gives better performance based upon the consistency results than EKF method. The future extension for this paper is BLDCM with Improvement of PF algorithm by introducing Monte Carlo Sampling along with the Bayesian Estimation which is integrated with the basic PF to improve the weak targets in PF algorithm results.

References

1. Applications of aero space technology report on "Brushless DC Motors" NASA technologies utilities publications prepared by Midwest research institute
2. https://ntrs.nasa.gov/search.jsp?R=19750007247
3. Y. Narendra Kumar, A. Venkatesh, Ch. Anusha, K. Vani, D. Siddarda, Estimation of speed based on flux oriented control of an induction motor drive model with reference adaptive system scheme. Trans. Eng. Sci. **2**(4) (2014)
4. A. Venkatesh, Research Scholar.: An improved adaptive SMO for speed estimation of sensorless Dsfoc induction motor drives and stability analysis using lyapunov theorem at low frequencies. Int. J. Eng. Res. Technol. (IJERT), **8**(09). ISSN: 2278-0181 (2019)
5. https://www.embitel.com/blog/embedded-blog/brushless-dc-motor-vs-pmsm-how-these-motors-and-motor-control-solutions-work
6. M. Siva Kumar, N. V. Anand, A novel approach for the design of controller for higher order discrete—time systems via its reduced model (IEEE, 2010)
7. A. Tollkuhn, F. Particke, J. Thielecke, Gaussian state estimation with non-Gaussian measurement noise, sensor data fusion: trends, solutions, applications (SDF) (2018)
8. Y. Xu, K. Xu, J. Wan, Z. Xiong, Y. Li, Research on particle filter tracking method based on Kalman filter, in *2nd IEEE Advanced Information Management, Communicates, Electronic and Automation Control Conference (IMCEC)* (IEEE, 2018)
9. F.L. Lewis, L. Xie, D. Popa, *Automation and Control Engineering—Optimal and Robust Estimation with an Introduction to Stochastic Control Theory*, 2nd edn. (CRC Press, 2007)
10. A. Ejlali, J. Soleimani, Sensorless vector control of 3-phase BLDC motor using a novel Extended Kalman (IEEE, 2012)
11. Y. Salih, A.S. Malik, 3D object tracking using three Kalman filters, in *Symposium on Computers & Informatics* (IEEE)
12. Y. Xu, K. Xu, Z. Xiong, Y. Li, Research on particle filter tracking method based on Kalman filter. IMCEC (IEEE, 2018)
13. Y. Xu, K. Xu, J. Wan, Z. Xiong, Y. Li, Research on Particle filter tracking method based on Kalman Filter. IMCEC (2018)
14. E. Suzdaleva, I. Nagy, L. Pavelkova, Bayesian filtering with discrete-valued state (IEEE, 2009)
15. B. Chen, L. Xing, J. Liang, N. Zheng, J.C. Príncipe, Steady-state mean-square error analysis for adaptive filtering under the maximum correntropy criterion. Sig. Process. Lett. **21**(7), 070–9908. IEEE (2014)

PWM Control Technique for Switched Reluctance Generator in Variable Speed Applications

Rishiraj Sarker⊙, Debaparna Sengupta, and Asim Datta⊙

1 Introduction

Switched reluctance generator (SRG) can be defined as a recent type of electronically commutated and controlled, brushless, rotating electrical machine. Switched reluctance machine (SRM) is proficient in operating as a motor as well as a generator by regulating its converter firing angles [1]. The constructional feature of this machine as a generator is same of a motor, as both does not require any permanent magnet or field windings on its rotor. However, there is a lot of significant differences in their design of the controller. SRG has attracted the researchers by its unique properties over the existing machines as listed below

- Low production cost due to absence of permanent magnet.
- Low inertia as there are no windings on the rotor.
- High-speed operation.
- Flexible operation in high temperature.

The aforementioned properties allow the SRGs to be adopted in various power system applications, i.e. aircraft system [2], hybrid electric vehicles [3] and wind-generation applications [4]. A 6/4 SRG is constructed for automotive applications and its performance is elaborated in [1]. A simulation modelling and experimental

R. Sarker (✉)
Jadavpur University, Kolkata 700032, India
e-mail: sarker.rishiraj88@gmail.com

D. Sengupta
Techno International New Town, Kolkata 700156, India
e-mail: imdebaparna@gmail.com

A. Datta
Mizoram University, Aizawl 796004, India
e-mail: asimdatta2012@gmail.com

© The Author(s), under exclusive license to Springer Nature Singapore Pte Ltd. 2021   339
J. Kumar and P. Jena (eds.), *Recent Advances in Power Electronics*
and Drives, Lecture Notes in Electrical Engineering 707,
https://doi.org/10.1007/978-981-15-8586-9_31

setup of SRG control technique are reported in [5], but the design is not suitable for high-speed operation. A gas turbine application-based SRG model is designed and its performance analysis is accomplished in [6]. A sensorless SRG control technique is presented in [7]. A 24/16 SRG has been utilized for automotive application. A suitable agreement is achieved between high starting torque and low-speed operation while attaining optimum operating condition, reported in [8]. A switched reluctance generator with an extra step-down converter is presented for the enhancement of the previously developed systems for variable speed wind energy applications. A simulated model encompassing the high-speed operation of SRG has been illustrated in [9]. A state of the art of SRG for both high- and low-speed applications is proposed and presented in [10, 11]. The control of SRGs has not been broadly investigated till now.

This paper discusses the electromechanics of SRG which consists of a brief discussion on the procedure of energy conversion and torque generation. The objective is to enlighten a perception on the control issues of SRG. Discussion on the control structure and its implementation is included which is built on the previously developed electromechanical concepts. The PWM control technique of SRG is verified with MATLAB/SIMULINK to verify its suitability in the variable speed applications.

2 Electromagnetic Energy Conversion in SRG

The objective is to analyze the characteristics of SRG with a brief explanation of the energy conversion procedure so that the control of the SRG can be discussed.

Figure 1 depicts the cross-sectional view of a three-phase SRG which consists of 6 rotor poles and 8 stator poles. The coils rounding the different poles of the stator are reciprocally interlinked to assemble the three-phase windings. The structural compactness enables the rotor to be used in high-speed operations due to the absence of permanent magnets on the rotor.

Fig. 1 Cross-sectional view of a conventional 6/4 switched reluctance generator

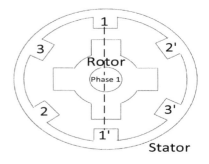

2.1 Principle of Torque Generation

The switched reluctance generator has a salient structure for a laminated ferromagnetic material-based rotor and a salient pole-mounted stator. The machine operates as a controlled stepper motor while working in motoring mode, but the machine does not have electromagnets (DC operated) on the rotor. This machine simply works on the force of attraction inbetween a ferromagnetic material (salient shaped rotor) and a rotating magnet (stator poles governed by pulses generated and instrumented in power electronic devices). An example of this type of machine is shown in Fig. 1. The two coils wound around the reverse stator poles are excited concurrently, and thus the magnetic flux is produced. The resultant torque rotates in the anticlockwise direction. Thus, torque is generated over a restricted area, based on the stator poles arc. In a simple 6/4 machine, as shown in Fig. 1, the coil inductance (L) fluctuates with the rotor position. Positive rotation is in the anticlockwise direction. To eradicate the negative torque impulse, the current must be zero whilst the poles are extricating, i.e. during the interval between A and K, as in Fig. 2.

The torque production cycle related to each of the current pulses is termed as stroke. Consequently, the generation of continuous one-directional torque entails more than a single phase, such that the gap between the torque waveform generated by a single phase is occupied by the current flowing in other phases. Usually, there is a single stroke/rotor pole pitch in each phase, and the current in other phases are flowing over only a segment of the rotor pole pitch. The current and inductance waveforms infer a saw-tooth waveform of flux-linkage ($\Psi = Li$).

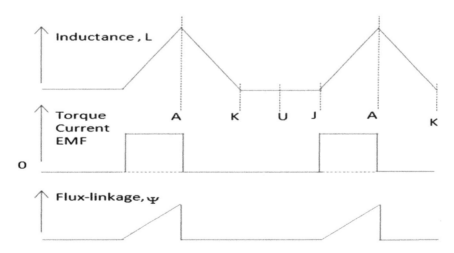

Fig. 2 Torque production in SRG

2.2 Torque Equation

In order to avoid the complexity, the fringing flux effect around the poles is neglected, assuming that all the flux cross the air gap in a circular direction. Mutual coupling between two phases is usually very low, thus neglected. The equation of voltage for single phase is [6]

$$
\begin{aligned}
v &= Ri + \frac{d\Psi}{dt} \\
&= Ri + L\frac{di}{dt} + i\frac{dL}{dt} \\
&= Ri + L\frac{di}{dt} + \omega_m\frac{dL}{d\theta}
\end{aligned}
\tag{1}
$$

where v, i, Ψ, R, L, θ and ω_m are termed as the terminal voltage, current/phase, flux-linkage (Volt-seconds), resistance/phase, inductance/phase, angular position of the rotor and the angular velocity of rotor (rad/s), respectively.

The back EMF (e) is expressed as

$$
e = \omega_m \cdot \frac{dL}{d\theta}
\tag{2}
$$

The instant electrical power (vi) is expressed as

$$
vi = Ri^2 + iL\frac{di}{dt} + \omega_m i^2 \frac{dL}{d\theta}
\tag{3}
$$

At any instant, the rate of change of stored magnetic energy is given by

$$
\frac{d\left(\frac{1}{2}Li^2\right)}{dt} = \frac{1}{2}i^2\frac{dL}{dt} + Li\frac{di}{dt} = \frac{1}{2}i^2\omega_m\frac{dL}{d\theta} + Li\frac{di}{dt}
\tag{4}
$$

As per the law of conservation of energy $P = T_e\omega_m$, where P and T_e are the instantaneous mechanical energy and electromagnetic torque.

Thus, rearranging (2) and (3)

$$
P = T_e\omega_m = vi - Ri^2 - \frac{d\left(\frac{1}{2}Li^2\right)}{dt}
$$

$$
\text{and } T_e = \frac{1}{2}i^2\frac{dL}{d\theta}
\tag{5}
$$

where the slope of the inductance graph in Fig. 2 is depicted by $\frac{dL}{d\theta}$.

The firing angles are chosen in a manner that $\frac{dL}{d\theta} < 0$ for generating mode of operation. The back emf is negative (generating), while $\frac{dL}{d\theta} > 0$ depicts that the back emf is positive (motoring).

3 SRG Control Strategies

Opinionated on the electromechanics of the SRG and the principle of energy conversion studied in the preceding sections, the control of SRG is discussed henceforth. The study starts with an analysis of a conventional SRG controller architecture. The different methods of controlling the SRG are discussed later on. Finally, the discussion ends with a specific controller implementation.

Figure 3 shows a general SRG control architecture. The DC voltage (V_{DC}) is given back and equated with the reference voltage (V_{DC}^*). The controller adjusts the variables, i.e. the switching on/switching off angles (θ_{on}/θ_{off}) and reference point of current (I_{HI}). The controller can utilize the input parameters, i.e. rotor speed (ω) and bus voltage (V_{DC}). The SRG, which characterizes the convener and the machine in Fig. 3, produces current (I_o).

3.1 High Energy Conversion Control Method

To attain the maximum energy conversion, an SRG system is regulated with the DC bus voltage according to the rotor speed. As i^2 is positive, the torque does not vary towards the current direction. The motor winding voltage is $v = V_s$ (the supply voltage is termed as V_s) when Q_1 and Q_2 are simultaneously kept on at a time. In the same way, $v = -V_s$ when Q_1 and Q_2 are simultaneously off at a time.

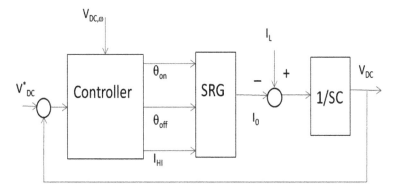

Fig. 3 Block diagram of basic SRG control technique

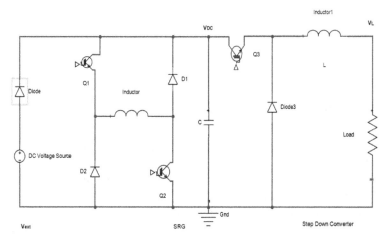

Fig. 4 SRG converter

Controller helps to maintain the DC bus voltage (V_{DC}) across C in a perfect flat-topped shape. The load voltage (V_L) is controlled at a fixed value with a load voltage controller by chopping action provided by Q_3 (Fig. 4).

3.2 Proposed PWM Control Method

By changing the applied voltage, PWM control method can be used to adjust the current to a user-defined value. Generally, the generator is expected to be operated at a very high speed.

In a single PWM period, the excitation current refers to the area *A1* and the generated current refers to area *A2* in Fig. 5. It is assumed that the duty cycle (α) is fixed for a single switching period and the current ripple due to PWM is insignificant compared to its average peak value (i_p) [8];

$$A_1 = D_1 t_{PWM} i_P$$
$$A_2 = D_2 t_{PWM} i_P \tag{6}$$

where $D_1 = \alpha$, $D_2 = (1 - \alpha)$ and t_{PWM} is the time period of each PWM pulses ($t_{PWM} = 1/f_{PWM}$). The total current in a single period of PWM ($I_{o,PWM}$) is

$$I_{o,PWM} = A_2 - A_1 = (D_2 - D_1) t_{PWM} i_P \tag{7}$$

The preliminary excitation current and the generated current [area specified as (1) and (2) in Fig. 5] are ignored in the calculation. The proportion of the excitation energy and the generated energy ε with the duty cycle is expressed as

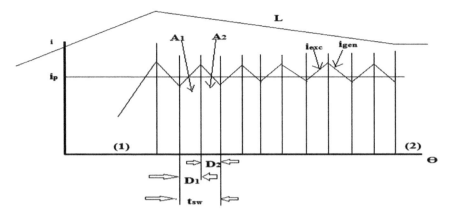

Fig. 5 Excitation area of PWM strategy

$$\varepsilon = D_1/(D_2 - D_1) \tag{8}$$

$$= d/(1 - 2\alpha) \tag{9}$$

Each of the three-phase circuit equation is articulated as follows:

$$V_{DC} = L\frac{di_{exc}}{dt} - e \tag{10}$$

where the coil resistance is ignored and the back EMF (e) is expressed as

$$e = \omega i L\frac{dl}{dt} \tag{11}$$

where ω is the rotor speed. At any condition, $V_{DC} > e$ must be satisfied else the total current will be increased throughout the drifting period. The increased current at the excitation period is

$$L\frac{di_{exc}}{dt} = V_{DC} + e > 0 \tag{12}$$

It is assumed that the variation in inductance (L) of a PWM period is minor and can be neglected to maintain the constant current over each of the PWM periods,

$$\frac{di_{exc}}{dt}D_1 t_{PWM} = \frac{di_{gen}}{dt}D_2 t_{PWM}[V_{DC} + e]D_1 = [V_{DC} - e]D_2 \tag{13}$$

By reshuffling the above Eq. (13)

$$\alpha = \frac{1}{2}\left(1 - \frac{e}{V_{DC}}\right) \tag{14}$$

From Eq. (9), if $\alpha \geq 0.333$ then $\varepsilon \geq 1$. To generate the total energy larger than the excitation energy, the condition of $\alpha < 0.333$ must be fulfilled. Substituting this condition into Eq. (14), we get the condition of $e \geq 0.333V_{DC}$. Hence, the condition to generate total power more than the power used for the excitation with PWM control is

$$0.333V_{DC} < e < V_{DC} \tag{15}$$

$$0.333\omega_B < e < \omega_B \tag{16}$$

where the base speed is ω_B. The result shows that to generate a total power more than the excitation power with the PWM control, the speed of the prime-mover must be higher than one-third of the base speed (speed range of 1:3).

4   Implementation of the Proposed Controller

Figure 6 shows a schematic diagram of the implementation steps of the generator controller. The controller consists of three main parts: microcontroller device, commutation panel and input/output (I/O) terminal. The commutation panel includes an field programmable gate array (FPGA), receiving terminal of encoder signal and a digital/analog converter. The commutator panel of the FPGA device specifies the converter's switching states from the incoming encoder signals and sends the information from the microcontroller about the switching angle. The commutation signals are sent to the input/output board. The digital/analog converter is used to detect

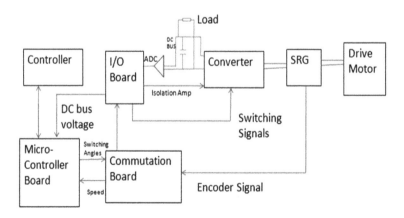

Fig. 6 Schematic block diagram of the proposed SRG controller

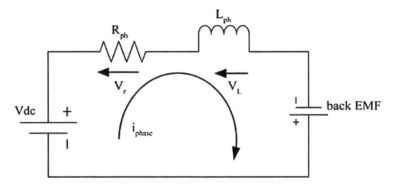

Fig. 7 Equivalent circuit for SRG single-phase structure

and regulate the overcurrent. The input/output board sends the commutation signals through optical transmitters to the power converter. The DC bus voltage is fed back through an isolation amplifier and an analog/digital converter.

4.1 Simulation Model

Output power of SRG may vary depending upon several variables, i.e. excitation current (I_{exc}), excitation voltage (V_{exc}), rotor position, rotor speed, etc. The above-mentioned parameters can be regulated by achieving a higher proficiency in the SRG application. Thus, bringing out the finest parameters to generate an optimum performance is a challenge to the researchers. Hence, inbuilt block diagrams and special functions of MATLAB/SIMULINK have been utilized to realize an effective PWM control model of the SRG.

The SRG is displayed as a combination of series resistor and inductor as illustrated in Fig. 7. Each phase of the SRG is modelled in Simulink based on (1). The model is then combined with quite a few modules such as power converter, controller block and positioning sensor. Each of individual modules has been tested separately and assembled to arrange a complete system as shown in Fig. 8.

4.2 Results and Discussion

The maximum and minimum phase inductances used for the simulated 3 phase 6/4 SRG model are 50 mH and 10 mH, respectively. A machine with a higher number of rotor poles will involve more excitation. The current and torque profile of a 3 phase 6/4 SRG have been realized and presented in Fig. 9a, b.

Further, (1) can be re-written as

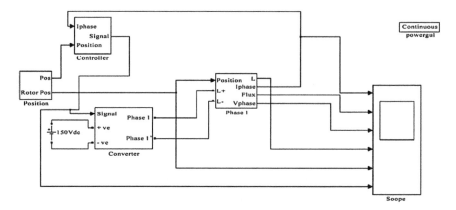

Fig. 8 MATLAB/SIMULINK-based model of SRG

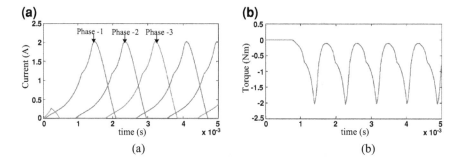

Fig. 9 a Current and **b** torque profile of a three-phase 6/4 SRG

$$i = \frac{\pm Vt}{L + \left(R + \omega\left(\pm \frac{dL}{d\theta}\right)\right)t} \tag{17}$$

According to (17), the current steadily increases corresponding to the voltage and position of θ_{on} and the slope of the inductance. The current (i) then starts increasing as the rotor shifts from its allied position till the back EMF is greater than V_{DC} and it slowly reaches to zero. The amplitude of the current depends on the magnitude of the back EMF and the stored energy, as presented in Fig. 10.

It is visible that, if θ_{on} is placed before its allied position, the current profile looks like a catenary curve, as in Fig. 10d. On the other hand, if θ_{off} is kept constant, while θ_{on} is shifted with the decreasing inductance profile, the current reduces as shown in Fig. 10a–c.

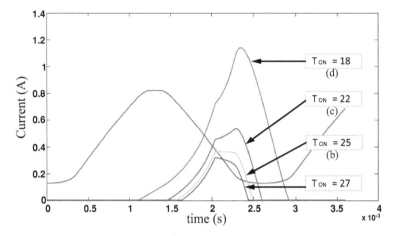

Fig. 10 Current variation in terms of changing inductance

5 Conclusions

This paper encompasses the PWM control strategy of SRG to enhance the existing research work. The mathematical formulations and theoretical analysis hold a good resemblance with the simulated results of the current–torque characteristics of a three-phase 6/4 SRG. FPGA platform is chosen to realize the effective results of the proposed control scheme. However, various changes in the parameters have been done for an in-depth analysis of the machine. This would result in a higher expenditure and would be laborious if the changes were to be achieved during the experimental phase. Hence, this work enlightens a new direction to the researchers to analyze the complete system before instrumenting the experimental prototype. Additional investigation is obligatory to fill the economic gap with the aid of quantitative and qualitative software to enhance its performance and acceptance. Further, the projected PWM control strategy can be implemented in the design of a switched reluctance motor-driven hybrid electric vehicle, SRM operated photovoltaic system and aircraft system.

References

1. F. Faradjizadeh, R. Tavakoli, E. Afjei, Accumulator capacitor converter for a switched reluctance generator. IEEE Trans. Power Electron. **33**(1), 501–512 (2018)
2. V. Valdivia, R. Todd, F.J. Bryan, A. Barrado, A. Lázaro, A.J. Forsyth, Behavioral modeling of a switched reluctance generator for aircraft power systems. IEEE Trans. Industr. Electron. **61**(6), 2690–2699 (2014)
3. F. Yi, W. Cai, Modeling, control, and seamless transition of the bidirectional battery-driven switched reluctance motor/generator drive based on integrated multiport power converter for electric vehicle applications. IEEE Trans. Power Electron. **31**(10), 7099–7111 (2016)

4. A. Sunan, F. Kucuk, F. Goto, H.J. Guo, O. Ichinokura, Three-phase full-bridge converter controlled permanent magnet reluctance generator for small-scale wind energy conversion systems. IEEE Trans. Energy Convers. **29**(3), 585–593 (2014)
5. G.P. Viajante, D.A. Andrade, A.W.F.V. Silveira, M.A.A. Freitas, L.C. Gomes, V. R. Bernardeli, Output DC voltage control strategy for switched reluctance generator, in *International Conference on Renewable Energies and Power Quality (ICREPQ'12)* (Santiago de Compostela, Spain, 2012)
6. M. Besbes, M. Gabsi, E. Hoang, M. Lecrivain, B. Grioni, C. Plasse, SRM design for starter-alternator system, in *Proceedings of ICEM 2000* (2000), pp. 1931–1935
7. E. Echenique, J. Dixon, R. Cárdenas, R. Peña, Sensorless control for a switched reluctance wind generator, based on current slopes and neural networks. IEEE Trans. Industr. Electron. **56**(3), 817–825 (2009)
8. D.A. Torrey, Switched reluctance generators and their control. IEEE Trans. Industr. Electron. **49**(1), 3–14 (2002)
9. V.S.D.C. Teixeira, A.B. Moreira, A.R. Filho, Methodology for the electromagnetic design of the axial-flux C-core switched reluctance generator. IEEE Access **6**, 65463–65473 (2018)
10. A. Arifin, I. Al-Bahadly, S.C. Mukhopadhyay, State of the art of switched reluctance generator. Energy and Power Eng. **4**, 447–458 (2012)
11. C.Y. Ho, J.C. Wang, K.W. Hu, C.M. Liaw, Development and operation control of a switched-reluctance motor driven flywheel. IEEE Trans. Power Electron. **34**(1), 526–537 (2019)

Genetic Algorithm Optimized Direct Torque Control of Mathematically Modeled Induction Motor Drive Using PI and Sliding Mode Controller

Abha Pragati, Bibhu Prasad Ganthia, and Bibhu Prasad Panigrahi

1 Introduction

Speed controller design significantly affects electric drive output. Because of their basic structure, PI speed controllers are commonly used in industrial applications. However, due to continuous variance of machine parameters, model uncertainties, nonlinear dynamics, and external device disturbances, fixed-gain PI controllers are often unable to enhance the necessary output of the function. Continuous adaptation of the controller parameters therefore becomes desirable when high output from the drive system is needed [1, 2]. Genetic Algorithms (GAs) are adaptive search techniques focused on a biological definition of "survival of the fittest." It can provide an inexpensive and reliable way to optimize applications by looking for a global minimum without having a cost-function derivative [3]. Thus, GA can be added to tuning the PI controller gains to ensure maximum control output under nominal operating conditions [4]. Another solution to this issue, however, is to fully replace the PI controller with adaptive control structures such as self-tuning and Artificial Intelligence techniques, Model Reference Adaptive Control (MRAC), and Sliding Mode Control (SMC) [5]. The sliding mode control strategy has demonstrated robustness against motor parameter uncertainty and dynamic modeling, insensitivity to external load disturbance, stability and rapid dynamic response among these different designs proposed [6, 7]. It is therefore found to be very efficient in regulating systems of electric drives. Big chattering of torque at a steady state can be called the key downside for such a control scheme [8, 9]. One way to boost the efficiency of the sliding mode controller is to link it with Genetic Algorithm (GA) to create a controller for GA Sliding Mode (GA-SM).

A. Pragati (✉) · B. P. Ganthia · B. P. Panigrahi
Electrical Engineering, IGIT, Sarang, Odisha Dhenkanal, India
e-mail: abhapragati026@gmail.com

Fig. 1 Modern variable
speed IM drive system

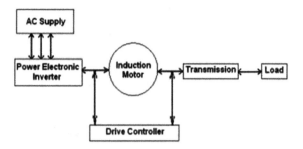

2 Induction Motor (IM) Drive

The IM is perhaps the most commonly used AC engine, as it provides many advantages compared to other engines. Two forms of 3-phase IMs exist. They are Wound rotor IM and Squirrel-cage IM [10, 11]. Both machines are identical from an electrical point of view, except that the previous has permanently shorted rotor winding terminals within the system. In the context of a wound-rotor system, the 3-phase rotor winding terminals are externally available for the operation [12, 13]. The project discusses only the speed strength of Squirrel-cage IM. Figure 1 displays the block diagram for a typical modern IM drive device with variable speed. The IM is directly or indirectly related to the charge (through gears). The power converter regulates the transfer of power from an ac supply toward motor by actively regulating the voltage semiconductor switches [14, 15].

Primary reasons for this complexity are the use of variable frequency, harmonically configured electrical systems for converters and complicated dynamics of ac systems, differences in system parameters, and complexities in managing feedback signals in the existence of transient harmonics [16]. The IM is superior because they are machines that are stable and reliable to their counterparts, because they have functionality such as repairs-free in operations [17].

3 Direct Torque Control Technique

This switching-based approach has been described as the approach that is simple and feasible to meet those needs. DTC is one of the most excellent and efficient methods for regulating control induction [18]. This approach emphasizes differentiated torque value and the stator flux monitoring and is now one of the vital systems commonly observed control methods with the purpose of regulating torque and flux efficiently [19]. Figure 2 displays the typical block diagram for conventional DTC scheme.

The diagram of the DTC block arrangement is displayed in Fig. 3 and the frame of reference for the stator and rotor axis. The reference torque can be measured using a PI unit using the variation in instantaneous speed and reference speed. Selection of this reference speed strengthens the torque and flux control dynamic response [19].

Fig. 2 Conventional DTC scheme

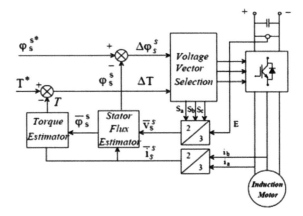

Fig. 3 Controller block diagram

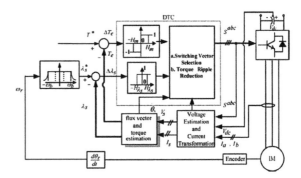

4 Genetic Algorithm

GA is a global stochastic tool for the optimizing of adaptive search based on natural selection frameworks. GA has been widely acknowledged as a major and successful tool for resolving optimization problems [20]. Compared to other techniques of optimization, such as simulating annealing and random search methods, GA is effective in the avoidance of local value of minima value which is a typical feature of the nonlinear models. The GA architecture is illustrated in Fig. 4 and PI controller implementation of the genetic algorithm well into the sliding control mode algorithm is shown in Fig. 5.

The detail parameters for the genetic algorithm are shown in Table1. This represents the iterations of the genetic algorithm defining the conditions for the controller operations.

Fig. 4 GA architecture

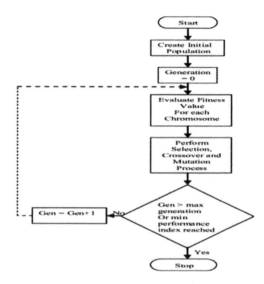

Fig. 5 Genetic algorithm in
PI-SMC-based DTC

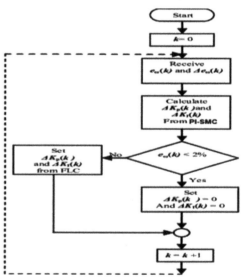

Table 1 Parameters for
genetic algorithm

Parameters	Value	Attributes	Strategy
Generations	20	Selection	Stochastic
Chromosomes	16	Crossover	Double Point
Genes	4	Crossover Probability	0.8
Chromosome length	40bit	Mutation Rate	0.07

Fig. 6 Sliding mode control phasor

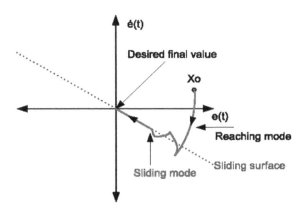

5 Sliding Mode Controller

This highly advanced technique is widely recognized for its robustness against internal uncertainties, uncertainties due to nonlinearity in parametric variations and anomalies that were excluded in the modeling. SMC procedure is applicable for enhancing DTC for IMs highlighted in this study [21].

Such strategies enhance the efficiency of the steady state and preserve the benefits of intermittent state and the torque and flux are resilient against system parameter variations. It presents models and experimental findings to demonstrate the feasibility of the proposed strategy [22]. Fig. 6 shows the concept of sliding mode technique in phasor diagram form where the desired value and reaching mode to the stability are presented [23].

6 Simulink Model and Result

The direct torque control design PI-DTC and SMC-DTC framework and the flux depending on the sliding mode of a speed-driven induction motor are shown in Fig. 7. The model is cascade order for electromagnetic torque control, square flux norm, and speed control. So there are sliding control method and genetic algorithms that are introduced to adjust the torque, flux, and speed in the control structure.

6.1 DTC Using Conventional PI Controller

The results from conventional PI controller are shown in Fig. 8 which represents the result for the electromagnetic torque; Fig. 9 shows the speed obtained from the rotor

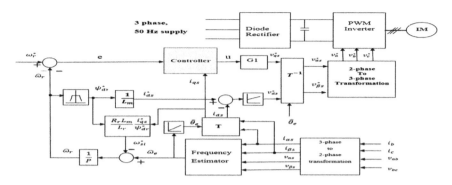

Fig. 7 Controller design and implementation in IM drive

Fig. 8 EM torque versus time(sec)

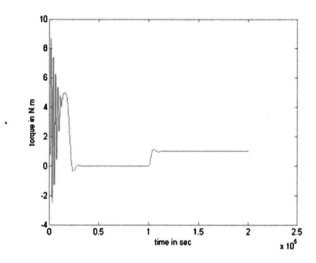

Fig. 9 Rotor speed (Nr) versus time(sec)

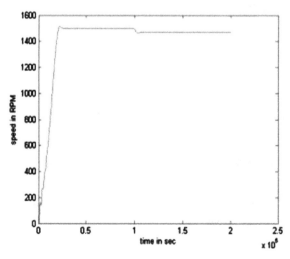

Fig. 10 Stator flux trajectory of d-axis and q-axis

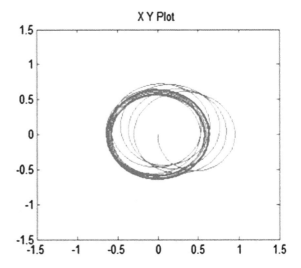

for the system. Figure 10 highlights the trajectory for the stator flux. The frame of reference for the model is stationary form.

6.2 DTC Using PI-Genetic Algorithm

The results using conventional PI-GA controller are shown in Fig. 11 which represents the result for the electromagnetic torque, Fig. 12 shows the rotor speed for the system, and the Fig. 13 highlights the trajectory for the stator flux in stationary reference frame.

6.3 DTC Using PI-Sliding Mode Controller-Genetic Algorithm

The proposed frame for the direct torque control architecture and the flux from the stator frame based on the sliding mode of a speed-controlled IMD (induction motor) (PI-SMC-GA-DTC) are shown in Fig. 14. Figure 15 represents SVM for DTC scheme of IM using GA in association with the proposed technique.

The proposed technique gives better results with respect to all the uncertainties that arise due to the speed variations represented in the following figures shown above. Figure 15 represents torque due to electromagnetic vs time, Fig. 16 shows Rotor speed vs Time, 18 shows d-axis voltage of DTC scheme of IM using PI with Nonlinearity, 19 shows q-axis voltage of DTC scheme of IM using PI with Nonlinearity, 20 represents d-axis voltage of DTC scheme of IM using SMC with Nonlinearity, 21 represents q-axis voltage of DTC scheme of IM using SMC with

Fig. 11 EM torque versus
time(sec)

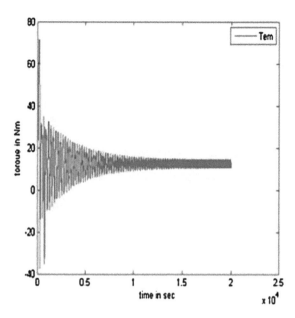

Fig. 12 Rotor speed versus
time(sec)

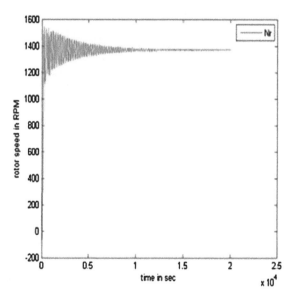

Nonlinearity, 22 demonstrates the d-axis voltage of the IM DTC strategy using PI-SMC-GA with nonlinearity of speed variations, 23 indicates the q-axis voltage of the IM DTC strategy employing nonlinearity PI-SMC-GA as well as Fig. 24 illustrates the d-axis graph and q-axis trajectory flux in the stationary reference frame in the research stated in modeling approach.

Fig. 13 Stator flux
trajectory of d-axis and
q-axis

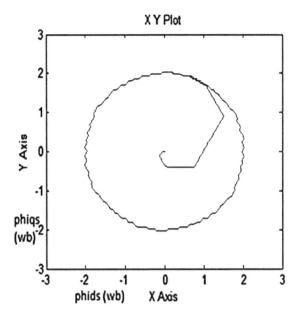

Fig. 14 Block diagram of
the proposed technique

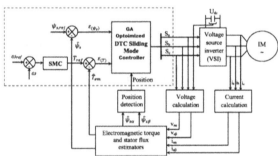

Fig. 15 SVM for DTC
scheme of IM using GA

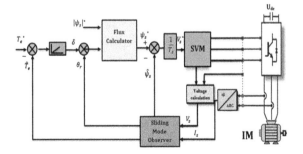

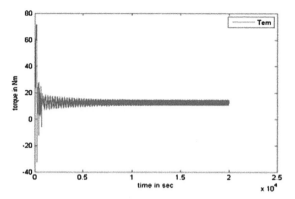

Fig. 16 Electromagnetic Torque versus time(sec)

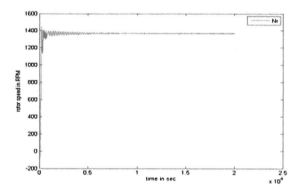

Fig. 17 Rotor speed versus time(sec)

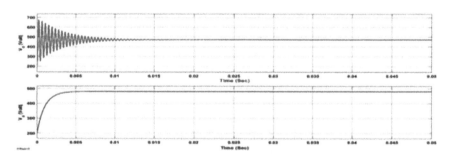

Fig. 18 d-axis voltage of DTC scheme of IM using PI with nonlinearity.

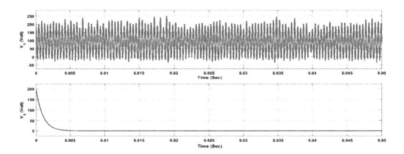

Fig. 19 q-axis voltage of DTC scheme of IM using PI with nonlinearity.

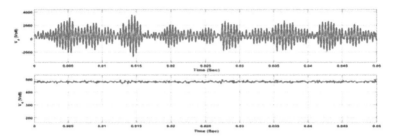

Fig. 20 d-axis voltage of DTC scheme of IM using SMC with nonlinearity.

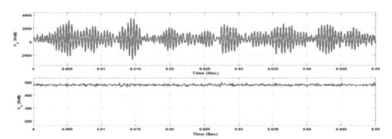

Fig. 21 q-axis voltage of DTC scheme of IM using SMC with nonlinearity.

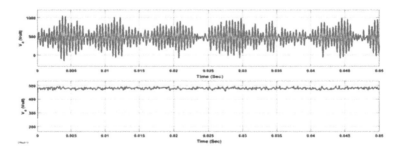

Fig. 22 d-axis voltage of DTC model of IM using PI-SMC-GA with nonlinearity.

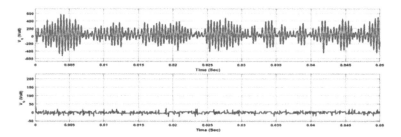

Fig. 23 q-axis voltage of DTC scheme of IM using PI-SMC-GA with nonlinearity.

Fig. 24 Stator flux
trajectory of d-axis and
q-axis

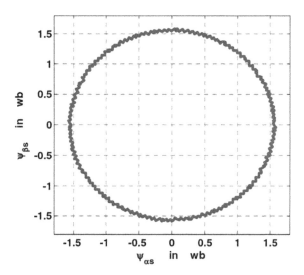

7   Result Analysis and Discussion

The cost function for the controller is to determine that every generation's individuals can be selected as the Integral Time of Absolute Error (ITAE). This cost function can be written in mathematical expression as follows:

$$\text{ITAE} = \int_0^t t|exp(t)|\text{d}t$$

GA is looking for the optimum set value of the conventional PI controller gains during the search process which minimizes ITAE cost function. This function is set as the criterion of evolution of the GA with positive and negative errors calculations. The comparison between controllers at different operating conditions is shown in Table 2.

Table 2 Comparison of controllers with the proposed technique at various operating conditions

		Conventional PI	PI-GA	PI-GA-SMC (Proposed technique)
Speed 60 rad/s at rated load torque of 30%	**ITAE**	0.08165	0.09514	0.09832
Reference speed of 120 rad/s at Nominal Parameters. Phase load torque applied at t = 2 s from 30 to 90% rated load		0.32	0.229	0.232
Reference speed of 60 rad/s and rated load torque of 30%		0.01182	0.04266	0.04319
Reference speed 120 rad/s at nominal parameters and load torque rated 30%		0.6061	0.5199	0.5102
Applied with 120% uncertainty in motor inertia (J) 60 rad/s reference speed at 30% level load		0.3258	0.3025	0.3005
Speed variation from 60 to 250 rad/s at a level load of 30% and nominal parameters		12.78	9.49	9.05
Reversal of speed value from 60 to − 60 rad/s at 30% average load with 60% uncertainty in engine inertia (J)		9.45	5.28	5.21
Reference speed 150 rad/s at nominal parameters and load torque rated 50%		0.6158	0.5210	0.5100

The Simulink results compared with the proposed technique are categorized according to following operating conditions: Speed 60 rad/s at rated load torque of 30%, Reference speed of 120 rad/s at Nominal Parameters. Phase load torque applied at t = 2 s from 30 to 90% rated load, Reference speed of 60 rad/s and rated load torque of 30%, Reference speed 120 rad/s at nominal parameters and load torque rated 30%, Applied with 120% uncertainty in motor inertia (J) 60 rad/s reference speed at 30% level load, Speed variation (Nr) from 60 to 250 rad/s at a level load of 30% and nominal parameters, Reversal of speed value from 60 to −60 rad/s at 30% average load with 60% uncertainty in engine inertia (J), and Reference speed 150 rad/s at nominal parameters and load torque rated 50%. A phase shift in the load torque from 30% changed to 90% of its actual value is at time t = 2 s where the motor is operating at 120 rad/s to analyze the disturbance value of rejection potential of the proposed schemes. When the change in load torque is applied, PI-GA shows a low capacity for disturbance with the steady-state error of 0.15% while GA-SMC shows zero sensitive to sudden change in load. PI-SMC-GA-based system has the faster response toward the speed variations and achieves faster steady state.

In Table 3, it is concluded that the proposed technique can play a vital role in DTC of induction motor drive with faster response toward the speed variation.

Table 3 Comparison between controllers at different operating conditions

	DTC	SVM DTC	PI SMC DTC	PI GA DTC	Proposed technique
Torque & flux ripple	More	Less	Average	Average	Average
Torque response	Fast	Fast	Fast	Faster	Faster
Switching frequency	Vary	Constant	Constant	Constant	Constant
THD	High Distortions	Low Distortions	Low Distortions	Less Distortions	Less Distortions
Switching loss	More	Less	Average	Average	Less
Sensitivity	No	Yes	No	No	No
Complexity	Simple	Simple	More	More	More
Speed Regulation	Slow	Fast	Good	Very Good	Very Good
Precession	Less	Average	Average	Average	More
Computation time	Low	Average	High	Average	High
Controller Regulation	Hysteresis	PI Controlled	SMC Controller	GA and PI Controller	PI + SMC + GA + DTC Controller

8 Conclusion

This paper highlights the implementation of GA-genetic algorithm characteristics using sliding mode control technique with conventional PI controller for the direct torque control of IMD; induction motor drive system. This follows research paper the conventional controller technique compared with the proposed PI-SMC-GA control technique and it gives superior efficiency on speed variations and faster response toward steady-state operation. The DTC drive is the electromagnetic torque command based on the Lyapunov principle for GA-SM control. PI-GA demonstrates better output under nominal driving conditions while GA-SM reveals robustness over variance in stator resistance, inertia instability, and crassness of stability to the. The model provides better efficient and smooth operating process with faster steady-state operations. In future, more adaptive techniques can be implemented for better steady-state operations with higher constraints.

References

1. M. Hajian, G.R. Arab Markadeh, J. Soltani, S. Hoseinnia, Energy optimized sliding-mode control of sensorless induction motor drives. Energy Convers. Manage. **50**, 2296–2306 (2009)
2. A. Chikhi, M. Djarallah, K. Chikh, A comparative study of field-oriented control and direct-torque control of induction motors using an adaptive flux observer. Serb. J. Electr. Eng. **7**, 41–55 (2010)
3. R. Abdelli, D. Rekioua, T. Rekioua, Performances improvements and torque ripple minimization for VSI fed induction machine with direct control torque. ISA Trans. **50**, 213–219 (2011)
4. P.R. Tripathy, B.P. Panigrahi, Simulation studies on switching table based DTC and fuzzy rule based DTC for three-phase squirrel cage induction motor. ETASR-Eng., Technol. & Appl. Sci. Res. **2**(1) (2012)
5. M.V. Kazemia, M. Moradib, R.V. Kazemic, Minimization of powers ripple of direct power controlled DFIG by fuzzy controller and improved discrete space vector modulation. Electr. Power Syst. Res. **89**, 23–30 (2012)
6. C. Saravanan, J. Sathiswar, S. Raja, Performance of three phase induction motor using modified stator winding. Int. J. Comput. Appl. **46**, 1–4 (2012)
7. R. Trabelsi, A. Khedher, M.F. Mimouni, M'sahli, F. , Backstepping control for an induction motor using an adaptive sliding rotor-flux observer. Electr. Power Syst. Res. **93**, 1–15 (2012)
8. I.M. Alsofyani, N.R.N. Idris, A review on sensorless techniques for sustainable reliablity and efficient variable frequency drives of induction motors. Renew. Sustain. Energy Rev. **24**, 111–121 (2013)
9. A. Derouich, A. Lagrioui, Real-time simulation and analysis of the induction machine performances operating at flux constant. Int. J. Adv. Comput. Sci. Appl. **5**, 59–64 (2014)
10. L. Amezquita-Brooks, E. Liceaga-Castro, J. Liceaga-Castro, C.E. Ugalde-Loo, Flux-torque cross-coupling analysis of FOC schemes: Novel perturbation rejection characteristics. ISA Trans. **52**, 446–461 (2015)
11. N. Goel, R.N. Patel, S. Chacko, Torque ripple reduction of DTC IM drive using artificial intelligence, in *International Conference on Electrical Power and Energy Systems* (2016)
12. H.M.D. Habbi, H.J. Ajeel, I.A. Inaam, Speed Control of induction motor using PI and V/F scalar vector controllers. Int. J. Comput. Appl. **151**, 36–43 (2016)
13. B.P. Ganthia, P.K. Rana, S.A. Pattanai, Space vector pulse width modulation fed direct torque control of induction motor drive using matlab-simulink, in *Proceedings of the 3rd International Conference on Electrical, Electronics, Engineering Trends, Communication, Optimization and Sciences (EEECOS), Tadepalligudem, India* (2016), pp. 1–2
14. N. El Ouanjli, A. Derouich, A. El Ghzizal, A. Chebabhi, M. Taoussi, A comparative study between FOC and DTC controls of the doubly fed induction motor (DFIM), in *International Conference on Electrical and Information Technologies* (IEEE, 2017)
15. N. El Ouanjli, A. Derouich, A. El Gzizal, Y. El Mourabet, B. Bossoufi, M. Taoussi, Contribution to the performance improvement of doubly fed induction machine functioning in motor mode by the DTC control. Int. J. Power Electron. Drive Syst. **8**, 1117–1127 (2017)
16. A. Ammarn, A. Bourek, A. Benakcha, Nonlinear SVM-DTC for induction motor drive using input-output feedback linearization and high order sliding mode control. ISA Trans. **67**, 428–442 (2017)
17. S. Boubzizi, H. Abid, M. Chaabane, Comparative study of three types of controllers for DFIG in wind energy conversion system. Prot. Control Mod. Power Syst. **3**(1), 21 (2018)
18. F. Wang, Z. Zhang, X. Mei, J. Rodríguez, R. Kennel, Advanced control strategies of induction machine: Field oriented control, direct torque control and model predictive control. Energies **11**(1), 120 (2018)
19. B. Ganthia, S. Sahu, S. Biswal, A. Abhisekh, S. Kumar Barik, Genetic algorithm based direct torque control of VSI fed induction motor drive using MATLAB simulation. Int. J. Adv. Trends Comput. Sci. Eng. **8**(5), 2359–2369 (2019)

20. M.M. Ali, W. Xu, M.F. Elmorshedy, Y. Liu, S.M. Allam, M. Dong, Sliding mode speed regulation of linear induction motors based on direct thrust control with space-vector modulation strategy, in *2019 22nd International Conference on Electrical Machines and Systems (ICEMS)* (Harbin, China, 2019), pp. 1–6

21. S. Jnayah, A. Khedher, Sensorless direct torque control of induction motor using sliding mode flux observer, in *2019 19th International Conference on Sciences and Techniques of Automatic Control and Computer Engineering (STA)* (Sousse, Tunisia, 2019), pp. 536–541

22. S. Krim, S. Gdaim, A. Mtibaa, M. Faouzi Mimouni, FPGA-based real-time implementation of a direct torque control with second-order sliding mode control and input–output feedback linearisation for an induction motor drive. IET Electr. Power Appl. **14**(3), 480–491 (2020)

23. W. Wang, H. Lin, H. Yang, W. Liu, S. Lyu, Second-order sliding mode-based direct torque control of variable-flux memory machine. IEEE Access **8**, 34981–34992 (2020)

24. B.P. Ganthia, A. Pritam, K. Rout, S. Singhsamant, J. Nayak, Study of AGC in two-area hydro-thermal power system, in *Advances in Power Systems and Energy Management* ed. by A. Garg, A. Bhoi, P. Sanjeevikumar, K. Kamani, Lecture Notes in Electrical Engineering, vol. 436 (Springer, Singapore, 2018)

25. B.P. Ganthia, R. Pradhan, S. Das, S. Ganthia, Analytical study of MPPT based PV system using fuzzy logic controller, in *2017 International Conference on Energy, Communication, Data Analytics and Soft Computing (ICECDS)* (Chennai, India, 2017), pp. 3266–3269

26. B.P. Ganthia, P.K. Rana, T. Patra, R. Pradhan, R. Sahu, Design and analysis of gravitational search algorithm based TCSC controller in power system. Mater. Today: Proc. **5**(1), pp. 841–847 (2018). Part 1, ISSN 2214–7853. https://doi.org/10.1016/j.matpr.2017.11.155

27. B.P. Ganthia, A. Abhisikta, D. Pradhan, A. Pradhan, A variable structured TCSC controller for power system stability enhancement. Mater. Today: Proc. **5**(1), 665–672 (2018). Part 1, ISSN 2214–7853. https://doi.org/10.1016/j.matpr.2017.11.131

28. B.P. Ganthia, S. Mohanty, P.K. Rana, P.K. Sahu, Compensation of voltage sag using DVR with PI controller, in *Electrical Electronics and Optimization Techniques (ICEEOT) International Conference* (2016), pp. 2138–2142

29. B.P. Ganthia, K. Rout, Deregulated power system based study of agc using pid and fuzzy logic controller Int. J. of Adv. Res. **4**, 847–855. www.journalijar.co

30. I. Takahashi, T. Noguchi, A new quick-response and high-efficiency control strategy of an induction motor. IEEE Trans. Ind. Appl. **IA-22**(5) (1986)

Implementation of Modified Five-Level Inverter Using a Reduced Number of Switches

Sumit Kumar Rai[ID], **S. L. Shimi, and Lini Mathew**

1 Introduction

The multilevel power converter structure can be employed as an alternative in high-power and medium-voltage situations. A multilevel converter can not only achieve high power rating but also allows the use of renewable sources of energy. Renewable energy sources such as photovoltaic, fuel cells, and wind can be interfaced with an MLI system for high-power applications with ease. Different major multilevel structures have been applied in industrial applications such as Cascaded H-bridges multilevel inverter with separate dc sources, diode clamped multilevel inverter, and Capacitor clamped multilevel inverter. Recent studies show that topologies of three levels are quite suitable solutions for low-voltage applications whereby medium or high switching frequency is used [2]. When the switching frequency is more than 10 kHz [1], the effectiveness of the classical three-level neutral-point-clamped (NPC) converter [3] may also be more helpful than the traditional two-level inverter. Over the years, the literature has recorded many different multilevel converter topologies [3–8]. High-power medium-voltage motor drives fed from multilevel inverters are increasingly being used in various applications in the industry [9, 10]. Compared to conventional two-level voltage-source inverters, multilevel inverters produce voltage waveforms with low harmonic distortion at relatively low switching frequencies. Also, many of the problems encountered in two-level inverters like high dv/dt, high

S. K. Rai (✉) · S. L. Shimi · L. Mathew
Department of Electrical Engineering, National Institute of Technical Teachers Training and Research (NITTTR), Chandigarh, India
e-mail: raisumit38@gmail.com

S. L. Shimi
e-mail: shimi.reji@gmail.com

L. Mathew
e-mail: lenimathew@yahoo.com

© The Author(s), under exclusive license to Springer Nature Singapore Pte Ltd. 2021
J. Kumar and P. Jena (eds.), *Recent Advances in Power Electronics and Drives*, Lecture Notes in Electrical Engineering 707,
https://doi.org/10.1007/978-981-15-8586-9_33

electromagnetic interference, high voltage stress on the systems, high switching losses, output filter requirements can be significantly mitigated by using multilevel inverters. These advantages of the multilevel power conversion have introduced high-level research in this area, leading to the introduction of several new multilevel inverter topologies and related pulse width modulation (PWM) techniques [11, 12].

The paper is designed as follows: Sect. 1 highlights the theoretical concept of the multilevel inverter. The proposed five-level inverter topology and the SHE-PWM switching technique are described in Sect. 2. Section 3 discusses the simulation studies to obtain current and voltage waveforms. The hardware implementation of the modified 5-level with six switch inverter designs and their result are described in Sect. 4. Section 5 provides hardware analysis and discussion of the overall study. Lastly, the conclusions are presented in Sect. 6.

2 Proposed Topology

The topology shown in Fig. 1 consists of six switches and one diode with two DC sources to obtain single-phase five-level voltage waveform. The conventional cascaded H-bridge requires two CHB to obtain a single-phase five-level voltage waveform, in which a total of eight switches with two DC sources. Thus, it increases the power loss and circuit complexity of the circuit. Therefore, an improved topology with better electrical performance of five-level inverter like efficiency, total harmonics distortion (THD), Staircase waveform quality, common-mode voltage, and input current. This topology operates in both symmetrical and unsymmetrical cases. If S_1 or S_2 fails, then this topology works as a three-level voltage waveform, and even if the voltage source (V_1) and S_1 fail to operate then also obtain a three-level voltage waveform. This topology operates five-level and three-level multilevel inverter depending upon the situation of a circuit.

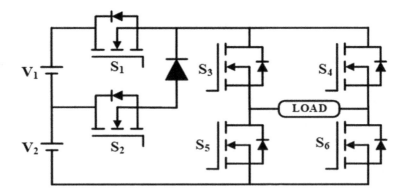

Fig. 1 The circuit model of five-level inverter topology

In this paper, first of all, topology of the circuit model has been developed in MATLAB/SIMULINK, which yields a five-level voltage waveform circuit model with reduced switches. The voltage waveforms at different THD have been obtained by varying the switching angle of gate driver circuit. The switching angle is calculated to eliminate lower order selective harmonics with the help of a genetic algorithm (GA) and give the gate pulse according to selective harmonics elimination, pulse width modulation (SHE-PWM) technique. The gate pulse is given to different switches using Arduino Microcontroller. The gate driver circuit aims to isolate the topology with the power circuit and controller circuit by using very few components, which improves topology performance and reduced circuit complexity. The MOSFET used as a switch because it is best for medium voltage, power and also operates in a wider range of frequency. The main characteristic of the MOSFET is 'on' and 'off' state switching losses is less but more voltage drop and high-power losses during running condition, this is also a drawback of MOSFET. The complete setup is arranged and designed to validate the results, which is shown in this paper.

3 MATLAB/Simulation Results

In MATLAB/SIMULINK, MOSFET-based 5-level inverter has been designed using the SimPowerSystems toolbox. A snubber circuit consisting of a series RC circuit is connected in parallel with each power switch to provide protection. The gate pulse is generated in MATLAB on the basis of the SHE-PWM technique. Here, six gate pulses are generated for each switch and are shown in Fig. 2 for a complete cycle.

The voltage and current waveforms of five-level inverter obtained from the MATLAB/SIMULINK model are shown in Fig. 3. The parameter set in MATLAB/Simulink of 24 and 36 V with RL load ($R = 250\ \Omega$, $L = 20$ mH) and discrete solver.

The harmonics of the 5th order are eliminated by using the SHE algorithm.

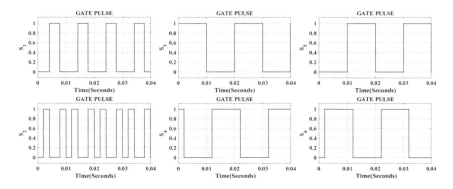

Fig. 2 Gate pulses generate using MATLAB/Simulink to trigger MOSFET

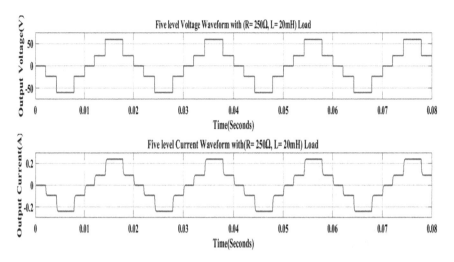

Fig. 3 Voltage and current waveform of five-level inverter in MATLAB/Simulink

The frequency spectrum of phase voltage obtained using FFT analysis is shown in Fig. 4. It is observed that the minimum value of phase voltage THD is 28.87% at 0.9 modulation index and also observed the current THD of the single-phase 5-level inverter with motor (RL) is 21.84%. The current THD is less than voltage THD because of the reason that the circulating current presence in circuit which act as a filter.

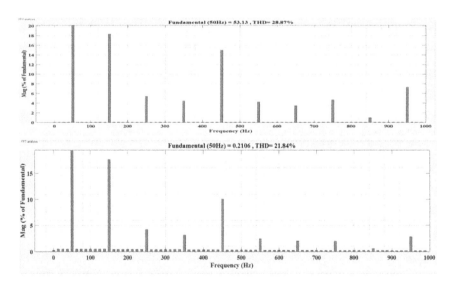

Fig. 4 THD of voltage with RL load

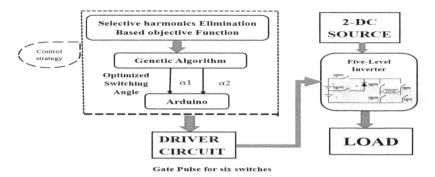

Fig. 5 Complete flow chart of control strategy of modified CHB five-level inverter

If the topology works as a three-level inverter, then voltage stress increases for the switch and also increases to 35.18% we observed in MATLAB/SIMULINK.

4 Flow Chart of The Proposed Five-Level Topology

Figure 5 shows the block diagram of control strategy used in modified CHB topology for eliminating selective lower order harmonics in output wave form. The gate pulse for five-level inverter power switching is obtained using SHE control strategy. The optimized switching angles $\alpha1$, $\alpha2$ are obtained using SHE techanique. Thus, to solve SHE-based transcendental equation of the five-level inverter using Genetic algorithm, this signal is used as distributed gate pulses for six switches where α is the firing-angle of the power switch.

5 Hardware Design

5.1 Driver Circuit

The driver circuit consists of Arduino microcontroller, resistances, optocouplers, and LEDs. The LED is used to indicate the signal presence which communicates between the power circuit and the driver circuit. The complete setup of the driver circuit is shown in Fig. 6. The internal connection of the driver circuit is shown in the PCB layout, which is designed in Easy EDA software as shown in Fig. 8b.

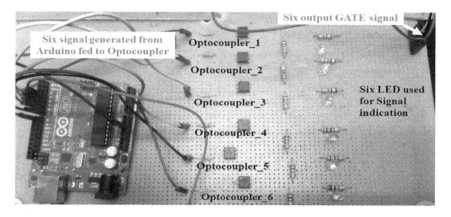

Fig. 6 The hardware setup of driver circuit with Arduino microcontroller

5.2 Power Circuit

The power circuit is designed with the help of six MOSFET switches and one diode. Here, we used heatsink for protection against overheating. IRFZ44N is used as a N-channel MOSFET which has medium power rating, The MOSFET is mostly used at broad frequencies ranging from kHz to hundreds of kHz as switching instruments. An advantage of a MOSFET as a switching device is the low-power consumption required for gate drive. Figure 7 shows the hardware setup of power circuit with MOSFET, heatsink, diode, and load. Here, in power circuit resistive load is shown, but during experimentation an induction motor as RL load has been used.

The complete hardware setup which consists of two batteries each of 12 V to generate 24 V for one source (V_1) and 36 V taken from DC power supply for the second source (V_2), Arduino controller for generating pulses for driver circuit on the basis of programming, the electrical circuit consisting of the power circuit and driver circuit, using power quality analyzer (FLUKE 438) and YOKOGAWA DLM2024

Fig. 7 Hardware setup of power circuit with six MOSFET with heatsink and one diode

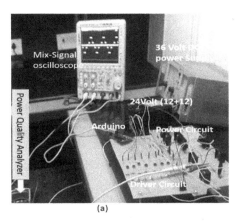 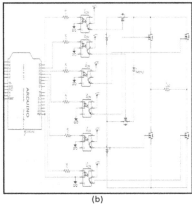

| (a) | (b) |

Fig. 8 **a** The complete hardware setup of the proposed topology, **b** PCB layout of the proposed topology

model are shown in Fig. 8a. The PCB layout of the complete hardware setup to assemble all components in a single board for final hardware design and used for practical application is shown in Fig. 8b.

6 Hardware Result and Discussion

Figure 9 shows the five-level voltage waveform when two DC sources of 24 and 36 V are connected with power circuit and driver circuit connected with each gate terminal of power circuit.

This voltage waveform obtained across RL load using power quality analyzer (PQA) shows power factor and displacement power factor as unity, the active and

 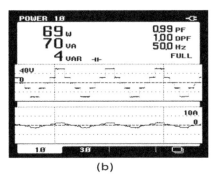

| (a) | (b) |

Fig. 9 Shows active, reactive, and apparent power with current, voltage waveform in PQA

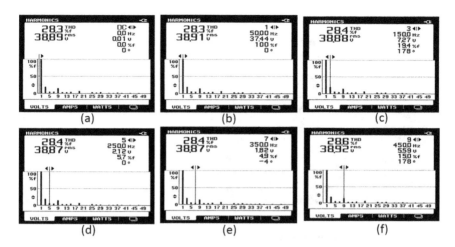

Fig. 10 Harmonics spectrum of phase voltage up to 9th order at 0.9 modulation index

reactive power is 69 W, 4 Var, respectively, at 50 Hz frequency with 38.98 V rms voltage and 1.72 A current.

6.1 Voltage THD up to 9th Order in PQA in RL Load

The harmonics of 0th, 1st, 3rd, 5th, 7th, and 9th order with corresponding amplitude values as obtained in PQA are shown in Fig. 10a–f. The THD of 1st and 3rd order harmonics of 38.91 V rms are 28.3,19.4%, respectively, in RL load and also THD of 5th, 7th, and 9th of 38.87 V rms are 5.70, 4.90, and 15.0%, respectively. It is observed from figure that 3rd harmonics is introduced because of the magnetizing current under no-load condition. The 3rd harmonics introduced by this system is of the value of 19.4% but can be eliminated in three-phase system. The experimental results of the targeted 1st, 3rd, 5th, 7th harmonics were found to be higher than the simulation results because of the significantly high value of hardware sampling time and all physical parameters considered here.

6.2 Current THD up to 9th Order in PQA in RL Load

The harmonics of 0th, 1st, 3rd, 5th, 7th, and 9th order current with corresponding amplitude are shown in Fig. 11a–f. The THD of 1st and 3rd harmonics at 1.69 A rms are 25.9%, 18.9%, respectively, and the THD of 5th, 7th and 9th of 1.69 A rms are 5.2, 4.4, 13.5%, respectively.

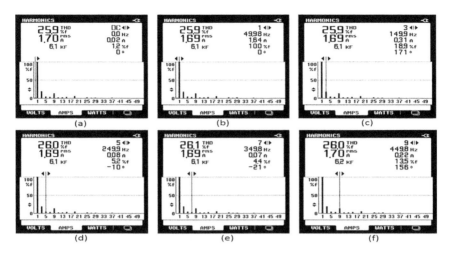

Fig. 11 Harmonics spectrum of phase current up to 9th order at 0.9 modulation index

The summarized table of phase voltage and phase current and their corresponding amplitude of THD for the proposed topology with RL load ($R = 250\ \Omega$, $L = 20$ mH) is shown in Table 1.

Figure 10b shows the THD of voltage at fundamental frequency, i.e., 50 Hz as 28.3% at 38.91 RMS voltage. This is the same result obtained in MATLAB/Simulink using FFT analysis (Refer Fig. 4a) and is 28.87%, hence verified the hardware and software results. In Fig. 11b, the current THD is shown as 25.9% at 1.69 A current. Referring Fig. 4b, the current THD obtained using MATLAB/Simulink was 21.84%. The current THD value is less in Simulink result compared to hardware because in hardware some losses occur and all physical characteristics have to be considered. Hence, all software simulation result and hardware result are verified.

Table 1 Output voltage, current, and corresponding their amplitude of THD

Order of harmonics	Voltage in (%)	Current in (%)	Magnitude rms (V)	Magnitude rms (A)
0	0	1.2	0.01	0.02
1	100	100	37.44	1.64
3	19.4	18.9	7.27	0.031
5	5.70	5.20	2.12	0.08
7	4.90	4.40	1.82	0.07
9	15.0	13.5	5.59	0.22
THD	28.3	25.9	38.91	1.70

6.3 Result Comparison

For 5-level inverter with two switches and one H-bridge phase voltage by using of
24 V, 36 V two dc input, the minimum THD value for GA-SHE algorithm has been
obtained at 0.9 value of modulation index. The theoretical output fundamental phase
voltage has been synthesized at a frequency of 50 Hz is given by Eq. (1) and is given
by:

$$V_{(peak)} = M_i \left(4 * \frac{Vdc}{\pi} \right) * \sin(60^\circ)$$

$$= 0.9 \left(4 * \frac{60}{\pi} \right) * 0.8660 = 59.53 \text{ V} \tag{1}$$

Theoretical value $(V)_{rms} = 59.53/\sqrt{2} = 42.100$ V.
Hardware calculated value $(V)_{rms} = 38.98$ V.
Error in RMS Voltage $= 42.10–38.98 = 3.12$ V
The voltage drop in hardware is 3.19 V due to physical characteristics of switching.

7 Conclusion

In this paper, a single-phase modified five-level unsymmetrical CHB inverter with six
switches and two dc sources has been presented. The major advantages of this type of
topology in contrary to conventional topologies are reduced number of devices and
simple control. This reduction in switches has lowered the size, complexity, cost, and
losses, very low ripple current can be present, resulting in a volume reduction of a
filter. If switching frequency is higher, which increased the switching losses, but the
resulting reduction in THD has immensely improved the inverter performance. As a
result, increased switching losses can be neglected safely. The simulated results are
validated on the experimental setup and found correct. The validation is tested on a
70 W prototype which was designed and built, and check prototype with an induction
motor type load. In the future, this work can be expanded to focus on applying this
design in a real-life standalone, grid-connected PV system, and industries in the field
of low-voltage and high-power applications.

References

1. M. Schweizer, I. Lizama, T. Friedli, J.W. Kolar, Comparison of the chip area usage of 2-level and
   3-level voltage source converter topologies, in *Proceedings of the 36th Annual IEEE IECON*
   (2010), pp. 391–396

2. M. Schweizer, I. Lizama, T. Friedli, J.W. Kolar, Comparison of the chip area usage of 2-level and 3-level voltage source converter topologies, in *Proceedings of the 36th Annual IEEE IECON*, pp. 391–396
3. R. Teichmann, S. Bernet, A comparison of three-level converters versus two-level converters for low-voltage drives, traction, and utility applications. IEEE Trans. Ind. Appl. **41**(3), 855–865 (2005)
4. A. Nabae, I. Takahashi, H. Akagi, A new neutral-point-clamped PWM inverter. IEEE Trans. Ind. Appl. **17**(5), 518–523 (1981)
5. R.H. Baker, L.H. Bannister, Electric Power Converter. U.S. Patent 3 867 643 (1975)
6. T.A. Meynard, H. Foch, Multilevel conversion: High voltage chopper and voltage source inverter, in *Proceedings of the 23rd Annual IEEE PESC*, vol. 1, (1992), pp. 397–403
7. T. Bruckner, S. Bernet, H. Güldner, The active NPC converter and its loss-balancing control. IEEE Trans. Ind. Electron. **52**(3), 855–868 (2005)
8. P. Barbosa et al., Active neutral-point-clamped multilevel converters, in *Proceedings of the Power Electron. Spec. Conf., Recife, Brazil* (2005), pp. 2296–2301
9. J. Rodriguez, S. Bernet, B. Wu, J.O. Pontt, S. Kouro, Multilevel voltage-source-converter topologies for industrial medium-voltage drives. IEEE Trans. Ind. Electron. **54**(6), 2930–2945 (2007)
10. N. Hatti, K. Hasegawa, H. Akagi, A 6.6-kV transformer less motor drive using a five-level diode-clamped PWM inverter for energy savings of pumps and blowers. IEEE Trans. Power Electron. **24**(3), 796–803 (2009)
11. S. Kouro, M. Malinowski, K. Gopakumar, J. Pou, L.G. Franquelo, B. Wu, J. Rodríguez, M.A. Pérez, J.I. Leon, Recent advances and industrial applications of multilevel converters. IEEE Trans. Ind. Electron. **57**(8), 2553–2580 (2010)
12. H. Abu-Rub, J. Holtz, J. Rodriguez, G. Baoming, Medium-voltage multilevel converters—State of the art, challenges, and requirements in industrial applications. IEEE Trans. Ind. Electron. **57**(8), 2581–2596 (2010)

Fuzzy Logic Controller-Based BLDC Motor Drive

Kishore Kumar Pedapenki, Jayendra Kumar, and Anumeha

1 Introduction

BLDC motor is an ideal motor for applications of low and medium power because of its high torque/inertia ratio, high efficiency, high energy density, wide range of speed control, low maintenance requirement. It is a three-phase synchronous motor with permanent magnets on the rotor and three phase windings on the stator. It is also known as an electronically switched motor because there are no mechanical brushes and commutator assembly, instead an electronic commutation is used based on the position of the rotor sensed by the Hall-effect position sensor. Applications can be found in a broad range of household appliances, office automation I and industrial I instruments, ventilation, automotive, air conditioning and many more.

Most widely used controller for speed control of any motor drive is proportional-integral (PI) controller. Due to its robust performance over a wide range of operating conditions and functional simplicity, the PI controller is extremely popular. Using PI controller, motor drive is controlled because it is used to reach speed accurately and decreases steady state error. Because of speed response of the drive is sluggish using this controller which is due to integral portion of the controller. Since PI controller doesn't have the ability to predict the future errors of the system, therefore, it cannot eliminate steady state oscillations and reduces settling time. Fuzzy logic has rapidly become one of the most successful of today's technology for developing sophisticated

K. K. Pedapenki (✉)
Jain Deemed-to-be University, Bangalore, India
e-mail: iitr.kis@gmail.com

J. Kumar
National Institute of Technology Jamshedpur, Jamshedpur, India
e-mail: jkumar.ece@nitjsr.ac.in

Anumeha
Government Women's Polytechnic Jamshedpur, Jamshedpur, India

© The Author(s), under exclusive license to Springer Nature Singapore Pte Ltd. 2021   379
J. Kumar and P. Jena (eds.), *Recent Advances in Power Electronics
and Drives*, Lecture Notes in Electrical Engineering 707,
https://doi.org/10.1007/978-981-15-8586-9_34

control systems. Several studies show, both in simulations and experimental results, that fuzzy logic control yields superior results with respect to those obtained by conventional control algorithms. Thus, in industrial applications, the fuzzy logic control has become an attractive solution in controlling the electrical motor drives. Fuzzy logic controller [1] is one of the promising controllers [2] in shunt active power filters [3] also. The hardware setup based on the fuzzy logic was also developed in this area of shunt active power filter [4]. In this paper, a simulation model with the fuzzy logic controller with zeta converter is presented using MATLAB/Simulink.

2   Zeta Converter

The power factor corrected (PFC) zeta converter [5–7] is a fourth-order DC-DC converter that works in both discontinuous and continuous current conduction mode as shown in Fig. 1. It operates as a buck boost feature that is not inverted output voltage.

Steady State Analysis [8]

Using inductor [9, 10] volt-sec balance equation, the relationship between input and output voltage can be obtained as shown in Fig. 2a, b.

Let the switch Q is ON for a period of DT in the total period of T.

Fig. 1 Wave forms of Zeta PFC converter

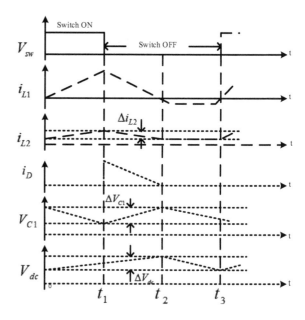

Fig. 2 **a** Voltage across inductor L_1 [21]. **b** Voltage across inductor L_2 [22]

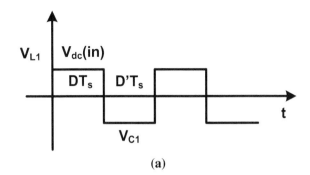

(a)

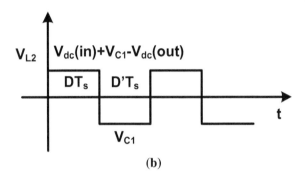

(b)

Inductor volt-sec balance equation:

$$V_{dc(in)}DT_s - V_{C1}D^1T_s = 0 \tag{1}$$

$$V_{dc(in)}DT_s = V_{C1}D^1T_s \tag{2}$$

$$V_{C1} = \frac{V_{dc(in)}D}{D^1} \tag{3}$$

From Fig. 4b,

$$(V_{in(dc)} + V_{C1} - V_{dc(out)})DT_s - V_{dc(out)}D^1T_s = 0 \tag{4}$$

$$(V_{in(dc)} + V_{C1} - V_{dc(out)})DT_s = V_{dc(out)}D^1T_s$$

$$V_{C1} = \frac{V_{dc(out)}}{D} - V_{dc(in)} \tag{5}$$

From the Eqs. (5) and (6)

$$\frac{V_{dc(out)}}{D} - V_S = \frac{V_{dc(in)} D}{D^1} \tag{6}$$

$$\frac{V_{dc(out)}}{D} = V_{dc(in)} + \frac{V_{dc(in)} D}{D^1} \tag{7}$$

$$V_{dc(out)} = \frac{V_{dc(in)} D}{(1 - D)} \tag{8}$$

$$\frac{D}{1 - D} = \frac{V_0}{V_{in}} = \frac{I_{in}}{I_0} \tag{9}$$

The duty cycle of the zeta converter is given by

$$D = \frac{V_{dc(out)}}{V_{dc(out)} + V_{dc(in)}} \tag{10}$$

3 Fuzzy Logic Controller (FLC)

Fuzzy logic control (FLC) [11] is a control algorithm based on a linguistic control strategy which tries to account the human knowledge about how to control a system without requiring a mathematical model. Input and output are non-fuzzy values. The block diagram of the FLC is shown in Fig. 3 and the calculated value of parameter expression used in this work is in Table 1.

The Fuzzy Logic Controller [12] has two input signals, i.e. the error E and the change in error CE [13]. The change in error is related to the derivative of error (d*E*/d*t*). The controller observes the pattern of the speed loop error signal and correspondingly updates the output, so that actual speed matches the command speed reference [14].

Fuzzification: Fuzzification is the controller's first block, which converts each piece of input data into membership degrees through a lookup into one or more membership functions.

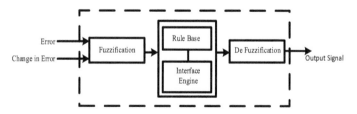

Fig. 3 Block diagram of FLC

Table 1 Parameter expressions and their values used in this work

S. no.	Parameter expression	Calculated values
1	$L_1 = \dfrac{V_s^2}{2P_{max}f_s}\left(\dfrac{V_{dc\,max}}{V_{dc\,max}+\sqrt{2}V_s}\right)$	1.29 mH
2	$C_1 = \dfrac{P_{max}}{\eta\left(\sqrt{2}V_s+V_{dc\,max}\right)^2 f_s}$	423.06 nF
3	$L_o = \dfrac{V_s^2}{P_{max}}\dfrac{V_{dc\,max}}{\lambda f_s(\sqrt{2}V_s)}\left(\dfrac{V_{dc\,max}}{V_{dc\,max}+\sqrt{2}V_s}\right)$	3.89 mH
4	$C_{dc} = \dfrac{P_{min}}{2\omega k V_{dc\,min}^2}$	5 mF
5	$C_{f\,max} = \dfrac{I_m}{\omega_L V_m}\tan\theta$	378 nF
6	$L_f = \dfrac{1}{4\Pi^2 f_c^2 C_f}$	1.03 mH

Table 2 Rule base of FLC

E/CE	NH	NM	NL	Z	PL	PM	PH
NH	NH	NH	NH	NH	NM	NL	Z
NM	NH	NH	NH	NM	NL	Z	PL
NL	NH	NH	NM	NL	Z	PL	PM
Z	NH	NM	NL	Z	PL	PM	PH
PL	NM	NL	Z	PL	PM	PH	PH
PM	NL	Z	PL	PM	PH	PH	PH
PH	Z	PL	PM	PH	PH	PH	PH

Rule ibase: The rules may use multiple variables in both the status and the rule conclusion. Therefore, the controllers can be applied to both single-input single-output (SISO) issues and multi-input multi-output (MIMO) issues. The rule base used in this work is in Table 2.

Membership ifunctions:

Every element in the universe of discourse is a member of a fuzzy set to some grade, perhaps even zero. The grade of membership for all its members describes a fuzzy set, such as neg. A grade of membership functions is allocated in fuzzy sets components, so the shift from membership to non-membership is more gradual than abrupt.

Defuzzification: [15]

Defuzzification is the process of converting a fuzzified output into a single crisp value with respect to a fuzzy set. The defuzzified value in fuzzy logic controller (FLC) represents the action to be taken in controlling the process.

4 Method and Result Analysis

This motor control drive consists of diode bridge rectifier, dc filter, Zeta converter, voltage sourced inverter, BLDC motor [16, 17]. Zeta converter [18] is fed from the rectifier output through dc filter. The single-phase AC supply is fed to the rectifier to convert AC to DC. The rectifier output, i.e. DC consist of harmonics when it is converted from AC to DC. So, DC filter can be used to limit the harmonic content and provide filtered dc voltage which is given as input to the zeta converter. As discussed in earlier section, Zeta converter [19] is well suited for power factor improvement [20] compared to boost converter and having buck boost capability.

Zeta converter is used to control the dc link voltage which is given as input to the VSI fed BLDC motor. The dc link voltage is controlled by varying the switching pulses of Zeta converter, which are generated by the using Fuzzy Logic Controller. The controlled dc link voltage from zeta converter is used to drive the BLDC motor through voltage sourced inverter. The Position tracking of the rotor is converted into respective emf signal using Hall sensors. A Hall sensor is used to give gating signals to voltage source inverter with respect to rotor position tracking. This way Hall sensor is used to control the gate pulses from the position tracking of the BLDC motor drive. The VSI is operated at the fundamental frequency switching is accomplished by the electronic commutation of the BLDC motor. The same is shown in Fig. 4 above.

The speed of the BLDC motor drive is compared with the reference speed. The error speed signal is given to Fuzzy Logic Controller with rule base shown in Table 2. The processing signal of FLC is compared with the sawtooth signal to get the switching pulses to the switch of Zeta converter. MOSFET is used for switch in

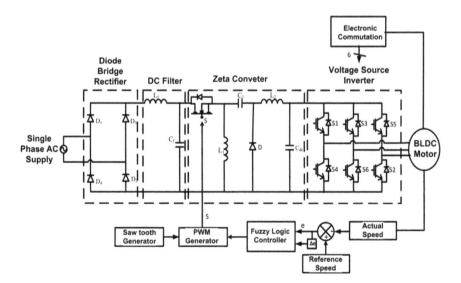

Fig. 4 Proposed Zeta PFC converter-based BLDC motor drive with fuzzy logic controller

the Zeta converter because of it is having a capability of higher switching frequency. By controlling the duty cycle of switch, DC link voltage of the zeta converter is controlled. By controlling the DC link voltage of the Zeta converter, speed of the BLDC motor is controlled. Figures 5, 6, 7 and 8 show the DC link voltage, speed of the BLDC motor, stator current and electromagnetic torque respectively using PI controller, whereas the same parameters using fuzzy logic controller (FLC) have been shown in Figs. 9, 10, 11 and 12 The comparison of the PI and FLC is shown in Table 3.

Fig. 5 DC link voltage (V) using PI controller

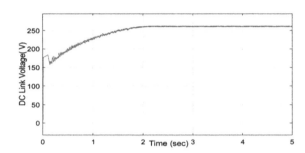

Fig. 6 Speed response using PI controller

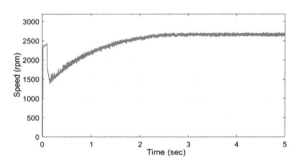

Fig. 7 Stator current using PI controller

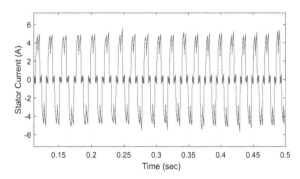

Fig. 8 Electromagnetic
torque using PI controller

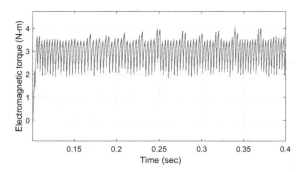

Fig. 9 DC link voltage (V)
using FLC

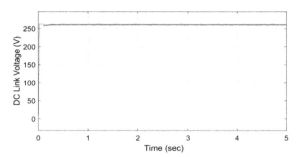

Fig. 10 Speed response
using FLC

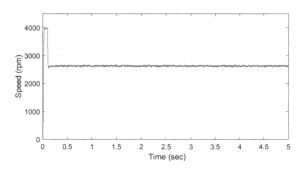

Fig. 11 Stator currents
using fuzzy logic controller

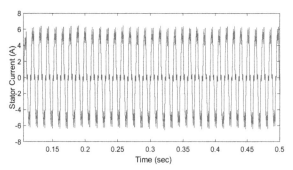

Fig. 12 Electromagnetic torque using fuzzy logic controller

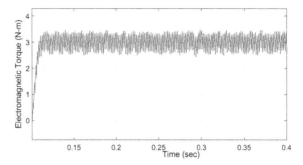

Table 3 Comparison of parameters obtained: PI and fuzzy logic controller

Parameters	PI controller	Fuzzy logic controller
Power factor	0.918	0.986
Settling time(in sec)	2.1	0.2
Rise time (in sec)	2.4	0.08
Voltage ripple (in V)	2.5	2.5

5 Conclusion

The proposed Zeta converter-based BLDC motor drive using Fuzzy Logic Controller has designed and simulated in MATLAB/Simulink. This work presents a comparative analysis of speed control of brushless DC motor (BLDC) drive fed with voltage source inverter (VSI) with PI Controller and Fuzzy Logic Controller. The main objective of this work is to control the speed response of BLDC motor drive which is achieved by Fuzzy Logic Controller. For the speed control of BLDC drive, zeta converter is preferred over boost converter. Because of the zeta converter has the capability of buck-boost and better power factor improvement is achieved for this proposed work. Speed response of BLDC drive using PI controller is having rise time higher and high settling time. So compared to PI controller-based BLDC motor, fuzzy logic controller-based BLDC motor given low rise time and faster speed response and improved Power Factor. The speed response of BLDC drive is compared by both controllers which are simulated in MATLAB/Simulink.

References

1. K.K. Pedapenki, S.P. Gupta, M.K. Pathak, Two controllers for shunt active power filter based on fuzzy logic, in *IEEE International Conference on Research in Computational Intelligence and Communication Networks (ICRCICN)* (IEEE, 2015), pp. 141–144
2. K.K. Pedapenki, S.P. Gupta, M.K. Patha, Comparison of fuzzy logic and neuro fuzzy controller for shunt active power filter, in *International Conference on Computational Intelligence and Communication Networks (CICN)* (IEEE, 2015), pp. 1247–1250

3. K.K. Pedapenki, G. Swathi,, Analysis of shunt active power filter with unit voltage template method, in *International Conference on Computation of Power, Energy Information and Communication (ICCPEIC)* (IEEE, 2018), pp. 749–753
4. K.K. Pedapenki, S.P. Gupta, M.K. Pathak, Shunt active power filter with MATLAB and dSPACE 1104 verification. Int. J. Appl. Eng. Res. **11**(6), 4085–4090 (2016)
5. B. Singh, V. Bist, Power quality improvements in a zeta converter for brushless DC motor drives. IET Sci. Meas. Technol. **9**(3), 351–361 (2015)
6. V. Bist, B. Singh, A brushless DC motor drive with power factor correction using isolated-zeta converter. IEEE Trans. Indus. Inf. **9**(2), 1–6 (2014)
7. S. Singh, B. Singh, G. Bhuvaneswari, V. Bist, Power factor corrected zeta converter based improved power quality switched mode power supply. IEEE Trans. Indus. Electron. **62**(9), 5422–5433 (2015)
8. R. Kumar, B. Singh, BLDC motor-driven solar PV array-fed water pumping system employing zeta converter. IEEE Trans. Ind. Appl. **52**(3), 2315–2322 (2016)
9. C.C. Shijith, S. Shijith, PFC and speed control of BLDC motor using zeta converter. Int. J. Adv. Res. Electr. Electron. Eng. **2**(2), 52–58 (2014)
10. M. Vimal, V. Sojan, Vector controlled PMSM drive with power factor correction using zeta converter, in *International Conference on Energy, Communication, Data Analytics and Soft Computing* (IEEE, 2017), pp. 289–295
11. T. Mathew, C. Ann Sam C, Closed loop control of BLDC motor using a fuzzy logic controller and single current sensor, in *International Conference on Advanced Computing and Communication Systems (ICCACC)* (IEEE, 2013), pp. 1–6
12. A. Maamoun, Y.M. Alsayed, A. Shaltout, Fuzzy logic based speed controller for permanent magnet synchronous motor drive, in *IEEE International Conference on Mechatronics and Automation* (IEEE, 2013), pp. 1518–1522
13. H.M. Kamel, H.M. Hasanien, H.E.A. Ibrahim, Speed control of permanent magnet synchronous motor using fuzzy logic controller, in *IEEE International Electric Machines and Drives Conference* (IEEE, 2009), pp. 1587–1591
14. G. Dewantoro, Y.L. Kuo, Robust speed-controlled permanent magnet synchronous motor drive using fuzzy logic controller, in *IEEE International Conference on Fuzzy Systems* (IEEE, 2011), pp. 884–888
15. C.W. Tao, J.S. Taur, Design of fuzzy-learning fuzzy controllers, in *IEEE International Conference on Fuzzy Systems Proceedings IEEE World Congress on Computational Intelligence*, (IEEE, 1998), pp. 416–421
16. K. Sreeram, Design of fuzzy logic controller for speed control of sensor less BLDC motor drive, in *International Conference on Control, Power Communication and Computing Technologies* (IEEE, 2018), pp. 18–24
17. N.N. Baharudin, S.M. Ayob, Brushless DC motor drive control using single input fuzzy PI controller, in *IEEE Conference on Energy Conversion (CENCON)* (IEEE, 2015), pp. 13–18
18. B. Singh, V. Bist, A single sensor-based PFC Zeta converter FED BLDC motor drive for fan applications, in *IEEE Fifth Power India Conference* (IEEE, 2012), pp. 1–6
19. S. Singh, B. Singh, Voltage controlled PFC Zeta converter based PMBLDCM drive for an air-conditioner, in *5th International Conference on Industrial and Information Systems* (IEEE, 2010), pp. 550–555
20. B. Singh, S. Singh, Isolated Zeta PFC converter-based voltage controlled PMBLDCM drive for air-conditioning application, in *India International Conference on Power Electronics*, (IEEE, 2011) pp. 1–5
21. E.G. Janardhan, *Special Electrical Machines* (PHI publications, 2007)
22. J. Jantzen, *Design of Fuzzy Controllers* (Wiley Press, Beijing, 1999)

Artificial Ant Colony Optimized Direct Torque Control of Mathematically Modeled Induction Motor Drive Using PI and Sliding Mode Controller

Bibhu Prasad Ganthia, Rosalin Pradhan, Rajashree Sahu, and Aditya Kumar Pati

1 Introduction

Speed controller design significantly affects electric drive output. Because of their basic structure, PI speed controllers are commonly used in industrial applications. However, due to continuous variance of machine parameters, model uncertainties, nonlinear dynamics, and external device disturbances, fixed-gain PI controllers are often unable to enhance the necessary output of the function. Continuous adaptation of the controller parameters therefore becomes desirable when high output from the drive system is needed [1, 2]. Genetic Algorithms (GAs) are adaptive search techniques focused on a biological definition of "survival of the fittest." It can provide an inexpensive and reliable way to optimize applications by looking for a global minimum without having a cost-function derivative [3]. Thus, GA can be added to tuning the PI controller gains to ensure maximum control output under nominal operating conditions [4]. Another solution to this issue, however, is to fully replace the PI controller with adaptive control structures such as self-tuning and Artificial Intelligence techniques, Model Reference Adaptive Control (MRAC), and Sliding Mode Control (SMC) [5]. The sliding mode control strategy has demonstrated robustness against motor parameter uncertainty and dynamic modeling, insensitivity to external load disturbance, stability, and rapid dynamic response among these different designs proposed [6, 7]. It is therefore found to be very efficient in regulating systems of electric drives. Big chattering of torque at a steady state can be called the key downside for such a control scheme [8, 9]. One way to boost the efficiency of the sliding mode controller is to link it with Genetic Algorithm (GA) to create a controller for GA Sliding Mode (GASM).

B. P. Ganthia (✉) · R. Pradhan · R. Sahu · A. K. Pati
Electrical Engineering, IGIT, Sarang, Dhenkanal, Odisha, India
e-mail: jb.bibhu@gmail.com

J. Kumar and P. Jena (eds.), *Recent Advances in Power Electronics and Drives*, Lecture Notes in Electrical Engineering 707,
https://doi.org/10.1007/978-981-15-8586-9_35

389

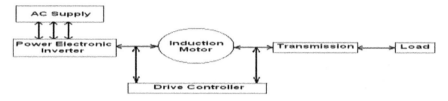

Fig. 1 Modern variable speed IM drive system

2 Induction Motor (IM) Drive

The IM is perhaps the most commonly used AC engine, as it provides many advantages compared to other engines. Two forms of 3-phase IMs exist. They are Wound rotor IM and Squirrel-cage IM [10, 11]. Both machines are identical from an electrical point of view, except that the previous has permanently shorted rotor winding terminals within the system. In the context of a wound-rotor system, the 3-phase rotor winding terminals are externally available for the operation [12, 13]. The project discusses only the speed strength of Squirrel-cage IM. Figure 1 displays the block diagram for a typical modern IM drive device with variable speed. The IM is directly or indirectly related to the charge (through gears). The power converter regulates the transfer of power form from an ac supply toward motor by actively regulating the voltage semiconductor switches [1, 14, 15].

Primary reasons for this complexity are the use of variable frequency, harmonically configured electrical systems for converters and complicated dynamics of ac systems, differences in system parameters and complexities in managing feedback signals in the existence of transient harmonics [16]. The IM is superior because they are machines that are stable and reliable to their counterparts, because they have functionality such as repairs-free in operations [17].

3 Direct Torque Control Technique

This switching-based approach has been described as the approach that is simple and feasible to meet those needs. DTC is one of the most excellent and efficient methods for regulating control induction. This approach emphasizes differentiated torque value and the stator flux monitoring and is now one of the vital system commonly observed control methods with the purpose of regulating torque and flux efficiently. Figure 2 displays the typical block diagram for conventional DTC scheme. The look-up switching table is taken according to the switching vectors [1, 17].

The diagram of the DTC block arrangement is displayed in Fig. 3 and the frame of reference for the stator and rotor axis. The reference torque can be measured using a PI unit using the variation in instantaneous speed and reference speed. Selection of this reference speed strengthens the torque and flux control dynamic response [5].

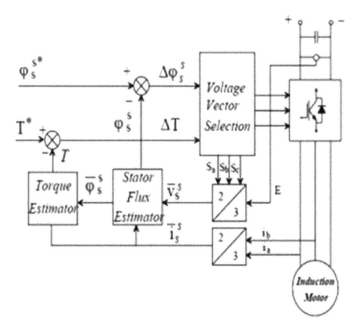

Fig. 2 Conventional DTC scheme

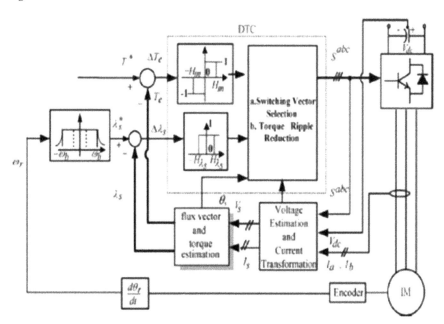

Fig. 3 Controller block diagram

4 Genetic Algorithm

GA is a global stochastic tool for the optimizing of adaptive search based on natural selection frameworks. GA has been widely acknowledged as a major and successful tool for resolving optimization problems. Compared to other techniques of optimization, such as simulating annealing and random search methods, GA is effective in the avoidance of local value of minima value which is a typical feature of the nonlinear models. The GA architecture is illustrated in Fig. 4 and PI controller implementation of the genetic algorithm well into the sliding control mode algorithm is shown in Fig. 5.

Fig. 4 GA architecture

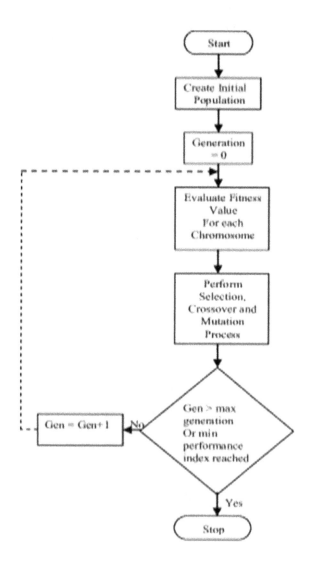

Fig. 5 Genetic algorithm
based DTC

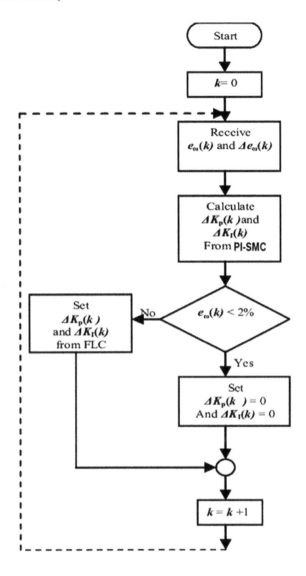

The detail parameters for the genetic algorithm are shown in Table 1. This represents the iterations of the genetic algorithm defining the conditions for the controller operations.

Table 1 Parameters for genetic algorithm

Parameters	Value	Attributes	Strategy
Generations	20	Selection	Stochastic
Chromosomes	16	Crossover	Double point
Genes	4	Crossover probability	0.8
Chromosome length	40 bit	Mutation rate	0.07

5 Artificial Ant Colony Algorithm

M. Dorigo in the late 1900s implemented the ACO technique influenced meta-heuristic solutions for the problem of combinatorial optimization. This algorithm is based on the actual actions of ant while searching for food source. It is obvious that there are large concentrations of pheromone on the shortest path, so more ants appear to choose it to fly. The ant colony algorithm comprises three major phases: initialization, development of ant response, processing of pheromone. The universal upgrading rule is enforced in the ant system in which all ants start their walk and pheromone gets accumulated and modified on all ends shown in the equation below:

$$\rho = \frac{[\tau ij(t)]\alpha[1/Tij]\beta}{\sum[\tau ij(t)]\alpha[1/Tij]\beta}$$

Tij is the time for the process, i is the tugboat by j. α and β are the coefficients.

The flow chart is the architecture of ACO technique as shown in Fig. 6 which is implemented for the finding specific torque for the control of induction motor drives. Figure 7 shows the operational control technique for the optimizing the references value with respect to the suitable value of torque control. The ACO minimizes the constraints of large no. of torques into limited numbers which can be directly used with the IM drive demand.

A collection of food source sites (*eb*) are generated randomly at the initialization of the algorithm. Let's consider *u*th food source in the population as

$$d_u = d_u, 1, d_u, 2, d_u, 3, \ldots, d_u, n \quad (9)$$

and every food source site is generated according to the Eq.

$$D_{uv} = d\min + rand(0, 1)(d\max - d\min)$$

where u refers to the size of food sites, $u = 1, 2, 3 \ldots Eb$, v means the parameters to be optimized, $v = 1, 2, 3 \ldots$ ncv d_u max & d_u min are the upper and lower dimensional boundaries u. The $f_i t_u$ amounts are determined after initialization of the food source sites. In Table 2, parameters for genetic algorithm are shown below.

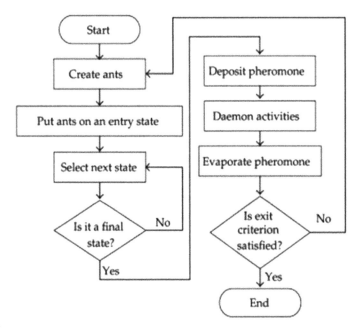

Fig. 6 ACO architecture

6 Sliding Mode Controller

This highly advanced technique is widely recognized for its robustness against internal uncertainties, uncertainties due to nonlinearity in parametric variations and anomalies that were excluded in the modeling. SMC procedure is applicable for enhancing DTC for IMs highlighted in this study.

Such strategies enhance the efficiency of the steady state and preserve the benefits of intermittent state and the torque and flux are resilient against system parameter variations. It presents models and experimental findings to demonstrate the feasibility of the proposed strategy. Fig. 8 shows the concept of sliding mode technique in phasor diagram form where the desired vale and reaching mode to the stability is presented.

7 Simulink Model and Result

The direct torque control design PI-DTC and SMC-DTC framework and the flux depending on the sliding mode of a speed-driven induction motor are shown in Fig. 9. The model is cascade order for electromagnetic torque control, square flux norm, and speed control. So sliding control method and genetic algorithms are introduced to adjust the torque, flux, and speed in the control structure.

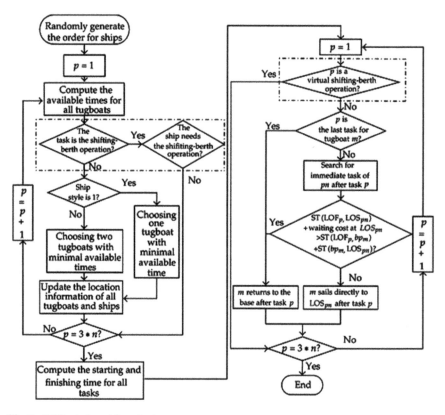

Fig. 7 ACO rule-based flowchart

Table 2 Parameters for genetic algorithm

Parameters	Value	Attributes	Strategy
Generations	20	Initialization	Search
Bandwidth	0.2	Construction of colony	Multi-points
Memory rate	0.95	Updating pheromone	1.0
No. of iteration	100	Tour rate	0.2

7.1 DTC Using Conventional PI Controller

The results from conventional PI controller are shown in above where Fig. 10 represents the result for the electromagnetic torque; Fig. 11 shows the speed obtained from the rotor for the system. Figure 12 highlights the trajectory for the stator flux. The frame of reference for the model is stationary form.

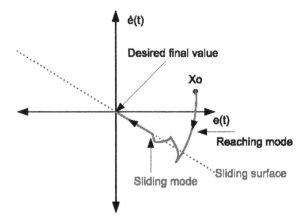

Fig. 8 Sliding mode control phasor

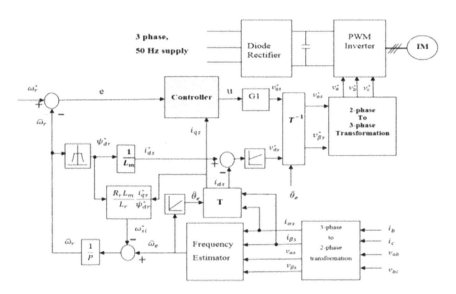

Fig. 9 Controller design and implementation in IM drive

7.2 DTC Using PI-Genetic Algorithm

The results using conventional PI-GA controller are shown in above where Fig. 13 represents the result for the electromagnetic torque, Fig. 14 shows the rotor speed for the system and the Fig. 15 highlights the trajectory for the stator flux in stationary reference frame.

Fig. 10 EM torque versus
time (sec)

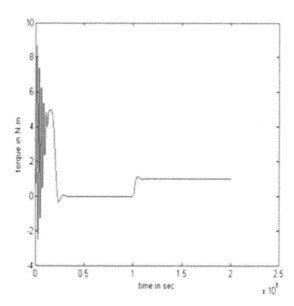

Fig. 11 Rotor speed (Nr)
versus time (sec)

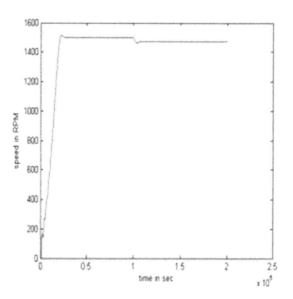

7.3 DTC Using PI-Sliding Mode Controller-ACO Technique

The proposed frame for the direct torque control architecture and the flux from the
stator frame based on the sliding mode of a speed-controlled IMD (induction motor)
(PI-SMC-ACO-DTC) are shown in Fig. 16. Figure 17 represents SVM for DTC
scheme of IM using ACO in association with the proposed technique.

Fig. 12 Stator flux trajectory of d-axis and q-axis

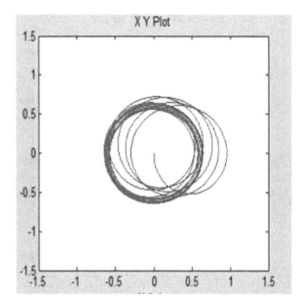

Fig. 13 EM torque versus time (sec)

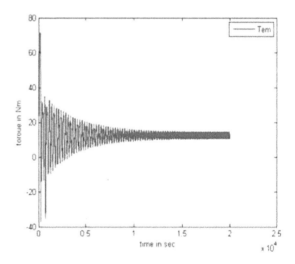

The proposed technique gives better results with respect to all the uncertainties that arise due to the speed variations represented in the following figures shown above. Figure 18 shows d-axis voltage of DTC scheme of IM using PI with Nonlinearity, Fig. 19 shows q-axis voltage of DTC scheme of IM using PI with Nonlinearity, Fig. 20 shows d-axis voltage of DTC scheme of IM using SMC with Nonlinearity, Fig. 21 shows q-axis voltage of DTC scheme of IM using SMC with Nonlinearity, Fig. 22 shows d-axis voltage of DTC model of IM using PI-SMC-ACO with Nonlinearity, Fig. 23 shows q-axis voltage of DTC scheme of IM using PI-SMC-ACO with

Fig. 14 Rotor speed versus
time (sec)

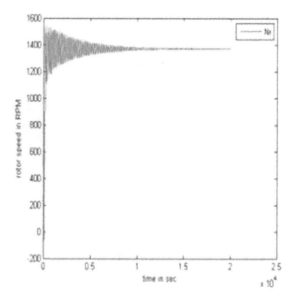

Fig. 15 Stator flux
trajectory of d-axis and
q-axis

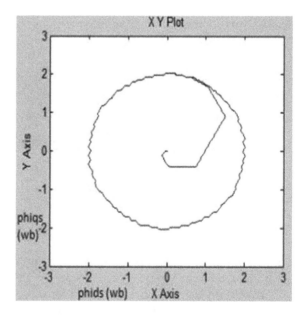

Nonlinearity, Fig. 24 shows Electromagnetic Torque versus time(sec), Fig. 25 shows
Rotor speed versus Time (sec), and Fig. 26 shows the Stator flux Trajectory of d-axis
and q-axis of the proposed technique.

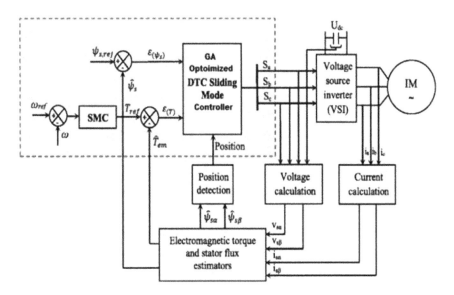

Fig. 16 Block diagram of the proposed technique

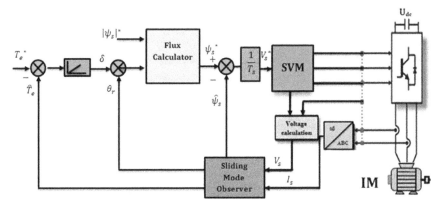

Fig. 17 SVM for DTC scheme of IM using ACO

8 Result Analysis and Discussion

The cost function for the controller is to determine that every generation's individuals can be selected as the Integral Time of Absolute Error (ITAE). This cost function can be written in mathematical expression as follows:

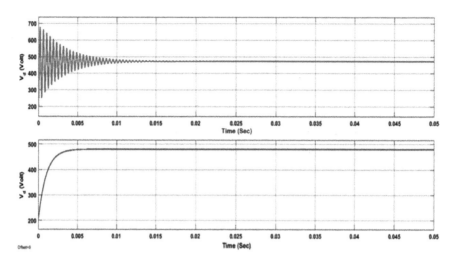

Fig. 18 d-axis voltage of DTC scheme of IM using PI with nonlinearity

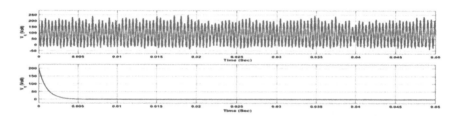

Fig. 19 q-axis voltage of DTC scheme of IM using PI with nonlinearity

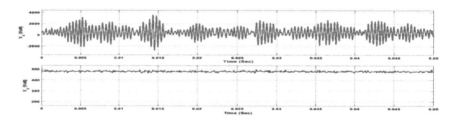

Fig. 20 d-axis voltage of DTC scheme of IM using SMC with nonlinearity

$$\text{ITAE} = \int_{0}^{t} t|exp(t)|dt$$

ACO is looking for the optimum set value of the conventional PI controller gains during the search process which minimizes ITAE cost function. This function is set as the criterion of evolution of the ACO with positive and negative errors calculations.

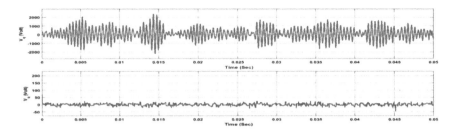

Fig. 21 q-axis voltage of DTC scheme of IM using SMC with nonlinearity

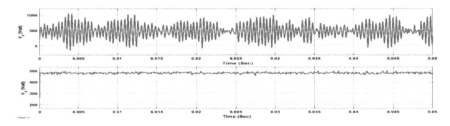

Fig. 22 d-axis voltage of DTC model of IM using PI-SMC-ACO with nonlinearity

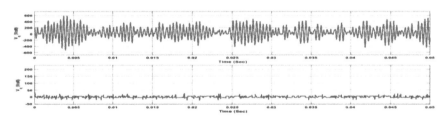

Fig. 23 q-axis voltage of DTC scheme of IM using PI-SMC-ACO with nonlinearity

Fig. 24 Electromagnetic
torque versus time (sec)

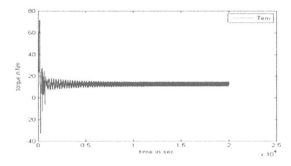

Fig. 25 Rotor speed versus
time (sec)

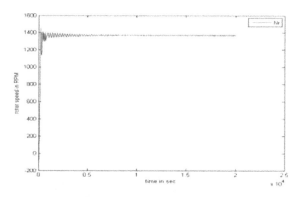

Fig. 26 Stator flux
Trajectory of d-axis and
q-axis

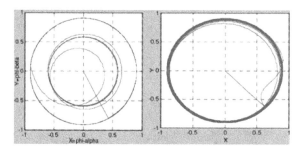

The comparison between controllers at different operating conditions is shown in
Table 3.

The Simulink results compared with the proposed technique are categorized
according to the following operating conditions: Speed 60 rad/s at rated load torque
of 30%, Reference speed of 120 rad/s at Nominal Parameters. Phase load torque
applied at t = 2 sec from 30 to 90% rated load, Reference speed of 60 rad/s and rated
load torque of 30%, Reference speed 120 rad/s at nominal parameters and load torque
rated 30%, Applied with 120% uncertainty in motor inertia (J) 60 rad/s reference
speed at 30% level load, Speed variation (Nr) from 60 rad/s to 250 rad/s at a level load
of 30% and nominal parameters, Reversal of speed value from 60 rad/s to −60 rad/s
at 30% average load with 60% uncertainty in engine inertia (J) and Reference speed
150 rad/s at nominal parameters and load torque rated 50%. A phase shift in the load
torque from 30% changed to 90% of its actual value is at time t = 2 s where the
motor is operating at 120 rad/s to analyze the disturbance value of rejection potential
of the proposed schemes. When the change in load torque is applied, PI-GA shows
a low capacity for disturbance with the steady-state error of 0.15% while GA-SMC
shows zero sensitive to sudden change in load. PI-SMC-ACO-based system has the
faster response toward the speed variations and achieve faster steady state.

From the result in Table 4, we get the results of different speed variations due to
different controllers and optimizing techniques, and Table 5 shows the comparison
between torque variations.

Table 3 Comparison of controllers with the proposed technique at various operating conditions

		Conventional PI	PI-GA	PI-ACO-SMC (proposed technique)
Speed 60 rad/s at rated load torque of 30%	**ITAE**	0.08165	0.09514	**0.08830**
Reference speed of 120 rad/s at Nominal Parameters. Phase load torque applied at t = 2 sec from 30 to 90% rated load		0.32	0.229	**0.201**
Reference speed of 60 rad/s and rated load torque of 30%		0.01182	0.04266	**0.04213**
Reference speed 120 rad/s at nominal parameters and load torque rated 30%		0.6061	0.5199	**0.5001**
Applied with 120% uncertainty in motor inertia (J) 60 rad/s reference speed at 30% level load		0.3258	0.3025	**0.2992**
Speed variation from 60 rad/s to 250 rad/s at a level load of 30% and nominal parameters		12.78	9.49	**9.003**
Reversal of speed value from 60 rad/s to −60 rad/s at 30% average load with 60% uncertainty in engine inertia (J)		9.45	5.28	**5.112**
Reference speed 150 rad/s at nominal parameters and load torque rated 50%		0.6158	0.5210	**0.4990**

Table 4 Comparison between controllers at different speed reversal

Parameters	SVM DTC	PI DTC	PI GA DTC	PI-SMC-ACO technique
Peak speed	511.56	551.05	515.51	513.34
K_i	14.30	9.65	15.98	25.42
K_p	0.93	0.18	0.92	1.12
Max Overshoot %	2.91	13.74	3.90	3.36
Settling time	0.18	0.17	0.13	0.10

Table 5 Comparison between controllers at different load torque

Parameters	SVM DTC	PI DTC	PI GA DTC	PI-ACO-SMC technique
Torque	1105.2	1008.3	1096.2	1005.72
K_i	2.2	9.95	15.10	29.22
K_p	1.2	1.10	0.75	2.10
Max Overshoot %	16.6896	3.1453	12.9048	2.8514
Settling time	0.12	0.09	0.007	0.009

Table 6 Comparison between controllers at different operating conditions

	SVM DTC	PI DTC	PI GA DTC	Proposed technique
Torque and flux ripple	Less	Average	Average	Average
Torque response	Fast	Fast	Faster	Faster
Switching frequency	Constant	Constant	Constant	Constant
THD	Low distortions	Low distortions	Less distortions	Less distortions
Switching loss	Less	Average	Average	Less
Sensitivity	Yes	No	No	No
Complexity	Simple	More	More	More
Speed Regulation	Fast	Good	Very good	Very good
Precession	Average	Average	Average	More
Computation time	Average	High	Average	High
Controller Regulation	PI controlled	SMC controller	GA and PI CONTROLLER	PI + SMC + ACO + DTC controller

In Table 6, it is concluded that the proposed technique can play a vital role in DTC of induction motor drive with faster response toward the speed variation.

9 Conclusion

This paper highlights the implementation of Artificial Ant Colony Algorithm (ACO) characteristics using sliding mode control technique with conventional PI controller for the direct torque control of IMD; induction motor drive system. This follows research paper the conventional controller technique compared with the proposed PI-SMC-ACO control technique and it gives superior efficiency on speed variations and faster response toward steady-state operation. The DTC drive is the electromagnetic torque command based on the Lyapunov principle for heuristic control. PI-GA demonstrates better output under nominal driving conditions while ACO-SM reveals robustness over variance in stator resistance, inertia instability, and crassness of stability. The model provides better efficient and smooth operating process with faster steady-state operations. In future, more adaptive techniques can be implemented for better steady-state operations with higher constraints.

References

1. B.P. Ganthia, P.K. Rana, S.A. Pattanai, Space vector pulse width modulation fed direct torque control of induction motor drive using matlab-simulink, in *Proceedings of the 3rd International Conference on Electrical, Electronics, Engineering Trends, Communication, Optimization and Sciences (EEECOS)* (Tadepalligudem, India, 1–2 June 2016)
2. N. El Ouanjli, A. Derouich, A. El Ghzizal, A. Chebabhi, M. Taoussi, (2017). A comparative study between FOC and DTC controls of the doubly fed induction motor (DFIM), in *International Conference on Electrical and Information Technologies: IEEE*
3. A. Ammarn, A. Bourek, A. Benakcha, Nonlinear SVM-DTC for induction motor drive using input-output feedback linearization and high order sliding mode control. ISA Trans. **67**, 428–442 (2017)
4. S. Boubzizi, H. Abid, M. Chaabane, Comparative study of three types of controllers for DFIG in wind energy conversion system. Protect. Control Modern Power Syst. **3**(1), 21 (2018)
5. F. Wang, Z. Zhang, X. Mei, J. Rodríguez, R. Kennel, Advanced control strategies of induction machine: field oriented control, direct torque control and model predictive control. Energies **11**(1), 120 (2018)
6. B. Ganthia, S. Sahu, S. Biswal, A. Abhisekh, S. Kumar Barik, Genetic algorithm based direct torque control of VSI fed induction motor drive using MATLAB simulation. Int. J. Adv. Trends Comput. Sci. Eng. **8**(5), 2359–2369 (2019)
7. M.M. Ali, W. Xu, M.F. Elmorshedy, Y. Liu, S.M. Allam, M. Dong, Sliding mode speed regulation of linear induction motors based on direct thrust control with space-vector modulation strategy, in *2019 22nd International Conference on Electrical Machines and Systems (ICEMS)* (Harbin, China, 2019), pp. 1–6.
8. S. Jnayah, A. Khedher, Sensorless direct torque control of induction motor using sliding mode flux observer, in *2019 19th International Conference on Sciences and Techniques of Automatic Control and Computer Engineering (STA)* (Sousse, Tunisia, 2019), pp. 536–541.
9. S. Krim, S. Gdaim, A. Mtibaa, M. Faouzi Mimouni, FPGA-based real-time implementation of a direct torque control with second-order sliding mode control and input–output feedback linearisation for an induction motor drive. IET Electric Power Appl. **14**(3), 480–491 (2020)
10. W. Wang, H. Lin, H. Yang, W. Liu, S. Lyu, Second-order sliding mode-based direct torque control of variable-flux memory machine. IEEE Access **8**, 34981–34992 (2020)
11. B.P. Ganthia, A. Pritam, K. Rout, S. Singhsamant, J. Nayak, Study of AGC in two-area hydrothermal power system, in *Advances in Power Systems and Energy Management*, ed. by A. Garg, A. Bhoi, P. Sanjeevikumar, K. Kamani. Lecture Notes in Electrical Engineering, vol. 436 (Springer, Singapore, 2018)
12. B.P. Ganthia, R. Pradhan, S. Das, S. Ganthia, Analytical study of MPPT based PV system using fuzzy logic controller, in *2017 International Conference on Energy, Communication, Data Analytics and Soft Computing (ICECDS)* (Chennai, India, 2017), pp. 3266–3269

13. B.P. Ganthia, P.K. Rana, T. Patra, R. Pradhan, R. Sahu, Design and analysis of gravitational search algorithm based TCSC controller in power system. Mater. Today Proc. **5**(1), Part 1, 841–847 (2018). ISSN 2214-7853. https://doi.org/10.1016/j.matpr.2017.11.155

14. B.P. Ganthia, A. Abhisikta, D. Pradhan, A. Pradhan, A variable structured TCSC controller for power system stability enhancement. Mater. Today Proc. **5**(1), Part 1, 665–672 (2018). ISSN 2214-7853. https://doi.org/10.1016/j.matpr.2017.11.131

15. B.P. Ganthia, S. Mohanty, P.K. Rana, P.K. Sahu, Compensation of voltage sag using dvr with pi controller, in *Electrical Electronics and Optimization Techniques (ICEEOT) International Conference* (2016), pp. 2138–2142

16. B.P. Ganthia, K. Rout, Deregulated power system based study of AGC using PID and fuzzy logic controller. Int. J. of Adv. Res. **4**, 847–855 (2016). www.journalijar.co

17. I. Takahashi, T. Noguchi, A new quick-response and high-efficiency control strategy of an induction motor. IEEE Trans. Industry Appl. **IA-22**(5) (1986)

Comparative Analysis of Power Factor Correction Converters for Different Topologies

Sanatan Kumar⦿, Devashish⦿, and Madhu Singh

1 Introduction

With the advancement in power electronics equipment, the integration of Power Factor Correction (PFC) into modern motor drives system has developed in the last few years. Today a number of PFC circuits can be built with different operating modes, each having its own set of challenges. As there are different types of load, so it is difficult to create suitable PFC for every load [1].

Harmonic elimination and power factor correction specifications as required by IEC 61000-3-2 stick apart among these developments as one of the most important crucial stage in electrical supply and designs in previous years. Despite increase in power ratings across all electrical appliances and expanding implementation including its harmonic elimination rules, better power supply designs are coming up with integrated power factor correction capability. Engineers are finding difficult to implement the required PFC measures and comply with other requirements of legislation such as power quality improvements, active mode reliability, and EMI constraints [2–3].

Our dedication to the use of many choices for topology and its components is expressed in the design guidance. In this paper, comparison in detail of various PFC implementation options while preserving the overall system requirements has been done.

Choices for power factor compensation approaches vary from varied range of active circuits to narrow range of passive circuits. The suitable solution might change subject to the change in level of power and other system details. The progress that has been made over the last few years in the discrete semiconductor systems and the

S. Kumar (✉) · Devashish · M. Singh
Department of Electrical Engineering, NIT Jamshedpur, Jamshedpur, Jharkhand, India
e-mail: 2015rsee00@nitjsr.ac.in

© The Author(s), under exclusive license to Springer Nature Singapore Pte Ltd. 2021   409
J. Kumar and P. Jena (eds.), *Recent Advances in Power Electronics and Drives*, Lecture Notes in Electrical Engineering 707,
https://doi.org/10.1007/978-981-15-8586-9_36

availability of cheaper controller ICs made the active PFC design more reasonable for a varied choice of uses [4]. It is important to observe at PFC designs in terms of real system price and execution while evaluating such designs.

Power converters typically use a diode correction to transform AC voltage into DC voltage accompanied by a bulk condenser. More than 60% of energy is projected to be consumed by some kind of electronics system by the end of the year 2020. Almost all of the appliances might have a front-end filter DC-DC converter. The capacitor filtration circuit connected to diode rectifier circuit, without any control circuit, draws pulsating currents from the distribution grid, which leads to low power and high harmonics content. These results in increased neutral currents in three-phase network, and heating of rotating motor. This makes some power conditioning. This is necessary, because there is a necessity for the power factor correction to limit the harmonics quality of line currents generated from electrical devices associated with electricity distribution grids [5]. Regulatory bodies across the globe have been brought to notice by the crisis. Governments strengthen rules, set new standards for lower harmonic components of current which eliminate their total harmonic current which may be produced. Consequently, the line current harmonics need to be reduced, and PFC correction and harmonic reduction circuits need to be introduced.

The independence from harmonics often minimizes interference from the same source in other applications. PFCs also have to meet with regulatory requirements in many of the current energy sources. The IEC 61000-3-2 must now be complied with in electrical equipment in Europe, Asia, and Japan [6].

Energy level issues are not a modern trend, Duffey et al. [7] originally proposed an improvement of the power factor for the distribution system back in the 1920s. But the Power Factor Correction (PFC) technique has become increasingly popular in the world of power electronics in the last few years. Lai et al. [8] and Singh et al. [9] defined the power factor and illustrated its significance for the field of power electronic converter.

The growth of DC-DC converter has improved enormously over the past two decades. A detailed analysis of the specifications, design characteristics, device selection, and specification for specific applications of Improved Power Quality Converters (IPQCs) is presented. The objective is to provide scholars, developers, and computer analyst continuing to work on switched-mode AC-DC converter, as suggested by Singh et al. [10] has a broad variety of IPQC application profile.

A boost PFC has been defined by Caneson et al. [11] in which the switch is switched on and off only twice per line. As a result, the losses in di/dt and dv/dt and conversion are smaller, sluggish recovery diodes can be used and heavy EMI filters are avoided. Alternatively, the output voltage may be controlled in a broad load range by simply regulating it. Salmon et al. [12] also implemented a PWM rectifier with a soft-switching unit power factor, which significantly increases the performance of the system by flipping the main switches into a switching system without any auxiliary switches. Thanks to a single-stage converter topology instead of a front-end rectifier followed by a boost converter, the conduction losses can also be greatly decreased. However, this has some pitfalls, as it requires a number of power circuit machines. The modern boost converters impressed primarily with the inductor volume, weight, and power unit losses; these factors affected the converter's size, power density, and performance. Zhang et al. [13] suggested the three-level boost converter within only one PFC, with a relatively small inductor and a low power system. It has provided high density, high efficiency, and lower costs.

SEPIC, Cuk, and ZETA converters have been proposed in [14–18] are used for large power applications. But these converters do not reflect the component stresses, the control of output voltage which is leading the distortion of output voltage and noise interference.

Depending on the configuration and specifications of the system, either solution can require the converters to work in DCM or CCM. However, in that article, active PFC converter operation for the AC mains unit power factor is provided.

2 Mathematical Description on Power Factor Correction

The description of power factor is given as proportion of the actual electrical power (defined under W) and the total electrical power is given as (VA): where actual electrical power is the total summation of the continuous product of voltage or current over a cycle and the total electrical power is the combination of current rms and the voltage rms.

The power ratio is unity when both current curve and voltage curve with respect to time are sinusoidal. All the three powers in electrical system and their relation are shown in Fig. 1. Time

$$S = V \times I(Volt - Amp) \tag{1}$$

$$P = S.Cos\phi(Watt) \tag{2}$$

$$Q = S.Sin\phi(\text{var}) \tag{3}$$

Fig. 1 Power triangle for electrical power supply

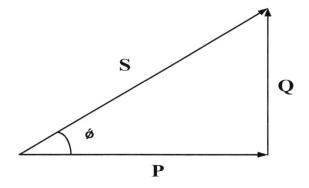

$$S = \sqrt{(P^2 + Q^2)}(VA) \tag{4}$$

Power factor also is introduced as a result of both the distortion factor as well as the displacement factor as specified in Eqn. 5, the distortion power factor K_θ is the cosine of just the angular position in between simple source reference current and the source reference voltage.

$$Power\_Factor = K_d K_\theta \tag{5}$$

$$K_d = \frac{I_{rms}}{I_r} \tag{6}$$

$$K_\theta = Cos\theta \tag{7}$$

The factor for distortion is defined for sinusoidal voltage only and can be expressed with the total harmonic distortion (THD):

$$DPF = \frac{1}{\sqrt{1 + THD^2}} \tag{8}$$

$$PF = \frac{Cos\phi}{1 + THD^2} \tag{9}$$

The overall power factor is far below to unit value as only the fundamental component provides true power so another harmonic is introduced to total power. This variance was defined as the effect of distortion, and is primarily responsible for the nonlinear portion. The general balance of actual power and apparent power is presented through

$$P_{in} = V_{inrms} \times I_{inrms} \times Cos\phi \times Cos\theta \tag{10}$$

where $cos\phi$ is indicated as the displacement factor between the source voltage and source current waveforms originating from the phase angle ϕ and where $cos\theta$ is the factor of distortion. Consequentially, in Fig. 2, power factor correction technique is not implemented for nonlinear load. Harmonic content is injected in the input AC current that's why the power factor of the given power source with the waveform is around 0.6. Harmonic spectrum analysis for different harmonic number is displayed in Fig. 4.

Figure 4 shows the waveform of pure input AC voltage and input current when PFC is implemented with an ideal power factor adjustment between non-linear load and source. It has a current waveform imitating the form as well as the phase of the waveform of the voltage. Notice that their present input harmonics are nearly zero. Harmonic spectrum analysis for different harmonic number is shown in Fig. 5.

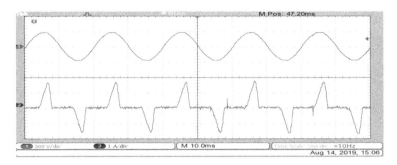

Fig. 2 Source voltage and current waveform without PFC converter

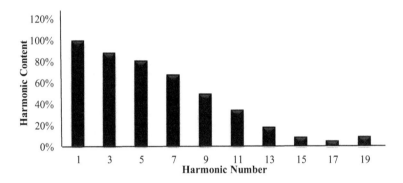

Fig. 3 Harmonic spectrum analysis of the source current waveform shown in Fig. 2

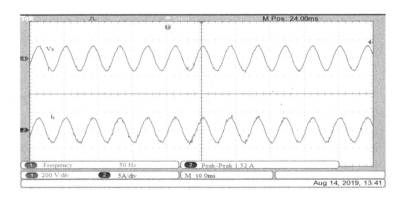

Fig. 4 Source voltage and current waveform with PFC converter

The goal is to increase the displacement factor and distortion factors by using a PFC stage for AC-DC converter in the main phase in directive to decrease the reactive power available from the supply. Acceptable obtained sinusoidal source current In

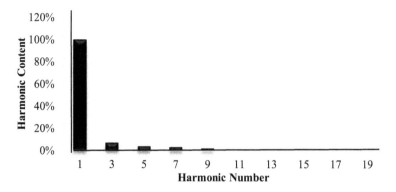

Fig. 5 Harmonic spectrum analysis of the current waveform in Fig. 4

phase with source voltage, the load must be strictly immune from the ac view. This results in better power output and better machine performance.

3 Types of Power Factor Correction

Two major PFC categories are generally available: active and passive PFC. Passive PFC includes basic reactive elements such as inductors and condensers to make up for line voltage and current displacement angle. This process is called correction of passive power factor. Active PFC employs switch and controller circuits for raising the power factor and minimizing the harmonic distortion. Both approaches have their own merits and demerits but a more modern approach appears to be active PFC.

Active PFC is used to mitigate input current distortion, often switch mode power supply (SMPS) are realized in it. Even if they are more complicated than passive PFC circuits, evolving integrated circuits and electronic components made an inexpensive option for active PFCs. Such power systems are capable of reaching a power factor of 99% and a THD of less than 5%. Responsive PFC topologies occur in both low and high-frequency operation.

The scale and weightiness of the passive solution are controversial at higher power levels. Figure 6 shows three distinct power supplies with current waveform with same input characteristics. As seen, the maximum current in passive PFC circuits remains thirty-three percent more than in active circuit maximum currents. Furthermore, although the harmonic rates of the level two that follows the IEC61000-3-2, certain recent regulations may fail to comply with the strict 0.9 PF criterion [19–20].

The market trends (higher copper and metallic core content costs and lower semi-conductor prices) have over the past years significantly tipped the difference in terms of active PFC in some of the most price-sensitive applications for consumer use. In combination with the effective action of active PFC circuits [21], it seems that this pattern will continue and make the designers available more advanced active PFC

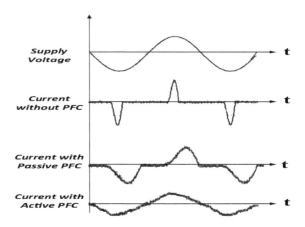

Fig. 6 Various input Current waveforms with different types of PFC converter

solutions. Therefore, the primary objective of PFC circuit intended for nonlinear load is present to diminish the harmonic value of line current.

3.1 Active Power Factor Improvement Approaches

Active power factor improvement provides a brighter and more effective way to control input side power factor when a power supply configuration is above hundred Watt and it is the preferred form of PFC. Active PFC consists of a high switching regulator which can produce a Quantitative power factor of more than 95% [22]. Effectively power factor correction on AC side of input voltage is automatically corrected and capable of a multiple range of input voltages. The additional costs associated with the increased complexity is needed when implementing Active PFC is a drawback. The following sections will discuss an active PFC circuit which has also improved features and but don't have several of the directly earlier mentioned disadvantages.

Active PFC Based on Boost Converter
The boost converter is widely used with single-phase full-wave diode bridge rectifier for optimizing power efficiency among the various current PFC topologies. It is a high-frequency active PFC system and is most often picked because the boost converter experiences lower transistor stresses than other DC-DC converters like Buck-boost, Cuk, SEPIC, etc.

There are mainly two modes in which the boost converter works that are the continuous mode of conduction and discontinuous mode of conduction, which is the most prevalent power factor improvement architecture. Through changing the switching frequency, the transfer mode power, also termed to as the Critical Conduction form or Barrier Conduction form, keeps the converters at the border stuck in between CMC as well as DMC. Figure 7 shows the active PFC circuit for Boost converter.

This architecture increases the voltage of the input side. Because the converter can function during the line-cycle, the crossover distortions in the source input current is zero. Across the envelope near source input voltage zero crossing the distortion is zero. Figure 8 shows the current and voltage wave pattern of a PFC circuit centered on a CCM boost converter. The converter switch current is continuous because the boost converter topology is connected in chain sequence through the input and the input power is not interrupted by the high-frequency operation of switch [23]. Therefore, the source current has high-frequency components of harmonics, leading to lower EMI and lower circuits for filtration. The output condenser restricts the turn-off voltage of the switch S_w to approximately the output voltage by diode D and consequently guards the switch.

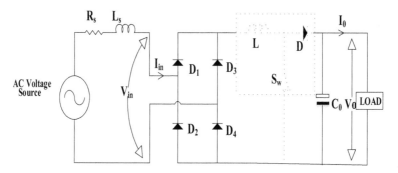

Fig. 7 Active PFC architecture centered on the Boost Converter

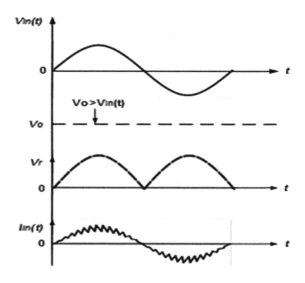

Fig. 8 Current and voltage wave pattern of a PFC circuit centered on a CCM boost converter

The regulator controls inductor current and tries to maintain it at a constant value so that it could not fall beyond the permissible limit. In comparison to CCM or CRM, the DCM converter functions in stable frequency and offers current interruptions. Comparison waveform is displayed in Fig. 9. It is seldom or never used because of high peak current and EMI linked to the DCM converter.

The hysteretic control variation is used for CRM converter for the lower limit of zero current on the other pointer. It remains an intrinsically stable variable frequency control technique, which prevents reverse recovery corrective losses. The on time remains the same for a certain constraint of input, but the off time is varied. As a consequence, when the source voltage is lowest instantaneously, and vice versa, the power converter has the highest switching frequency.

Active PFC Buck-Boost Converter

Ultimately, in the figure, you will see the boost buck PFC centered converter circuit with its waveforms Figs. 10 or 11. As even the name indicates, the buck-boost converter is accomplished of creating an inferior or advanced output voltage than the

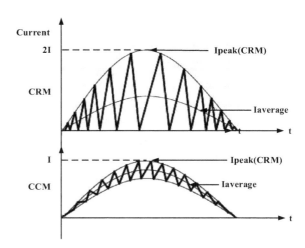

Fig. 9 Comparison of CCM and CRM operating configurations with inductor current

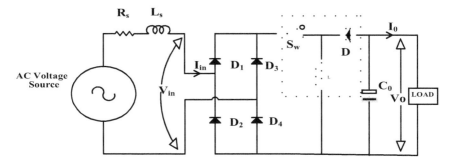

Fig. 10 Active PFC architecture centered as buck-boost converter

Fig. 11 Buck-Boost
converter centered PFC
circuit current and voltage
wave patterns

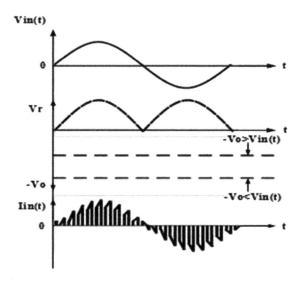

source voltage. It can be produced as the only cascade attachment between a buck-
and a boost converter [24]. As a result, in steady-state, presuming CCM process, its
voltage transfer ratio (output-to-input) is the product of the cascading ratio of the
attached buck as well as boost converters, supposing that they have a specific duty
ratio of D.

The input current takes no crossover distortions since the converter could work
during the line cycle. This does not distort the current line envelope across the zero-
crossing voltage. Furthermore, even if the inductor current, like that of the buck
converter, is constant, since the peak frequency switch S_w stops the source current
the converter's input switch is discontinuous [25]. Therefore, the input current is
important for high-frequency components which raise the requirement of filtering
and EMI.

The production of voltage for the boost-buck converter may be both superior and
lesser than the input voltage. In addition, the buck-boost converter can be served to
regulate DC voltage and to run the BLDC motor at low speeds, as compared to boost
PFC topology.

4   Conclusion

The PFC manufacturer has seen a large rise in the number of choices available in
the over few years. All this is because of the growing interest of IEC 61000-3-2 and
the vigorous competition among semiconductor providers. As PFC gets better and
more cost effective, end-users are going to receive maximum benefits. The through
skill of these IC controllers allows power supply manufacturers more opportunities
for execution.

Next, the technical context of the power factor correction being defined, such as the fundamentals of the energy usage and the quantity indication. Since the need for an improved power factor was extracted from the illustration of a diode bridge rectifier, PFC circuits were added. Two different topologies of converters are covered: boost and buck-boost converter.

Within this article, topics of power factor correction were discussed, which offered an in-depth knowledge of PFC circuits and extended power electronics understandings.

References

1. T.J.E. Miller, *Brushless Permanent Magnet and Reluctance Motor Drive* (Clarendon Press, Oxford, 1989)
2. A. McEachern, W.M. Grady, W.A. Moncrief, G.T. Heydt, M. McGranaghan, Revenue and harmonics: an evaluation of some proposed rate structures. IEEE Trans. Power Delivery **10**(1), 474–480 (1995)
3. I. Batarseh, Power Electronic Circuits (Wiley Inc. in press).
4. N. Mohan, T.M. Undeland, W.P. Robbins, *Power Electronics: Converters, Applications and Design* (Wiley, USA, 2003)
5. L. Dixon, Average current mode control of switching power supplies, in *Product & Applications Handbook, Unitrode Integrated Circuits Corporation, U140* (1993–1994_, pp. 9-457–9-470
6. Limits for Harmonic Current Emissions (Equipment Input Current $\leq$16 A per Phase). Int. Std. IEC 61000-3-2, 2
7. C.K. Duffey, R.P. Stratford, Update of harmonic standard IEEE-519: IEEE recommended practices and requirements for harmonic control in electric power systems. IEEE Trans. Indus. Appl. **25**(6), 1025–1034 (1989)
8. J.S. Lai, D. Hurst, T. Key, Switch-mode power supply power factor improvement via harmonic elimination methods, in *Conference Record of IEEE-APEC'91*, pp. 415–422.
9. B. Singh, S. Singh, A. Chandra, K. AI-Haddad, Comprehensive study of single-phase AC-DC power factor corrected converters with high-frequency isolation. IEEE Trans. Ind Inform. **7**(4), 540–555 (2011).
10. B. Singh, B.N. Singh, A. Chandra, K. Al-Haddad, A. Pandey, D.P. Kothari, A review of single-phase improved power quality AC-DC converters. IEEE Trans. Ind. Electron. **50**(5), 962–981 (2003)
11. C.A. Caneson, I. Barbi, Analysis and design of constant- frequency peak-current-controlled high-power-factor boost rectifier with slope compensation, in *Conference Record of IEEE-APEC'96*, pp. 807–813.
12. J.C. Salmon, Techniques for minimizing the input current distortion of current-controlled single-phase boost rectifiers. IEEE Trans. Power Electron. **8**(4), 509–520 (1993)
13. J. Yang, J. Zhang, X. Wu, Z. Qian, M. Xu, Performance comparison between buck and boost CRM PFC converter, in *Proc. IEEE Control Model. Power Electron. Conf.*, (2010) pp. 1–5.
14. S. Kumar, M. Singh, Improved power quality Cuk converter for variable speed BLDC motor drive. J. Eng. Appl. Sci. **13**(5), 1275–1285 (2018)
15. S. Kumar, M. Singh, A fuzzy logic controller based brushless DC motor using PFC Cuk converter. Int. J. Power Electron. Drive Syst. (IJPEDS) **10**(4), 1894–1905 (2019)
16. A.R. Prasad, P.D. Ziogas, S. Manias, A new active power factor correction method for single-phase buck-boost AC-DC converter, in *Conference Record of IEEE-APEC'92*, pp. 814–820.
17. C.W. Clark, F. Musavi, W. Eberle, Digital DCM detection and mixed conduction mode control for boost PFC converters. IEEE Trans. Power Electron. **29**(1), 347–355 (2014)

18. H.S. Youn, J.S. Park, K.B. Park, J.I. Baek, G.W. Moon, A digital predictive peak current control for power factor correction with low-input current distortion. IEEE Trans. Power Electron. **31**(1), 900–912 (2016)
19. IEEE Standards Association. 1076.3-2009, IEEE Standard VHDL Synthesis Packages (IEEE Press, New York, 2009).
20. J.P. Noon, D. Dalal, Practical design issues for PFC circuits, in *Conference Record of APEC'97*, pp. 51–58.
21. K.D. Gusseme, D.M. Van de Sype, A.P.M. Van den Bossche, J.A. Melkebeek, Input-current distortion of CCM boost PFC converters operated in DCM. IEEE Trans. Ind. Electron. **54**(2), 858–865 (2007)
22. K. Raggl, T. Nussbaumer, G. Doerig, J. Biela, J.W. Kolar, Comprehensive design and optimization of a high-power-density single-phase boost PFC. IEEE Trans. Ind. Electron. **56**(7), 2574–2587 (2009)
23. F. Yang, X. Ruan, Q. Ji, Z.H. Ye, Input differential-mode EMI of CRM boost PFC converter. IEEE Trans. Power Electron. **28**(3), 1177–1188 (2013)
24. B. Singh, S. Singh, Single-phase power factor controller topologies for permanent magnet brushless DC motor drives. IET Power Electron. **3**(2), 147–175 (2010)
25. Optimizing Inductor Design for Efficient PFC Stage, EDN Asia, 2010 November issue, https://www.endasia.com/article-27331.

Design and Development of a Low-Cost Grid Connected Solar Inverter for Maximum Solar Power Utilization

Muralidhar Nayak Bhukya, Manish Kumar, and V. Chandra Jagan Mohan

1 Introduction

Solar energy is much preferred due to abundant availability in nature and also easy to convert into electrical energy through Photovoltaic (PV) effect or via using a PV panel. Solar power generation mainly depends on available irradiance, temperature, and PV panel conversion efficiency [1]. Solar irradiance and temperature are inconsistence, hence the change in irradiance and temperature effects the power generation. In view of the above limitation, it is necessary to transmit maximum solar power to the load in any weather condition [2]. To transmit maximum solar power from source to load, Maximum Power Point (MPP) has to be tracked constantly with the help of MPPT algorithms. As discussed, the intensity of G and T changes for every clock second [3, 4]. Hence, every step change in G and T has its own MPP and operated at every change in MPP using MPPT scheme [5]. The extracted solar power is integrated into the grid using multilevel inverters.

Putri et al. [6] simulated PV system with INC-based MPPT scheme using buck-boost converter as an intermediate between source and load. The results are evident to show the better performance of INC compared with P&O, but still it fails to track the power path accurately and efficiently during dynamic weather conditions. Lashen and Adbel-Salam [7] enhanced the performance of conventional HC method by combining with ANFIS controller. The hybrid algorithm is tested using ropp and

M. N. Bhukya (✉) · M. Kumar
Department of Electrical Engineering, School of Engineering & Technology, Central University of Haryana, Mahendragarh, Haryana 123031, India
e-mail: murlidhar.nayak@cuh.ac.in

M. Kumar
e-mail: manish.kumar@cuh.ac.in

V. C. J. Mohan
Institute of Aeronautical Engineering, Hyderabad 500043, India

© The Author(s), under exclusive license to Springer Nature Singapore Pte Ltd. 2021   421
J. Kumar and P. Jena (eds.), *Recent Advances in Power Electronics and Drives*, Lecture Notes in Electrical Engineering 707,
https://doi.org/10.1007/978-981-15-8586-9_37

sinusoidal irradiance profiles. Lag communication between HC and ANFIS may misjudge duty ratio of the converter [8]. In [9, 10], intelligent MPPT controllers are presented to mitigate the drawbacks associated with conventional schemes. Applications of these intelligent controllers are limited to small scale PV systems. Boukenoui et al. [11] assessed the performance of conventional, improved, and soft computing-based MPPT techniques using d-Space controller. During low and high irradiance periods, fuzzy-based algorithm has better performance over conventional and improved schemes [12]. In [13], a new MPPT scheme based on Back-Stepping Sliding Mode technique is presented and the controller has acceptable dynamic response time during fast-changing weather conditions. At the same time, there exists a major percentage of tracking error, which forces the system to operate at a duty cycle nearer to the MPP value [14, 15].

Saxena et al. [16] designed and developed a solar system with a rating of 2.2 KW integrated into the grid. The proposed multioperated voltage source converter not only acts as a active filter but also compensates reactive power. Complex design of multifunctional control system and reconnection after power failure are the major issues. Chakravartula et al. [17] addressed and improved the efficiency of neutral point clamped grid-tied PV system using non-isolated full-bridge configuration [18]. Nimrod et al. [19] designed and analyzed the operation of 200 W transformerless grid-connected PV topology. Though the proposed system performance is not up to the benchmark in terms of eliminating the leakage current but overall cost, size, and efficiency are improved compared to the conventional topology. Alam et al. [20] incepted about electric vehicle to grid by developing an intelligent control technique for dynamic adjustment of charging and effective utilization of the batteries and hence improved functionality [21]. Khan and Xiao [22] maximized the solar power losses using a submodule integrated converter and examined under dynamic weather conditions and double frequency ripple.

In the proposed configuration, maximum amount of solar power is extracted using a new controller based on 2-Dimensional Lookup table [23]. A SIDO converter is employed to match low-level PV array voltage with grid DC bus voltage and also divides PV voltage into two individual voltages. The boosted PV voltage is fed to utility grid through SLI with reduced number of switches. Initially, using CS circuit, the two independent voltages are converted into three-level DC voltage and further fed to FB converter to develop seven-level AC voltage [24]. The RNG 500 D 500 Watt PV array is employed with a new MPPT controller based on 2DLT and SIDO converter incorporates a step-up converter.

2 SIDO Converter

The SIDO converter configuration consists of filter capacitor (C_f), inductor (L_f), two power electronic switches (S_A and S_B), and three diodes (Da, Db, and Dc). The dc-dc converter develops two output voltages (V_1 and V_2) given to CS circuit in the SLI. When the switch S_A is in on mode and S_B is in off mode as shown in Fig. 1a,

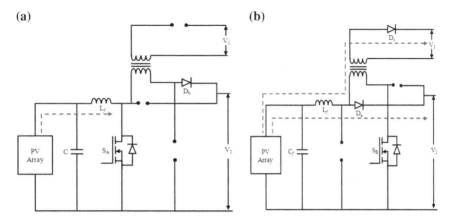

Fig. 1 **a** S_A switch on-state and **b** S_A switch off-state

inductor (L_f) stores energy from PV array. Output voltages of the converter are given to capacitors C_A and C_B in the CS circuit of the SLI. As the switching positions are reverted as shown in Fig. 1b, inductor stored energy and PV array charges the capacitors through Da and Dc. Voltage at both the capacitors are given as

$$C_A = \frac{1}{1 - \delta} V^I \tag{1}$$

$$C_B = \frac{1}{2(1 - \delta)} V^I \tag{2}$$

where δ is the duty ratio and V^I is the panel voltage.

3 MPPT Scheme

This paper proposes a new scheme based on 2DLT. The salient feature of the lookup table is that it gets required value directly from the memory, saves computational time which is generally involved in the traditional methods. Using RNG-500D 500 W mono-crystalline PV panel maximum voltage for different G and T is estimated and saved in the memory of the lookup table as shown in Table 1. With respect to G and T on panel surface, the proposed scheme fetches essential voltage value from the table. Using extrapolation and interpolation technique, it can calculate the required results that are not saved in the memory. Hence, the MPPT scheme is simple, low cost, and effective. The I-V and P-V curves of RNG-500D 500 W mono-crystalline PV panel are depicted in Fig. 2.

Table 1 Maximum voltage values of RNG-500D solar panel

Temp (°C)	G (W/m^2)				
	250	450	650	850	1000
16	95.2	95.23	97.02	97.48	97.6
20	92.48	95.04	95.65	95.97	95.7
24	90.27	92.62	93.23	93.24	94.27
28	87.79	90.01	90.55	91.37	91.1
32	85.56	81.19	88.01	87.25	88.59
36	85.55	86.65	86.43	86.35	85.85
40	81.21	83.61	83.64	81.64	84.72

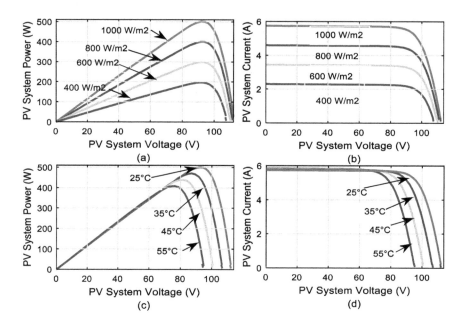

Fig. 2 P-V and I-V curves

4 Proposed Inverter (SLI) Configuration

4.1 Positive Half Cycle

Mode A: The CS circuit switches, i.e., S_{W1} and S_{W2} are in off mode. Hence, C_A starts discharging through D_1 as shown in Fig. 3a. The output voltage across CS circuit is equal to capacitor voltage $V_{dc}/3$. In the SLI, S_1 and S_4 switches are in on-state. At this point, inverter output voltage is equal to $V_{dc}/3$.

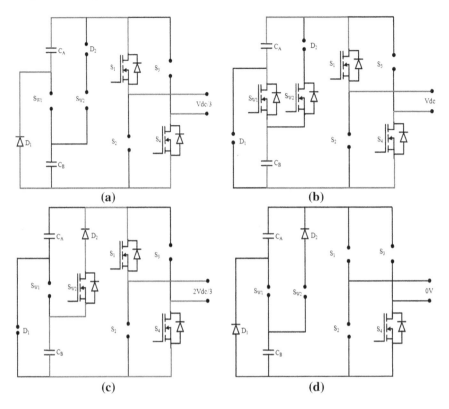

Fig. 3 Seven level inverter operating modes. **a** A, **b** B, **c** C, and **d** D

Mode B: One of the switches, i.e., S_{W1} is in of-state and S_{W2} is in on-state. Hence, CB capacitor starts discharging through the path D_2 and S_{W2} as depicted in Fig. 3b, the inverter output voltage at this stage is $2V_{dc}/3$.

Mode C: S_{W1} switch is turned to on-state which causes diode D_2 to reverse bias. Hence, the switching state of S_{W2} is ignorable. At this particular mode of operation, both capacitors start discharging in series and give an output voltage V_{dc} across the CS circuit as depicted in Fig. 3c.

Mode D: CS circuit switches are turned to off-state and only S_4 switch in the bridge converter is on as shown in Fig. 3d and output voltage is equal to zero. Therefore, the obtained voltages are $V_{dc}/3$, $2V_{dc}/3$, V_{dc}, and 0 V.

4.2 Negative Half Cycle

In the negative half cycle, apart from the switch position the rest of the operation is similar to the positive cycle and the obtained voltages are 0, $-V_{dc}/3$, $-2V_{dc}/3$, and

Table 2 Switching states of the power electronic switch

	Voltage level	S_{W1}	S_{W2}	S_1	S_2	S_3	S_4
Positive cycle	$UV < V_{dc}/3$	Of	Of	PWM	Of	Of	On
	$2V_{dc}/3 > UV > V_{dc}/3$	Of	PWM	On	Of	Of	On
	$UV > 2V_{dc}/3$	PWM	On	On	Of	Of	On
Negative cycle	$UV < V_{dc}/3$	Of	Of	Of	On	PWM	Of
	$2V_{dc}/3 > UV > V_{dc}/3$	Of	PWM	Of	On	On	Of
	$UV > 2V_{dc}/3$	PWM	On	Of	On	On	Of

$-V_{dc}$ and Table 2 depicts switching states. From the table, it is clear that only one switch is connected to PWM in all the voltage levels, which significantly reduces the power loss and switching losses. In A, C, D, E, G, and H modes of seven-level inverter operation, it uses only three switches in series to generate voltage levels. Whereas in traditional inverters it requires minimum four switches in series. Therefore, reduction in switches reduces conduction loss of the inverter.

5 Experimental Validation

DSP PIC 30F4011 microcontroller is employed to generate pulses to the inverter and converter configurations. Prototype technical details are mentioned in Table 3. PV simulator (N8937A) develops Power-Voltage and Current–Voltage curves, which are identical to RNG 500D 500 W PV module. PV voltage and current are fed to dc-dc SIDO converter. LV25-P and LA55-P Hall effect current and voltage sensors provided necessary scale down voltage to DSP PIC 30F4011 microcontroller to generate essential gate pulse required to the converter through 4081 PWM generator IC. PV parameters such as current, voltage, and power are depicted in Fig. 4a at STC (i.e., 1000 W/m^2 and 25 °C).

In continuation, converter output voltage V_1 and V_2 along with capacitor selection circuit, three-level dc output voltage experimental results are plotted in Fig. 4b. Figure 5a, b represents voltage and current of the proposed inverter from the results and it is evident that the output voltage and current have seven levels.

Table 3 Prototype parameters

Dc-Dc SIDO converter		Seven-level inverter	
Filter inductor	1 mH	Capacitors	1000 μF
Filter capacitor	10 μF	Filter inductor	1.8 mH
		PWM frequency	1550 Hz

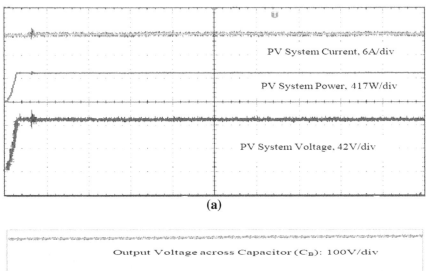

(a)

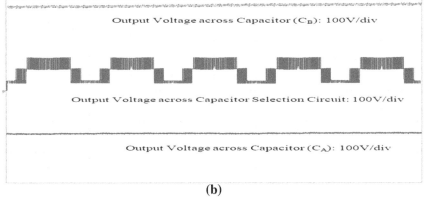

(b)

Fig. 4 Experimental results of **a** PV current, voltage and power **b** dc-dc SIDO converter and capacitor selection circuit

6 Conclusions

This paper puts forward a novel topology to harvest maximum amount of solar energy. The proposed configuration consists of a new controller based on 2DLT, SIDO converter, and a simple inverter having seven level in the output with reduced number of switches. The results provide enough evidence that the proposed configuration produces seven level in its output which are similar to sinusoidal shape with unity power factor to feed the utility. Therefore, the advantage of the proposed scheme is it harvests maximum solar energy in any weather condition and feeds the utility with low cost.

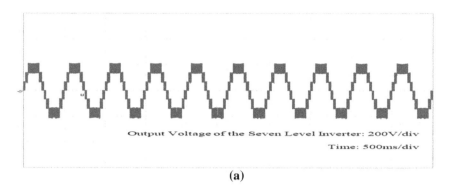

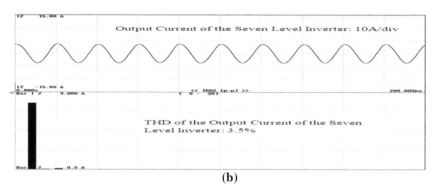

Fig. 5 Results of the seven-level inverter **a** voltage and **b** current with THD

References

1. S.B. Kjaer, J.K. Pedersen, F. Blaabjerg, A review of single-phase grid-connected inverters for photovoltaic modules. IEEE Trans. Indus. Appl. **41**(5), 1292–1306 (2005)
2. T. Hamed, J. Quaicoe, J. Benjamin, Power and frequency controllable multi level MHz inverter with soft switching, in *IEEE APE Conference & Expo* (2017), pp. 2576–2581. https://doi.org/10.1109/APEC.2017.7931061
3. A. Salem et al., Stand alone three phase symmetrical multi level inverter, in *IEEE Transportation Electrification Conference and Expo* (2015), pp. 1–6. https://doi.org/10.1109/INTEC.2015.7572459.
4. F.-S. Pai, R.-M. Chao, A new algorithm to photovoltaic power point tracking problems with quadratic maximization. IEEE Trans. Energy Convers. **25**(1), 262–264 (2010)
5. V.R. Kota, M.N. Bhukya, A simple and efficient MPPT scheme for PV module using 2-dimensional lookup table, in *IEEE Power and Energy Conference at Illinois (PECI)* (2016), pp. 1–7. https://doi.org/10.1109/PECI.2016.7459226
6. R. Putri, S. Wibowo, M. Rafi, Maximum power point tracking for photovoltaic using incremental conductance method. Energy Procedia **68**, 22–30 (2015)
7. M. Lashen, M. Adbel-Salam, Maximum power point tracking using hill-climbing and ANFIS technique for PV application. A review and a novel hybrid approach. Energy Convers. Manage. **171**, 1002–1019 (2018)

8. S.D. Al-Majidi, M.F. Abbod, H.S. Al-Raweshidy, A novel maximum power point tracking technique based on fuzzy logic for photovoltaic system. Int. J. Hydro. Energy **43**, 14158–14171 (2018)
9. K. Sundareswaran, P. Sankar, P.S.R. Nayak, S.P. Simon, S. Palani, Enhanced energy output from a PV system under partial shaded conditions through artificial bee colony. IEEE Trans. Sustain. Energy 1–12 (2014). https://doi.org/10.1109/TSTE.2014.2363521
10. S. Blaifi, S. Maulahoum, R. Benkercha, B. Taghezouit, A. Saim, M5P model tree based fast fuzzy maximum power point tracker. Solar Energy **163**, 405–424 (2018)
11. R. Boukenoui, M. Ghanes, J.-P. Barbot, R. Bradai, A. Melit, H. Salhi, Experimental assessment of maximum power point tracking methods for photovoltaic systems. Energy **132**, 324–340 (2017)
12. A.I.M. Ali, M.A. Sayed, E.E.M. Mohamed, Modified efficient perturb and observe maximum power point tracking technique for grid-tied PV system. Int. J. Electri. Power Energy Syst. **99**, 192–202 (2018)
13. C. Vimalarani, N. Kamaraj, B.B. Chitti, Improved method of maximum power point tracking of photovoltaic (PV) array using hybrid intelligent controller. Optik **168**, 403–415 (2018)
14. K. Dahech, M. Allouche, T. Damak, F. Tadeo, Back stepping sliding mode control for maximum power point tracking of a photovoltaic system. Electric Power Syst. Res. **143**, 182–188 (2017)
15. A. Anurag, N. Deshmukh, A. Maguluri, S. Anand, Integrated DC-DC converter based grid connected transformer less photovoltaic inverter with extended input voltage range. IEEE Trans. Power Electron. **33**, 8322–8330 (2018)
16. N. Saxena, I. Hussain, B. Sing, A.L. Vyas, Implementation of a grid integrated PV battery system for residential and electric vehicle application. IEEE Trans. Indust. Electron. **65**, 6592–6601 (2018)
17. C. Anandababu, B.G. Fernandes, Neutral point clamped MOSFET inverter with full bridge configuration for non isolated grid tied photovoltaic system. IEEE J. Emerg. Select. Topics Power Electron. **5**, 445–457 (2017)
18. M.N. Bhukya, V.R. Kota, D.S. Rani, A simple, efficient and novel standalone photovoltaic inverter configuration with reduced harmonic distortion. IEEE Access **7**, 43831–43845 (2019)
19. V. Nimrod, V. Jeziel, V. Juaquin, C. Hernandez, E. Vazquez, R. Osorio, Integrating two stages as a common mode transformer less photovoltaic converter. IEEE Trans. Indus. Electron. **64**, 7498–7507 (2017)
20. M.J.E. Alam, K.M. Muttagi, D. Sutanto, Effective utilization of available PEV battery capacity for mitigation of solar PV impact and grid support with integrated V2G functionality. IEEE Trans. Smart Grid **7**, 1562–1571 (2016)
21. V.R. Kota, M.N. Bhukya, A novel global MPP tracking scheme based on shading pattern identification using artificial neural networks for photovoltaic (PV) power generation during partial shaded condition. IET Renew. Power Generat. (2019) https://doi.org/10.1049/iet-rpg.2018.5142
22. O. Khan, W. Xiao, An efficient modeling technique to simulate and control sub module integrated PV system for single phase grid connection. IEEE Trans. Sustain. Energy **7**, 96–107 (2016)
23. M.N. Bhukya, V.R. Kota, A quick and effective MPPT scheme for solar power generation during dynamic weather and partial shaded conditions. Eng. Sci. Technol. Int. J. (2019) https://doi.org/10.1016/j.jestch.2019.01.015
24. M.N. Bhukya. V.R. Kota, DCA-TR-based MPP tracking scheme for photovoltaic power enhancement under dynamic weather conditions. Electri. Eng. **100**(4), 2383–2396 (2018)

UPFC Modelling for Augmentation of Voltage Stability and Reduction of Active Power Losses

Lalit Kumar, Khushboo Verma, and Sanjay Kumar

1 Introduction

Transmission systems of present power network are reliably crashed into increasingly focused on level because of the ascent in load requests and limitations of developing new lines. The result of such an accentuated framework would be gaining instability trailed by interference because of a fault. Use of FACTS devices has ended up being an exceptionally viable strategy to cut down the pressure of the power system grid and consequently issues in better usage of concerned accessories in power framework without bargaining the appropriate margin of stability [1]. Position of FACTS controllers play a crucial role in congestion management, and FACTS controllers used for the problem of reactive power compensation can be solved. Flexible AC transmission system (FACTS) was first brought into light by Hingorani [2] in 1988. Power flow control method has been presented by Gothana and Heydt [3]. A brief search for the optimal position of FACTS controllers is discussed by Lie and Dang [4]. This paper deals with improvement in operational performance of power system using UPFC FACTS devices. The simulation is executed using PSAT in the MATLAB. The area for the situation of UPFC is purposefully utilising an index termed as network branch index (NBI) [5]. Voltage stability improvement is assessed by extension power flow method present in [6]. UPFC is used with the optimal variables which are reducing

L. Kumar · K. Verma (✉) · S. Kumar
Department of Electrical Enginnering, National Institute of Technology Jamshedpur, Jamshedpur, India
e-mail: vkhushboo11@gmail.com

L. Kumar
e-mail: 2019rsee005@nitjsr.ac.in

S. Kumar
e-mail: sanjay.ee@nitjsr.ac.in

© The Author(s), under exclusive license to Springer Nature Singapore Pte Ltd. 2021 431
J. Kumar and P. Jena (eds.), *Recent Advances in Power Electronics and Drives*, Lecture Notes in Electrical Engineering 707,
https://doi.org/10.1007/978-981-15-8586-9_38

the real power loss and the system running cost [7]. Various optimization methods are explained in [8]. Increase in the loadability with FACTS devices is described in [9]. In [10], author describes the reactive power compensation. In this paper, voltage crash point using the load flow jacobian matrix is presented [11]. The authors in [12] presents gravitational search algorithm (GSA)-based enhancement strategy for the ideal coordination of FACTS controller with the current imaginary power sources present in an associated power grid. Basic principle of the UPFC is described in [13]. Power system analysis toolbox (PSAT) gives graphical consolidate of power flow analysis, and PSAT toolbox detects the most precise node [14]. In this paper, [15] shows how to the improve voltage stability using shunt devices. All the parameters are concluded for voltage collapse using the PSAT MATLAB and description of weak node with the help of voltage stability index [16].

2 Problem Formulation

2.1 Improvement of Bus Voltage

It is very essential to manage normal voltage profile constantly. The first objective is the minimization of the voltage deviation expressed as follows:

$$M_2 = \sum_{i=1}^{n1} [U_a - U_{specified}] \tag{1}$$

where $U_{specified}$ is the magnitude of bus voltage and nl is the total bus no. U_a is the voltage magnitude at a_{th} bus.

2.2 Reduction of Real Power Loss

Because of transmission loss, redistribution of reactive power in transmission network occurs firstly. So real power loss reduction changes the active power generated by slack bus.

Real power loss minimization is mathematically expressed as follows:

$$M_L = P_L \sum_{j=1}^{nl} G_{ij} [U_i^2 + U_j^2 - 2U_i U_j \cos(\beta_{ij})] \tag{2}$$

where

nl no. of transmission lines

G_{ij} Conductance of ith and jth bus in transmission line

U_i bus ith voltage
U_j bus jth voltage
B_{ij} Power angle at bus ith and jth
P_L Real power loss.

Equality limits: Imaginary and real power equation for nl bus system.

$$Q''_{gi} - Q''_{di} - U'_i \sum_{j=1}^{nl} U'_j [G'_{ij} \sin \varphi + B'_{ij} \cos \varphi] = 0 \tag{3}$$

where $i = 1, 2, 3 \ldots nl$

$$P''_{gi} - P''_{di} - U'_i \sum_{i=1}^{nl} U'_j [G'_{ij} \cos \varphi + B'_{ij} \sin \varphi] = 0 \tag{4}$$

where $j = 1, 2, 3 \ldots nl$

where $\varphi = \beta_i - \beta_j$

nl Total bus number.
P'_{gi} and P'_{di} is real power generation and ith bus demand, respectively
Q'_{di} and Q'_{gi} is and VAR demand and VAR Generation of the ith bus
G'_{ij} represents transfer conductance of ith and jth bus
B'_{ij} represents the transfer susceptance of ith and jth bus

Inequality limits: Maximum and minimum limit must be defined for the generator voltage magnitude and reactive power

$$P_{gx,\min} \leq P_{gx} \leq P_{gx,\max}, \; x = 1, 2, \ldots M_{PV} \tag{5}$$

Generator bus voltage and reactive power restraints are express as

$$Q_{gx,\min} \leq Q_{gx} \leq Q_{gx,\max}, \; x = 1, 2, 3, \ldots M_{PV} \tag{6}$$

$$U_{gx,\min} \leq U_{gx} \leq U_{gx,\max}, \; x = 1, 2, 3 \ldots M \tag{7}$$

Taps limits of Transformer: maximum and minimum limit

$$t_{x,\min} \leq t_x \leq t_{x,\max}, \; x = 1, 2, 3 \ldots \tag{8}$$

M_{PV} PV buses locations M buses locations

3   UPFC Circuit Modelling

The regular UPFC arrangement is as appeared in Fig. 1. The circuital model of
the UPFC is acquired from the STATCOM and SSSC. The model representation
fundamentally comprises shunt and series converters. A voltage source converter is
existing in branch of series type whose principle work is to infuse a voltage, with
variable extent and edge, in arrangement by means of a transformer. Thus, it has a
capacity to trade real power with the lines of transmission network. Be that as it may,
the shunt part of the converter is answerable for the compensation of any active power
provided furthermore, drawn by the series type branch just as the losses. The shunt
type converter arrangement has a solid capacity to freely trade a imaginary power
with the framework by means of the transformer through which it is associated with
the network.

From Fig. 1. The source voltage can be represented as

$$E_{sh} = V_{sh}(\cos\delta_{sh} + j\sin\delta_{sh}) \tag{9}$$

$$E_{se} = V_{se}(\cos\delta_{se} + j\sin\delta_{se}) \tag{10}$$

where δ_{sh} presents the voltage source of controllable phase angle and V_{sh} presents
the magnitude at the shunt type of converter area.

The viability of this methodology is exhibited in the area that is subsequent to
utilising an example of 6 bus and 14 bus system as a case study.

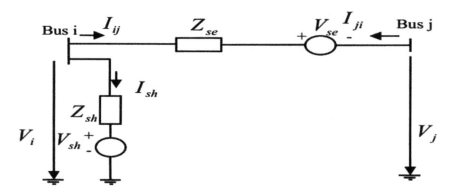

Fig. 1   UPFC configuration model

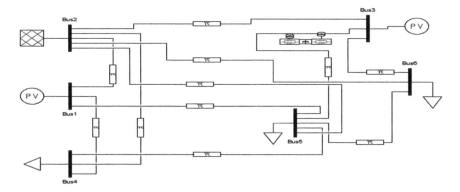

Fig. 2 A 6 bus modified system recreated in PSAT after the establishment of UPFC

4 Methodology

4.1 Psat

As the name suggests, PSAT effectively analyses the power system with the advantages of toolbox which is anything but difficult to get to. PSAT additionally empowers the capacities identified with various power flows locked in. When contrasted to other MATLAB toolboxes, PSAT gives wide scope of highlights that puts forth power system investigation accomplished with not so much attempt but rather more precise. The use of PSAT for the displaying of the test bus framework and its examination appeared as designing of test bus system.

4.2 IEEE 6 and 14 Test Bus System

The test system being analysed is 6 Bus Test System which involves of one slack bus, 2 generators at 2, 3 and remaining nodes are supposed to be load buses at 4, 5 and 6. It has 9 lines. IEEE-14 bus comprises one slack bus, 4 generators and 9 load buses. Utilising PSAT the model of IEEE 6 and 14 test bus system which was made is shown in Figs. 2 and 3.

5 Results and Discussion

The authors presented the bus voltage profile and real power loss before and after establishment of UPFC. The simulations are executed using PSAT in the MATLAB environment. Figures 2 and 3 show the single line diagram, and Figs. 4 and 5 shows

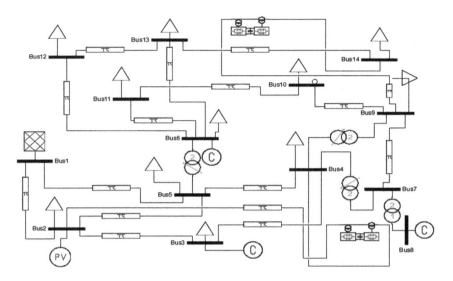

Fig. 3 A modified 14 bus system recreated in PSAT after the incorporating of UPFC

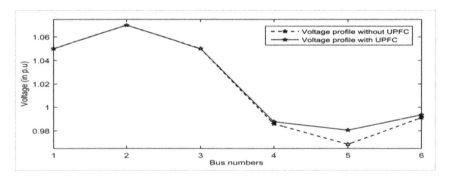

Fig. 4 Bus voltage due to absence and presence of UPFC for IEEE 6 bus system

the bus voltage due to absence and presence of UPFC for IEEE 6 and 14 bus network, respectively. It is detected that the bus voltage is low at node 6, 4 and 5 for 6 bus network and node 9, 10, 14 for 14 bus networks. Table 1 presents Bus voltage magnitude, and Table 2 presents active power losses due to absence and presence of UPFC for IEEE 6 bus system. Table 3 presents Bus voltage magnitude, and Table 4 presents real power losses due to absence and presence of UPFC for 14 bus system, respectively. It is clear that after using UPFC voltage magnitude increases and real power losses reduces. And in Table 5, the authors present the comparative analysis of total active power loss with and without using UPFC in both test bus system and also show the reduction of real power loss.

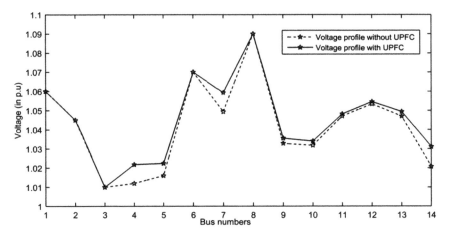

Fig. 5 Bus voltage due to absence and presence of UPFC in IEEE-14 test bus system

Table 1 Bus voltage due to absence and presence of UPFC for IEEE 6 bus system

Bus No.	01	02	03	04	05	06
Bus voltage due to absence of UPFC (pu)	1.05	1.07	1.05	0.9859	0.9685	0.9912
Bus voltage due to presence UPFC (pu)	1.05	1.07	1.05	0.9877	0.9806	0.9936

Table 2 Active power loss value due to absence and presence of UPFC for IEEE 6 bus

From node to node	Active power loss value due to absence of UPFC (pu)	Active power loss value due to presence of UPFC (pu)	From node to node	Active power loss value due to absence of UPFC (pu)	Active power loss value due to presence of UPFC (pu)
2–3	0.00106	0.00093	2–4	0.02326	0.02254
6–3	0.00957	0.00926	2–1	0.00142	0.00137
2–5	0.01199	0.01079	1–4	0.01013	0.00988
4–5	0.00128	0.00124	1–5	0.01360	0.01271
3–5	0.01178	0	2–6	0.01415	0.01353
5–6	0.00051	0.00018			

6 Conclusion

For the aforesaid system, Newton–Raphson algorithm is used for the minimization of active power loss in PSAT MATLAB software. We have presented two test system IEEE 6 bus and 14 bus system. In this proposed work, shunt series FACTS controller UPFC is placed in weak bus. The results got to signify a critical upgrade in the voltage outline of the test bus systems just as convincing decrement in the losses of lines after incorporating of UPFC. The result acquired by without UPFC is compared

Table 3 Bus voltage due to absence and presence of UPFC for IEEE-14 bus system

Bus no	Bus voltage due to absence of UPFC (pu)	Bus voltage due to presence UPFC (pu)	Bus no	Bus voltage due to absence of UPFC (pu)	Bus voltage due to presence UPFC (pu)
1	1.06	1.06	8	1.09	1.09
2	1.045	1.045	9	1.033	1.036
3	1.01	1.01	10	1.032	1.034
4	1.012	1.0218	11	1.047	1.0482
5	1.016	1.0224	12	1.053	1.0546
6	1.07	1.07	13	1.047	1.0494
7	1.049	1.053	14	1.0207	1.0311

Table 4 Active power loss value due to absence and presence of UPFC for IEEE-14 test bus system

From node to node	Active power loss value due to absence of UPFC (pu)	Active power loss value due to presence of UPFC (pu)	From node to node	Active power loss value due to absence of UPFC (pu)	Active power loss value due to presence of UPFC (pu)
1–2	0.04311	0.04201	4–9	0	0
2–3	0.02337	0.02108	7–9	0	0
2–4	0.01672	0	9–10	0.00006	0.00005
1–5	0.02771	0.02505	6–11	0.00114	0.00107
2–5	0.00921	0.00702	6–12	0.0008	0.0007
3–4	0.00397	0.00376	6–13	0.00248	0.00212
4–5	0.00478	0.00389	9–14	0.00091	0
5–6	0	0	10–11	0.00046	0.00041
4–7	0	0	12–13	0.00011	0.00007
7–8	0	0	13–14	0.00099	0.00055

Table 5 Comparison of total active power losses due to absence and presence of UPFC

Bus system	Active power loss value due to absence of UPFC (pu)	Active power loss value due to presence of UPFC (pu)	Active power loss reduction in %
IEEE 6 bus system	0.09875	0.08243	16.53
IEEE-14 bus system	0.13581	0.11046	18.67

with result acquired from with UPFC. It is clear that after using UPFC active power losses reduces to 15–20% and enhances the voltage stability.

References

1. K. Dwivedi, S. Vadhera, Reactive power sustainability and voltage stability with different FACTS devices using PSAT, in *2019 6th International Conference on Signal Processing and Integrated Networks (SPIN)* (2019), pp. 248–253
2. N. Hingorani, Flexible AC transmission. IEEE Spectr. **30**(4), 40–45 (1993)
3. D.J. Gotham, G.T. Heydt, Power flow control and power flow studies for system with FACTS devices. IEEE Trans. Power Syst. **19**(1), 60–65 (1998)
4. T.T. Lie, W. Deng, Optimal flexible AC transmission systems (FACTS) devices allocation. Int. J. Electr. Power Energy Syst. **19**(2), 125–134 (1999)
5. A.S. Alayande, O.U. Omeje, C.O.A. Awosope, T.O. Akinbulire, F.N. Okafor, On the enhancement of power system operational performance through UPFC: a topological-based approach, in *2019 IEEE PES/IAS PowerAfrica, Abuja, Nigeria* (2019), pp. 499–503
6. B. Bhattacharyya, V.K. Gupta, S. Kumar, UPFC with series and shunt FACTS controllers for the economic operation of a power system. Ain Shams Eng. J. **5**(3), 775–787 (2014)
7. M.A. Kamarposhti, M. Alinezhad, H. Lesani, N. Talebi, Comparison of SVC, STATCOM, TCSC, and UPFC controllers for static voltage stability evaluated by continuation power flow method, in *2008 IEEE Canada Electric Power Conference* (IEEE, 2008), pp. 1–8
8. B. Bhattacharyya, S. Kumar, Approach for the solution of transmission congestion with multi-type FACTS devices. IET Gener. Trans. Distrib. **10**(11), 2802–2809 (2016)
9. V.K. Gupta, S. Kumar, B. Bhattacharyya, Enhancement of power system loadability with FACTS devices. J. Inst. Eng. (India): Ser. B, **95**(2), 113–120 (2017)
10. A.R. Phadke, S.K. Bansal, K.R. Niazi, A comparison of voltage stability indices for placing shunt FACTS controllers, in *First International conference on Emerging Trends in Engineering and Technology* (2008), pp. 939–944
11. F. Karbalaei, H. Soleymani, S. Afsharnia, A comparison of voltage collapse proximity indicators, in *IPEC* (2010), pp. 429–432
12. L. Kumar, S. Kumar, S. Kumar Gupta, B.K. Raw, Optimal location of FACTS devices for loadability enhancement using gravitational search algorithm, in *2019 IEEE 5th International Conference for Convergence in Technology (I2CT)* (Bombay, India, 2019), pp. 1–5
13. R. Fuerte-Esquivel, E. Acha, H. Ambriz-Perez, A comprehensive Newton-Raphson UPFC model for the quadratic power flow solution of practical power networks. IEEE Trans. Power Syst. **15**(1), 102–109 (2000)
14. P.R. Sharma, R. Kr. Ahuja, S. Vashisth, V. Hudda, Computation of sensitive node for IEEE-14 bus system subjected to load variation. Int. J. Innov. Res. Electr. Electron. Instrum. Control Eng. **2**(6), 1603–1606 (2014)
15. A.F. Mohamad Nor, M. Sulaiman, A. Fazliana, A. Kadir, R. Omar, Voltage instability analysis for electrical power system using voltage stability margin and modal analysis. Indones. J. Electr. Eng. Comput. Sci. **3**(3) 655–662 (2016)
16. L. Kumar, B.K. Raw, S.K. Gupta, S. Kumar, Voltage stability enhancement using shunt devices and identification of weak bus through voltage stability indices, in *2019 4th International Conference on Recent Trends on Electronics, Information, Communication & Technology (RTEICT)* (Bangalore, India, 2019), pp. 247–251

FPGA-Based Implementation of Backstepping Controller for Three-Phase Shunt Active Power Filter Interfacing Solar Photovoltaic System to Distribution Grid

V. N. Jayasankar and U. Vinatha

1 Introduction

Globally, the load demand is rising exponentially due to the increase in population, urbanization, industrialization, and the growth in standard of living. To meet the increasing load demand, optimal utilization of the locally available renewable energy resources is necessary. Solar energy is the most abundant renewable resource all over the world. Grid connection of solar photovoltaic (PV) system presents the idea of storage-less solar power generation. In a grid-connected solar PV system, the excess power from the solar system after meeting the load demand is injected to the grid [1]. Nowadays, the proliferation of power electronic components is making both domestic and industrial loads non-linear and, therefore, power quality problems are introduced in distribution grid. Active filter with grid interfacing feature is a cost-effective way to alleviate distribution level power quality issues.

The control scheme of active filters consists of three stages namely reference current calculation, voltage control at dc link, and current control of voltage source inverter (VSI). In the literature, several control techniques for reference current calculation are reported so far. Instantaneous power theory (pq theory) algorithm is the most popular reference current calculation technique [2, 3]. To improve the performance during grid voltage unbalance and distortions, modified pq theory using self-tuning filter (STF) is proposed in this paper by the authors. Fundamental current and voltage detection is an important step in reference current calculation. Conventionally, the fundamental components of current and voltage are determined using low-pass filters, which introduces considerable phase-lags. This phase-lag can be eliminated by replacing low-pass filters with STF [4–7]. The control techniques for dc link voltage available in literature are PI, fuzzy, sliding-PI, Particle swam optimization-PI, etc [3, 8–11]. A backstepping controller (BSC) for the control of

V. N. Jayasankar (✉)
The National Institute of Engineering, Mysuru, Karnataka, India
e-mail: jayasankarvn@nie.ac.in

U. Vinatha
National Institute of Technology Karnataka, Surathkal, Mangalore, India

© The Author(s), under exclusive license to Springer Nature Singapore Pte Ltd. 2021
J. Kumar and P. Jena (eds.), *Recent Advances in Power Electronics and Drives*, Lecture Notes in Electrical Engineering 707,
https://doi.org/10.1007/978-981-15-8586-9_39

dc link voltage is proposed in this paper for improved performance under different system conditions. Among many current control techniques available in literature, hysteresis control is selected, because it is simple, easy to implement, and has good dynamic performance [12–14].

Real-time implementation of control algorithms is achieved using digital signal processing (DSP) cards, microcontrollers, FPGA, etc. Compared to DSP and microcontrollers, FPGA has the advantages such as short execution time, adaptability, higher accuracy, easiness to implement complex algorithm structures, low design cost, and high sampling period. FPGA is an array of logic elements, which are hardware programmable, and consist of large number of configurable inputs and outputs. The inherently optimized RAM and parallel processing capability of FPGA offer reduced execution time and rapid prototyping compared to DSPs and microcontrollers [15].

The final goals of authors are to implement the backstepping theory-based control of dc link voltage, harmonic current mitigation using STF-based instantaneous power theory and hysteresis control using a single on-chip FPGA. The proposed control algorithm is realized using Xilinx Zynq 7000 FPGA (XC7Z020-clg484-1).

2   Materials and Methods

This section deals with the details of system configuration and the control scheme. The details of hardware configuration, modeling of different components of the system, and the theoretical background of the control scheme are discussed in following subsections.

2.1   System Configuration

The solar PV system interfaced to grid is shown in Fig. 1. Different components of the system are solar PV system, Three-phase four-leg inverter, dc link capacitor, and ac filters.

2.2   Control Scheme

The control system block diagram is shown in Fig. 2. The reference currents for active filtering and dc link voltage control are calculated, and the sum of these two reference currents is given to hysteresis controller for generating pulses for the control of voltage source inverter. Following subsections discuss the calculation of reference currents in detail.

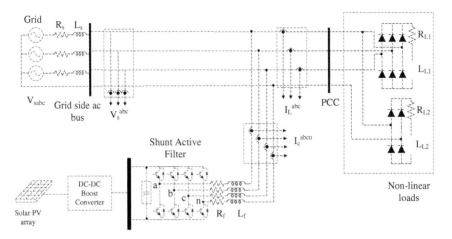

Fig. 1 Solar PV system interfaced to grid using a shunt active power filter

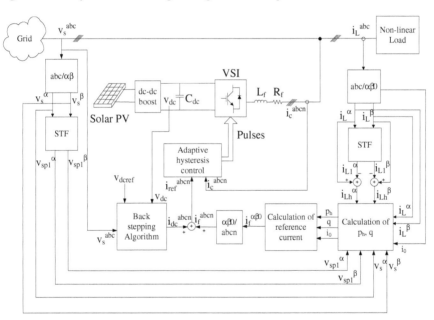

Fig. 2 Schematic diagram of control system

2.2.1 Reference Current Calculation for Harmonic Elimination

Modified pq theory is used for the reference current generation for harmonic elim-
ination. The grid voltages (v_s^a, v_s^b, v_s^c) and load currents (i_L^a, i_L^b, i_L^c) are transformed
into $\alpha - \beta$ frame as shown in (1) and (2), where v_s^α, v_s^β and i_L^α, i_L^β are voltages in
grid side and currents in load side in $\alpha - \beta$ frame. respectively [1, 16].

$$
\begin{bmatrix} v_s^\alpha \\ v_s^\beta \\ v_s^0 \end{bmatrix} = \sqrt{\frac{2}{3}} \begin{bmatrix} 1 & -\frac{1}{2} & -\frac{1}{2} \\ 0 & \frac{\sqrt{3}}{2} & -\frac{\sqrt{3}}{2} \\ \frac{1}{\sqrt{2}} & \frac{1}{\sqrt{2}} & \frac{1}{\sqrt{2}} \end{bmatrix} \begin{bmatrix} v_s^a \\ v_s^b \\ v_s^c \end{bmatrix} \tag{1}
$$

$$
\begin{bmatrix} i_L^\alpha \\ i_L^\beta \\ i_L^0 \end{bmatrix} = \sqrt{\frac{2}{3}} \begin{bmatrix} 1 & -\frac{1}{2} & -\frac{1}{2} \\ 0 & \frac{\sqrt{3}}{2} & -\frac{\sqrt{3}}{2} \\ \frac{1}{\sqrt{2}} & \frac{1}{\sqrt{2}} & \frac{1}{\sqrt{2}} \end{bmatrix} \begin{bmatrix} i_L^a \\ i_L^b \\ i_L^c \end{bmatrix} \tag{2}
$$

STF is used for detecting the positive sequence fundamental component of load
currents as well as grid voltages. STF can be represented by (3) and (4), where x^α
and x^β are the input signals to STF and x_1^α and x_1^β are the outputs. The constant K
controls the bandwidth and response time of STF [5–7].

$$
x_1^\alpha(s) = \frac{K}{s} \left[x^\alpha(s) - x_1^\alpha(s) \right] - \frac{\omega_1}{s} \times x_1^\beta(s) \tag{3}
$$

$$
x_1^\beta(s) = \frac{K}{s} \left[x^\beta(s) - x_1^\beta(s) \right] + \frac{\omega_1}{s} \times x_1^\alpha(s) \tag{4}
$$

Equation (5) represents the active power component of harmonic compensation
($p_h(t)$) and (6) represents reactive power component ($q(t)$), where i_{Lh}^α and i_{Lh}^β are
harmonic components of load current and v_{sp1}^α and v_{sp1}^β are the grid voltage fundamental components, respectively, in $\alpha - \beta$ frame. Equation (7) calculates the filter
currents with a phase difference of 180° to counter act the harmonic components.
The filter currents are transformed into abc frame using (8). Equation (9)computes
compensation currents for neutral.

$$
p_h(t) = \left(i_{Lh}^\alpha \times v_{sp1}^\alpha \right) + \left(i_{Lh}^\beta \times v_{sp1}^\beta \right) \tag{5}
$$

$$
q(t) = \left(i_L^\beta \times v_s^\alpha \right) - \left(i_L^\alpha \times v_s^\beta \right) \tag{6}
$$

$$
\begin{bmatrix} i_f^\alpha \\ i_f^\beta \\ i_f^0 \end{bmatrix} = \frac{-1}{(v_{sp1}^\alpha)^2 + (v_{sp1}^\beta)^2} \begin{bmatrix} v_{sp1}^\alpha & -v_{sp1}^\beta & 0 \\ v_{sp1}^\beta & v_{sp1}^\alpha & 0 \\ 0 & 0 & (v_{sp1}^\alpha)^2 + (v_{sp1}^\beta)^2 \end{bmatrix} \begin{bmatrix} p_h \\ q \\ i_0 \end{bmatrix} \tag{7}
$$

$$\begin{bmatrix} i_f^a \\ i_f^b \\ i_f^c \end{bmatrix} = \sqrt{\frac{2}{3}} \begin{bmatrix} 1 & 0 & \frac{1}{\sqrt{2}} \\ -\frac{1}{2} & \frac{\sqrt{3}}{2} & \frac{1}{\sqrt{2}} \\ -\frac{1}{2} & -\frac{\sqrt{3}}{2} & \frac{1}{\sqrt{2}} \end{bmatrix} \begin{bmatrix} i_f^\alpha \\ i_f^\beta \\ i_f^0 \end{bmatrix} \tag{8}$$

$$i_f^n = i_f^a + i_f^b + i_f^c \tag{9}$$

2.2.2 Reference Current Calculation for the Control of Dc Link Voltage

The solar PV system interfaced to grid is a dynamically changing system. For satisfactory performance, dc link voltage has to be controlled using an effective controller. Backstepping controller is used in this work, as it is a non-linear control algorithm with inherent dynamic stability [9, 17, 18]. The objective is to minimize the tracking error of voltage (z) as shown in (10), where x is the dc capacitor energy storage and the reference of x is marked as x^*.

$$z = x^* - x \tag{10}$$

The derivative of (10) can be written as shown in (11), where P_s is the power from grid to charge/discharge the capacitor, P_{Rdc} is the leakage losses of capacitor at dc link, and P_{sw} is the switching loss in the inverter. P_{sw} is treated as an unknown parameter.

$$\begin{aligned} \dot{z} &= \dot{x^*} - \dot{x} = \dot{x^*} - P_s + P_{Rdc} + P_{sw} \\ &= \dot{x^*} - P_s + \frac{2x}{C_{dc}R_{dc}} + P_{sw} \end{aligned} \tag{11}$$

A Lyapunov function as shown in (12) is introduced for designing a control law, where $\tilde{P}_{sw} = \hat{P}_{sw} - P_{sw}$ is the error in the estimation of switching loss (P_{sw}) and γ is a large positive constant.

$$V = \frac{1}{2}z^2 + \frac{1}{2\gamma}\tilde{P}_{sw}^{\,2} \tag{12}$$

Differentiating (12), we get (13).

$$\dot{V} = z\dot{z} + \frac{1}{\gamma}\tilde{P}_{sw}\dot{\tilde{P}}_{sw} = z\left[\dot{x^*} - P_s + \frac{2x}{C_{dc}R_{dc}} + P_{sw}\right] + \frac{1}{\gamma}\tilde{P}_{sw}\dot{\tilde{P}}_{sw} \tag{13}$$

Substituting $P_{sw} = \hat{P}_{sw} - \tilde{P}_{sw}$ in (13), we get (14).

$$\dot{V} = z\left[\dot{x^*} - P_s + \frac{2x}{C_{dc} \times R_{dc}} + \hat{P}_{sw}\right] + \tilde{P}_{sw}\left(\frac{1}{\gamma}\dot{\tilde{P}}_{sw} - z\right) \tag{14}$$

Since $\dot{V} <= 0$, $-cz = \tilde{P}_{sw} - z$, where $c \gg 0$. Equation (15) is the control law and (16) is the parameter adaptation law.

$$P_s = v_s^a i_{dc}^a + v_s^b i_{dc}^b + v_s^c i_{dc}^c + v_s^\beta i_{dc}^\beta = \dot{x^*} + \frac{2x}{C_{dc} \times R_{dc}} + \hat{P}_{sw} + cz \qquad (15)$$

$$\dot{\tilde{P}}_{sw} = \gamma z \qquad (16)$$

From the control law (15), the reference currents for dc voltage control can be derived as shown in (17) and (18).

$$\begin{bmatrix} i_{dc}^a \\ i_{dc}^b \\ i_{dc}^c \end{bmatrix} = \frac{1}{(v_s^a)^2 + (v_s^b)^2 + (v_s^c)^2} \begin{bmatrix} v_s^a(\dot{x^*} + \dfrac{2x}{C_{dc} \times R_{dc}} + \hat{P}_{sw} + cz) \\ v_s^b(\dot{x^*} + \dfrac{2x}{C_{dc} \times R_{dc}} + \hat{P}_{sw} + cz) \\ v_s^c(\dot{x^*} + \dfrac{2x}{C_{dc} \times R_{dc}} + \hat{P}_{sw} + cz) \end{bmatrix} \qquad (17)$$

$$i_{dc}^n = i_{dc}^a + i_{dc}^b + i_{dc}^c \qquad (18)$$

As per Lyapunov's theory, if $\dot{z}$ and $\dot{\tilde{P}}_{sw}$ need to have negative real value as shown in (19)for stable operation, which implies $c^2 - (4 \times \gamma) > 0$. The values of c and γ are selected by trial and error method using the inequality $c^2 - (4 \times \gamma) > 0$.

$$\begin{bmatrix} \dot{\tilde{P}}_{sw} \\ \dot{z} \end{bmatrix} = \begin{bmatrix} \gamma & 0 \\ -c & -1 \end{bmatrix} \begin{bmatrix} \tilde{P}_{sw} \\ z \end{bmatrix} \qquad (19)$$

The reference current for the inverter is the sum of reference currents for harmonic elimination and the reference currents for the control of dc link voltage. Hysteresis control can be used to generate pulses for controlling the voltage source inverter using the reference currents calculated.

2.3   FPGA-Based Design of Control Algorithm

The proposed algorithm is realized using FPGA with a modular structure as shown in Fig. 3. There are different modules such as analog to digital conversion (ADC) module, decode Xilinx-ADC (XADC) module, active filter control module, etc. Different control blocks like $\alpha - \beta$ to abc transformation, STF, instantaneous power calculation, reference current calculation, abc to $\alpha - \beta$ transformation, backstepping controller, and hysteresis controller are implemented using system generator blocks. IP cores are generated for synthesizing the algorithms in Xilinx Vivado design suite.

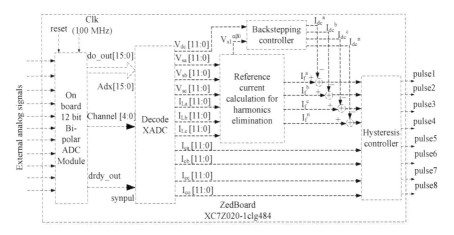

Fig. 3 Modular Design of FPGA Controller

The implemented modules are converted to verilog codes. Finally, the design synthesis, implementation and bit stream generation are done using Vivado design suite software.

3 Results and Discussion

Simulation results and FPGA implementation results are shown in this section.

3.1 Simulation Results

Dynamic model of solar PV system is simulated with solar cells of specification $V_{oc} = 21.1$ V and $I_{sc} = 3.8$ A. The solar PV system is connected to a three-phase, 415 V, 50 Hz distribution grid through a four-leg IGBT inverter. Grid resistance and inductance considered are 0.1 ohm, 0.01 mH, respectively. Capacitors of 4700 μF capacitors that are used are connected in series at the dc link. The dc link voltage reference is 700 V. The parameters of inductor filter are 0.1 ohm, 10 mH. A three-phase rectifier feeding a load of 100 Ω, 60 mH is considered. The backstepping controller parameters are c = 300 and γ = 20,000. Figures 4 and 5 show simulation results.

3.1.1 Case A: 50% Step Increment in Load

A 50% step increment of load is simulated by introducing an additional load of 100 Ω, 60 mH into the system at 0.1 s. In this case, solar power irradiation is considered

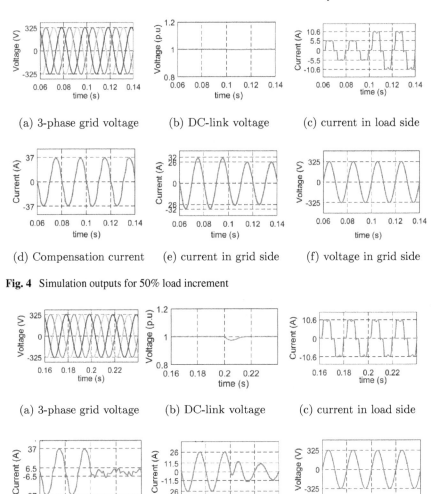

(a) 3-phase grid voltage   (b) DC-link voltage   (c) current in load side

(d) Compensation current   (e) current in grid side   (f) voltage in grid side

Fig. 4 Simulation outputs for 50% load increment

(a) 3-phase grid voltage   (b) DC-link voltage   (c) current in load side

(d) compensation current   (e) current in grid side   (f) voltage in grid side

Fig. 5 Simulation outputs when solar irradiation changes from 1 to 0 kW/m²

as constant at 1 kW/m². When the load demand is increased, power flow from solar system to load increases and the power injected to the grid decreases. From Fig. 4, it is observed that the magnitude of the current in grid side is decreased when the load is increased. Voltage at dc link is maintained constant at 1 pu throughout this case.

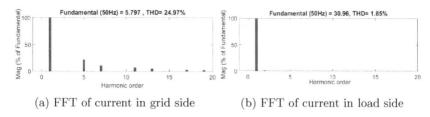

(a) FFT of current in grid side   (b) FFT of current in load side

Fig. 6 Harmonic spectrum when solar irradiation = 1 kW/m^2

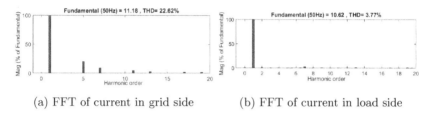

(a) FFT of current in grid side   (b) FFT of current in load side

Fig. 7 Harmonic spectrum, after load increment

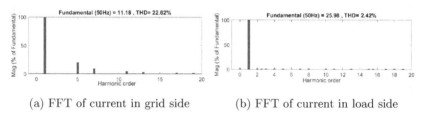

(a) FFT of current in grid side   (b) FFT of current in load side

Fig. 8 Harmonic spectrum when solar generation is absent

3.1.2 Case B: Solar Irradiation Changes from 1000 W/m^2 to Zero

In this case, a step change in solar irradiation from 1000 to 0 W/m^2 is introduced at 0.2 s. The additional power after meeting the load demand is injected to the grid. Therefore, the phase difference between current in grid side and voltage in grid side is 180° as shown in Fig. 5. When there is non-availability of solar power, the load demand is supplied by the grid. Under these conditions, the current in grid side and the voltage in grid side are in phase. However, a small dip in dc link voltage is observed during the dynamic change, returns to the steady-state value, 1 pu within 20 ms.

Figures 6, 7, and 8 show current in load side and current in grid side harmonic spectrum for different cases. In all these cases, the THDs of current in grid side are below 5%, which is below the limit set by IEEE standard of power quality (IEEE-519) [19].

3.2 FPGA Implementation and Hardware Co-Simulation Results

The entire control logic is implemented in system generator and then converted to IP cores. Vivado design suite is used for IP integrator-based block design. The HDL wrapper for the block design is generated, which generates the VHDL/Verilog code for the entire block algorithm. Test bench is created for the program, and behavioral simulations are carried out to verify the working of controller. After synthesis and implementation of the control algorithm in Vivado design suite, the bit stream is generated. The XC7Z020-clg484-1 FPGA Board in its JTAG mode is interfaced to the PC. Burning the bitstream generated into the FPGA is the last step of hardware implementation of control algorithm.

Hardware co-simulation is done with the power circuit of shunt active filter simulated in Matlab and controller hardware implemented in FPGA. Clock frequency of Zynq 7000 FPGA is 100 MHz (time period = 10 ns) and the Simulink simulation time period is taken as 1 μs. The hardware co-simulation results are observed in Xilinx Vivado design suite waveform viewer. The same dynamic cases considered in Matlab simulation are considered in hardware co-simulation also. The current and voltage waveforms for step increment in load and step change in solar irradiation are shown in Figs. 9 and 10, respectively. Table 1 shows the steady-state values of dc link voltage, current in load side and current in grid side observed in simulation as well as hardware co-simulation.

The results obtained in hardware co-simulation are the scaled down versions of actual voltages and currents. The waveforms observed are same as the Matlab simulation results. So the implementation of control system in FPGA is validated.

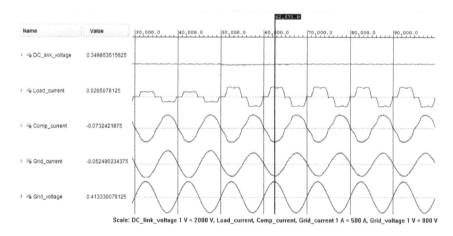

Fig. 9 Hardware co-simulation results for case a

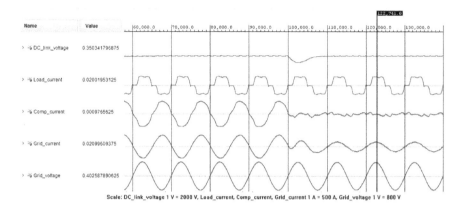

Fig. 10   Hardware co-simulation results for case b

Table 1   Comparison of simulation and hardware co-simulation results

Cases	System parameter	Steady state peak value		
		Simulation	Hardware co-simulation	
			Observed	Actual
Load = 100 Ω, 60 mH, Solar irradiation = 1 kW/m^2	DC link voltage (p.u)	1.0	0.3505	1.0
	Current in load side (A)	5.5	0.0110	5.5
	Current in grid side (A)	32	0.0642	32.1
After 50% load increment, Solar irradiation = 1 kW/m^2	DC link voltage (p.u)	1.0	0.3505	1.0
	Current in load side (A)	10.6	0.0212	10.6
	Current in grid side (A)	26	0.0522	26.1
When solar power generation is absent	DC link voltage (p.u)	1.0	0.3503	1.0
	Current in load side (A)	10.6	0.0212	10.6
	Current in grid side (A)	11.5	0.0230	11.5

4 Conclusion

The dynamic condition simulation results show that the controller performs well under different conditions considered. The controller is implemented in XC7Z020-clg484-1 FPGA and hardware co-simulation is done with implemented control system in FPGA, and the power circuit is simulated in Matlab/Simulink. The results obtained are similar to the simulation results, so the FPGA implementation is validated.

References

1. S. Mikkili, A. Padamati, PHC, id-iq and p-q control strategies for mitigation of current harmonics in three-phase three-wire shunt active filter with pi controller. Int. J. Emerg. Electric Power Syst. **18**(2), 1–29 (2017). https://doi.org/10.1515/ijeeps-2016-0055
2. H. Akagi, E. Watanabe, M. Aredes, *Shunt Active Filters*, chap. 4 (Wiley-Blackwell, 2017), pp. 111–236. https://doi.org/10.1002/9781119307181.ch4
3. H. Akagi, The state-of-the-art of active filters for power conditioning, in *Power Electronics and Applications, 2005 European Conference on* (2005), pp. 1–15
4. F. Nejabatkhah, Y.W. Li, B. Wu, Control strategies of three-phase distributed generation inverters for grid unbalanced voltage compensation. IEEE Trans. Power Electron. **31**(7), 5228–5241 (2016). https://doi.org/10.1109/TPEL.2015.2479601
5. M. Benchouia, I. Ghadbane, A. Golea, K. Srairi, M.E.H. Benbouzid, Implementation of adaptive fuzzy logic and pi controllers to regulate the dc bus voltage of shunt active power filter. Appl. Soft Comput. **28**, 125–131 (2015)
6. Y. Hoon, M.A. Mohd Radzi, M.K. Hassan, N.F. Mailah, A refined self-tuning filter-based instantaneous power theory algorithm for indirect current controlled three-level inverter-based shunt active power filters under non-sinusoidal source voltage conditions. Energies **10**, 3 (2017). https://doi.org/10.3390/en10030277
7. S. Biricik, S. Redif, C. Ozerdem, S.K. Khadem, M. Basu, Real-time control of shunt active power filter under distorted grid voltage and unbalanced load condition using self-tuning filter. IET Power Electron. **7**(7), 1895–1905 (2014). https://doi.org/10.1049/iet-pel.2013.0924
8. S. Mikkili, A.K. Panda, Instantaneous active and reactive power and current strategies for current harmonics cancellation in 3-ph 4wire shaf with both pi and fuzzy controllers. Energy Power Eng. **3**(03), 285–298 (2011)
9. A. Ghamri, T. Mahni, M. Benchouia, K. Srairi, A. Golea, Comparative study between different controllers used in three-phase four-wire shunt active filter. Energy Procedia **74**, 807–816 (2015)
10. R.L.D.A. Ribeiro, T.D.O.A. Rocha, R.M. de Sousa, E.C. dos Santos, A.M.N. Lima, A robust dc-link voltage control strategy to enhance the performance of shunt active power filters without harmonic detection schemes. IEEE Trans. Indus. Electron. **62**(2), 803–813 (2015). https://doi.org/10.1109/TIE.2014.2345329
11. T. Demirdelen, R.I. Kayaalp, M. Tümay, Pso-pi based dc link voltage control technique for shunt hybrid active power filter, in *2016 International Conference on Systems Informatics, Modelling and Simulation (SIMS)* (2016), pp. 97–102. https://doi.org/10.1109/SIMS.2016.18
12. F. Li, Y. Zou, W. Chen, J. Zhang, Comparison of current control techniques for single-phase voltage-source pwm rectifiers, in *2008 IEEE International Conference on Industrial Technology* (2008), pp. 1–4 . https://doi.org/10.1109/ICIT.2008.4608485
13. H. Vahedi, Y.R. Kukandeh, M.G. Kashani, A. Dankoob, A. Sheikholeslami, Comparison of adaptive and fixed-band hysteresis current control considering high frequency harmonics, in

2011 IEEE Applied Power Electronics Colloquium (IAPEC) (2011), pp. 185–188. https://doi. org/10.1109/IAPEC.2011.5779858

14. Y. Xiaojie, L. Yongdong, A shunt active power filter using dead-beat current control, in *IEEE 2002 28th Annual Conference of the Industrial Electronics Society. IECON 02*, vol. 1 (2002), pp. 633–637. https://doi.org/10.1109/IECON.2002.1187581

15. S. Charles, C. Vivekanandan, An efficient fpga based real-time implementation shunt active power filter for current harmonic elimination and reactive power compensation. J. Electr. Eng. Technol. **4**, 4 (2015). https://doi.org/10.5370/JEET.2015.10.4.1655

16. P. Karuppanan, K.K. Mahapatra, Active harmonic current compensation to enhance power quality. Int. J. Electr. Power Energy Syst. **62**, 144–151 (2014). https://doi.org/10.1016/j.ijepes. 2014.04.018

17. M. Salimi, J. Soltani, A. Zakipour, Experimental design of the adaptive backstepping control technique for single-phase shunt active power filters. IET Power Electron. **10**(8), 911–918 (2017). https://doi.org/10.1049/iet-pel.2016.0366

18. A. Mahfouz, S. Zaid, S. Saad, A. Hagras, Sensorless dc voltage control with backstepping design scheme for shunt active power filter. J. Electr. Eng. **15**, 303–312 (2015)

19. IEEE-519, IEEE recommended practice and requirements for harmonic control in electric power systems. IEEE Std 519-2014 (Revision of IEEE Std 519-1992) pp. 1–29 (2014). https:// doi.org/10.1109/IEEESTD.2014.6826459

Grounded FDNR Simulation Circuit Using VDDIBAs

Venkata Mohit Sai Chandra Kukunoori, Vinay Reddy Godhala, and Mayank Srivastava

1 Introduction

The concept of FDNR behaviour was very firstly introduced in [1]. This paper also described the transformation of any RLC passive network into FDNR-based network. The utilization of FDNR in high order filters is well understood. In the last three decades, the active simulation of FDNR element has been an alternative research domain and FDNR configuration based on different active building blocks (ABBs) has been presented by researches. The DVCCII-based realization reported in [2] requires matched passive elements. Similarly, the configuration discussed in [3] also suffer from matching constraints and also use of two resistors makes the circuit oversized. The grounded FDNR reported in [4] also employs two resistors based on OTRA and four passive elements also exhibit component matching limitation.

An op-Amp-based FDNR was reported in [5] which employs four op-Amp while another op-Amp-based FDNR [6] uses three op-Amp, seven resistors and two capacitors. The use of too many op-Amp and resistors makes these circuits not suitable for on-chip implementation.

The Second Generation Current Conveyor (CCII)-based FDNR circuits proposed in [7, 8] require meeting component matching conditions for lossless FDNR behaviour. The FDCCII-based FDNR configurations presented in [9, 10] are over-large as employed FDCCIIs are not fully utilized. All the represented circuits in

V. M. S. C. Kukunoori (✉) · V. R. Godhala · M. Srivastava
National Institute of Technology, Jamshedpur, Jharkhand, India
e-mail: kukunoori.mohit@gmail.com

V. R. Godhala
e-mail: gvinayreddy98@gmail.com

M. Srivastava
e-mail: mayank2780@gmail.com

© The Author(s), under exclusive license to Springer Nature Singapore Pte Ltd. 2021   455
J. Kumar and P. Jena (eds.), *Recent Advances in Power Electronics and Drives*, Lecture Notes in Electrical Engineering 707,
https://doi.org/10.1007/978-981-15-8586-9_40

[3–10] do not exhibit the facility of electronic tuning. Some electronically controllable FDNR simulators employing various active elements have been presented in [11–15].

Therefore, the aim of this research attempt is to add a new FDNR simulator in the available literature. The proposed circuit uses two VDDIBAs, two capacitors and one resistor. The VDDIBA is a modern ABB and no work has been reported so far describing the use of VDDIBA in FDNR realization. Hence, this work opens the door for researchers to pursue their research on this topic. The presented FDNR simulator enjoys advantages like electronic tuning, no requirements to meet matching conditions, use of less number of active and passive elements, low sensitivities of FDNR value with respect to circuit elements and exhibiting optimal behaviour even when using a non-ideal model of VDDIBAs.

2 VDDIBA

VDDIBA is an extension of VDBA, which is a popular active element [16]. VDDIBA is a voltage mode device with biasing current controllability. Hence, a very attractive active circuit concept was found. The single block symbolic representation of VDDIBA is illustrated in Fig. 1, and current–voltage relations among various terminals have been shown in Eq. 1–Eq. 3. The differential input voltage applied across the P and N terminals, get transferred to Z terminal in the form of current and diffused voltage across Z and V terminal is transferred to W terminal. The behaviour of VDDIBA can be realized by the circuit shown in Fig. 2, which employs two OTAs, single CFOA and one dual output OTA.

$$I_z = g_m(V_P - V_N) \tag{1}$$

$$V_W = V_V - V_Z \tag{2}$$

$$I_P = I_N = 0 \tag{3}$$

Fig. 1 Symbolic representation of VDDIBA

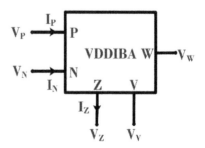

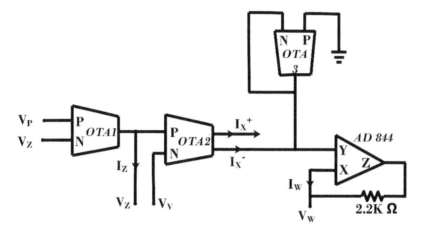

Fig. 2 VDDIBA realization using OTAs and CFOA

3 Proposed Configuration

The proposed FDNR circuit employing two VDDIBA has been shown in Fig. 3.

Through simple circuit analysis, the input impedance of the FDNR circuit given in Fig. 3 can be evaluated as follows:

$$Z_{IN} = \frac{1}{D_{eq}s^2} \tag{4}$$

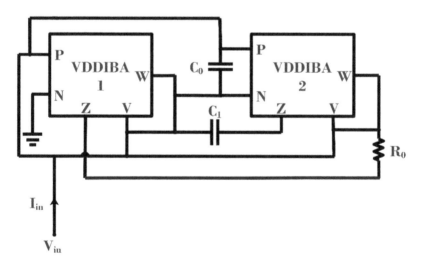

Fig. 3 Proposed FDNR simulator using VDDIBA active element

where

$$D_{eq} = \left(\frac{g_{m2}}{g_{m1} R_0 C_0 C_1} \right)$$ (5)

It can be noticed from Eq. 5 that the realized negative resistance value can be varied by g_{m1} and g_{m2} which confirms the availability of electronic tuning of the circuit. Also, no matching condition has been met to realize the FDNR function.

4 Non-ideal Analysis

On considering a non-ideal model of VDDIBA, the Eqs. (1–3) can be modified as follows:

$$I_Z = \alpha g_m (V_P - V_N)$$ (6)

$$V_W = \beta_V V_V - \beta_Z V_Z$$ (7)

$$I_P = I_N = 0$$ (8)

On considering a non-ideal model of VDDIBA, the Eqs. (6–8), the modified value of input impedance for the proposed circuit can be evaluated as

$$Z_{\text{non-ideal}} = \frac{1}{D_{\text{non-ideal}} s^2} + \frac{1}{C_{\text{non-ideal}} s}$$ (9)

where

$$D_{\text{non-ideal}} = \frac{\beta_{Z2} \alpha_2 g_{m2}}{C_0 C_1 \left[\alpha_1 g_{m1} R_0 + \beta_{V2} - \beta_{Z2} + \left(\frac{1-\beta_{V1}}{\beta_{Z1}} \right) \right]}$$ (10)

$$C_{\text{non-ideal}} = \frac{\beta_{Z1} - \beta_{Z2}}{C_1 \left[\alpha_1 g_{m1} R_0 + \beta_{V2} - \beta_{Z2} + \left(\frac{1-\beta_{V1}}{\beta_{Z1}} \right) \right]}$$ (11)

Therefore, it can be noticed from Eqs. 10 and 11 that on using a non-ideal model of employed VDDIBAs, the presented circuit does not simulate the behaviour of lossless FDNR. The impedance expression includes a lossy term along with the FDNR term. Therefore, it can be said that in a non-ideal environment the developed FDNR circuit simulates the working of a series $D - C$ network.

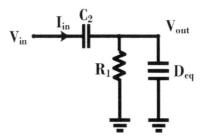

Fig. 4 Conventional RLC bandpass filter

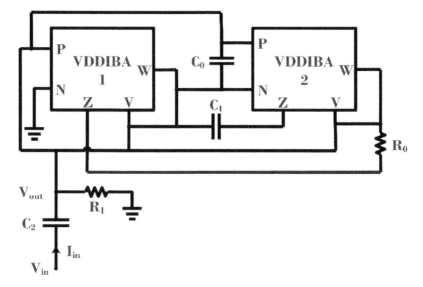

Fig. 5 Equivalent active realization of RLC bandpass filter using the proposed FDNR

5 Applications

To validate the behaviour of the presented FDNR simulator, a bandpass active filter
has been implemented. The conventional RLC bandpass filter is depicted in Fig. 4,
and its equivalent active realization using the proposed VDDIBA-based FDNR has
been demonstrated in Fig. 5.

6 Simulation Results

The behaviour of the presented FDNR simulator shown in Fig. 3 and the bandpass
filter depicted in Fig. 5 have been validated by executing PSPICE simulation. For

Fig. 6 CMOS model of
single output OTA

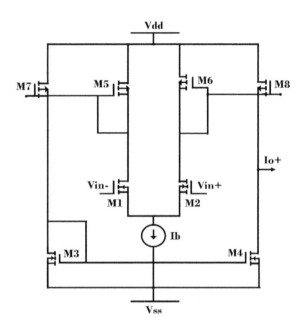

simulation purposes, the VDDIBA model illustrated in Fig. 2 has been taken into account. The single I/O OTA and dual output OTA shown in Fig. 2 can be implemented by CMOS models as shown in Figs. 6 and 7, respectively. The CFOA is implemented by SPICE model of ICAD844.

For the simulation of the circuit shown in Fig. 3, the supply voltage has been taken as 3 V with biasing currents of OTAs 50 μA. The values for used passive elements were selected as $C_0 = C_1 = 0.2, 0.3$ nF and 0.4 nF and $R_0 = 1$ kΩ. The PSPICE generated responses for FDNR circuit of Fig. 3 has been illustrated in Fig. 8. Figure 8 shows a variation in impedance magnitude with increasing frequency and it is clear from it that the presented FDNR behaves satisfactorily till 2.55 MHz. The electronic tuning for different values of biasing currents (20, 40 μA and 60 μA) has been depicted in Fig. 9.

The working of the developed bandpass filter (given in Fig. 5) also has been studied by PSPICE simulations with passive elements $C_0 = C_1 = 0.2$ nF, $C_2 = 0.1$ nF, $R_0 = 1$ kΩ, and $R_1 = 2.6$ kΩ. The frequency response of this filter has been given in Fig. 10.

7 Conclusion

A new VDDIBA-based grounded FDNR simulator has been reported in this work. The developed circuit employs two VDDIBAs, two capacitors and a single resistor. As there is no such FDNR circuit employing VDDIBA active element has been

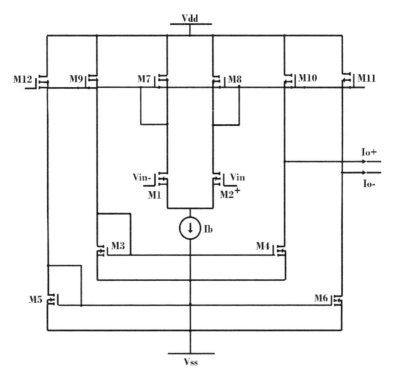

Fig. 7 CMOS model of dual output OTA

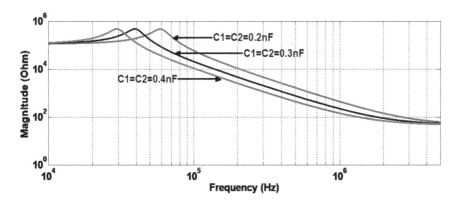

Fig. 8 FDNR response for different capacitance values

available in open literature hence, this work can be considered as novel work in FDNR realization. The working of VDDIBA block has been achieved by an OTA and CFOA-based circuit. The electronic tunability of realized FDNR has been demonstrated by varying the biasing current. The behaviour of the presented circuit has been checked

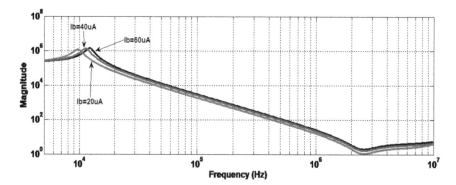

Fig. 9 Electronic tuning of realized FDNR

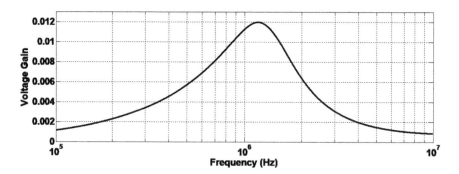

Fig. 10 The frequency response of filter shown in Fig. 5

by designing a bandpass filter using it. All the circuits are validated by PSPICE simulations.

References

1. A.M. Soliman, A new realization of the FDNC using the DVCCS/DVCVS. Int. J. Electron. Commun. (AEU) **33**(10), 423–424 (1979)
2. F. Kacar, H. Kuntman, New realization of FDNR and sixth order band pass filter application, in *2011 20th European Conference on Circuit Theory and Design (ECCTD)*, pp. 381–384 (2011)
3. A. Gupta, R. Senani, D.R. Bhaskar, A.K. Singh, OTRA-based grounded-FDNR and grounded-inductance simulators and their applications. Circuits Syst. Signal Process. **31**, 489–499 (2012)
4. R. Nandi, Novel active-R ideal frequency-dependent negative-resistance simulation. Proc. IEEE **67**(7), 1080 (1979)
5. R. Senani, D.R. Bhaskar, Versatile voltage-controlled impedance configuration. IEE Proc.-Circuits Devices Syst. **141** (5) (1994)
6. K. Pal, Novel F.D.N.C. simulation using current conveyors. Electron. Lett. **16**(16), 639–641 (1980)

7. R.I. Salawu, Realization of frequency dependent negative resistance. Microelectron. Reliabil. **20**, 853–857 (1980)
8. A.M. Soliman, Realisation of frequency-dependent negative-resistance circuits using two capacitors and a single current conveyor. Proc. IEE **125**(12), 1336–1337 (1978)
9. F. Kaçar, H. Kuntman, Novel electronically tunable FDNR simulator employing single FDCCII, in *2009, European Conference on Circuit Theory and Design* (Turkey, 2009), pp. 21–24
10. F. Kaçar, A. Yeşil, FDCCII-based FDNR simulator topologies. Int. J. Electron. **99**(2), 285–293 (2012)
11. M.T. Abuelma'Atti, N.A. Tasadduq, Electronically tunable capacitance multiplier and frequency-dependent negative-resistance simulator using the current-controlled current conveyor. Microelectron. J. **30**(9), 869–873 (1999)
12. M. Srivastava, P. Gupta, A. Roy, M.K. Verma, D. Prasad, New electronically tunable floating FDNR configuration employing grounded capacitances, in *Fifth IEEE International Conference on Signal Processing & Integrated Networks* (IEEE-SPIN 2018), pp. 754–759
13. P. Gupta, M. Srivastava, New frequency dependence negative conductance simulator employing VDTAs and grounded capacitances, in *Fifth IEEE International Conference on Signal Processing & Integrated Networks* (IEEE-SPIN 2018), pp. 275–279
14. M. Srivastava, A. Roy, R. Singh, New VDCC based electronically tunable grounded frequency dependent negative resistance simulator employing grounded passive elements, in *IEEE-RDCAPE-2017* (Noida)
15. M. Srivastava, A. Roy, R. Singh, New grounded FDNR simulation circuit with electronic control, in *IEEE-TELNET-2017* (Noida, India)
16. D. Biolek, R. Senani, V. Biolkova, Z. Kolka, Active elements for analog signal processing classification, review and new proposals. Radioengineering **17**(4), 15–32 (2008)

Designing a Series Active Power Filter for Mitigating Harmonics of a High Frequency Resonant Inverter

Rohan Sinha, Rahul Raman, Ananyo Bhattacharya, and Pradip Kumar Sadhu

1 Introduction

Modern state of the art electrical technology has lead to the increasing use of power electronic devices. Electrical sources like renewable resources generally call for the integration of power electronic devices so that the power can be utilized and tapped effectively. One such application is the use of heating equipments in industries to carry out brazing, welding, melting and many more. Although there are numerous methods of heating objects but, the use of an induction heater is the most preferred option [1, 2] because the benefits associated with induction heating provides an economical, cleaner and an energy efficient medium of heating. Also, the output energy of the induction equipment can be controlled by controlling the inverter thus, giving precise temperature control at the output. For effective induction heating, the inverter should switch at a very high frequency. This means that the output of the inverter produces an ac signal with a very high frequency. A high frequency ac signal generates high eddy current in the workpiece to be heated, this eddy current produces heat mainly due to Joule's law of heating, hysteresis losses in the work piece and an additional phenomenon skin effect is also associated with it. But the main problem associated with such a device is its non-linear nature. Being a non-linear load, it generates harmonics which has the tendency of following into the grid. This action

R. Sinha (✉)
Department of Electrical Engineering, Jorhat Engineering College, Jorhat, Assam, India
e-mail: sinharohan29@gmail.com

R. Raman · P. K. Sadhu
Department of Electrical Engineering, Indian Institute of Technology (ISM) Dhanbad, Dhanbad, India

A. Bhattacharya
Department of Electrical and Electronics Engineering, National Institute of Technology Jamshedpur, Jamshedpur, India

© The Author(s), under exclusive license to Springer Nature Singapore Pte Ltd. 2021   465
J. Kumar and P. Jena (eds.), *Recent Advances in Power Electronics and Drives*, Lecture Notes in Electrical Engineering 707,
https://doi.org/10.1007/978-981-15-8586-9_41

distorts the grid current leading to a poor power factor. Harmonics pose a serious issue associated with power quality, like damaging sensitive loads, malfunction of sensitive loads, heating of supply wire, transformer malfunction and many more. A poor power factor means that the grid is underutilized and is also uneconomical for the industry concerned. Therefore this situation demands for certain specialized circuit which can damp out these harmonics. Such circuits are called filters. It is a circuit which is installed near the non-linear devise to eliminate these distortions. Passive filters and active power filters are the two types of filter. The active power filters are further classified as shunt APF, Series APF, Unifies Power Quality Conditioners and Hybrid power filters [3, 4].

Here, the main non-linear load is the high frequency resonant inverter. The main harmonics which are generated are the high frequency harmonics and switching harmonics. To compensate these harmonics numerous works have been carried out. Raman et al. [5] gave a comparative study of the filters. They have implemented passive filter and semi active for compensation. Without filter the THD was 46.029%, with the passive filter 23.855%, still significantly high and with the semi-active filter the THD count is nearly 0%. But these filters have only been designed for a single phase application. Raman et al. [6] presented the use of a Vienna Rectifier for controlling harmonics. In this paper, the THD before the use of the PWM rectifier is 37.263% and after implementing the rectifier it reduced to almost 0%. But, a Vienna rectifier is usually direct current controlled which has the disadvantage of irregular control sequence. Palash et al. [7] present design of a passive filter along with a modified half-bridge inverter. Without the filter the value of THD was 44.99% and after using the passive filter it dropped to 17.90%, which is still a significant harmonic figure. Adding on, a half-bridge configuration is suitable for low power application only. These filtering techniques come with their own disadvantages. Passive filters have disadvantages associated such as resonance with source, high current flowing through passive devices due to resonance, being tuned filters they cannot mitigate harmonics in changing frequency conditions and many more. All of these characters compromise the filtering behavior of the circuits. The present papers deals with the mitigations of harmonics using the Series Active Power Filter (SeAPF), as the effectiveness of the SeAPF in the field of high frequency harmonic mitigation has not been evaluated.

Active power filters are power circuits which compensate the harmonics by actively generating the harmonics and injecting them at the point of installation. These generated signals are of equal amplitude but are 180° phase shifted. APFs have the added advantage of a good dynamic response, achieving steady state in a shorter duration of time and also adaptive schemes can be implemented using them in varying environmental conditions. Torre et al. [8] proposed their Active power filter for a single phase supply. Islam et al. [9] used an SeAPF based on the PQ theory, which is a computationally heavy algorithm and prone to voltage distortion. Pinto et al. [10] proposed a transformerless SeAPF, but it has isolated single phase VSI for each phase along with isolated power supply. This paper deals with designing an SeAPF with the control circuit of the filter and designing the induction heating inverter. It basically uses the DQ theory which is computationally less heavy and immune to grid voltage distortion. Also, an external power supply for the filter is not required in this configuration because it is equipped with a voltage balancing algorithm. The paper has been divided into parts which deal with designing the resonance circuit, the control circuit generating the reference voltage, the simulation results of the inverter with and without the filter and the comparative total harmonic distortion (THD).

2 Resonant Inverter

The application for which the inverter is used is for induction heating, a heating technology which is used both in domestic and industrial heating processes. Since the associated technology demands for a high frequency ac signal output, the inverter is called a High Frequency Resonant Inverter. When resonance is used in the inverter technique, the output power transferred to the load will be maximum. This is because when a circuit resonates, the inductive impedance cancels out the capacitive impedance leading to a condition where the load turns out to be purely resistive in nature. Also, on attaining resonance the power electronic switches can be operated at ZVS or ZCS [1] condition. A ZVS or ZCS scheme is usually employed during a high switching modulation scheme because on turning off or on the semiconductor switches at these conditions, the power loss associated will be at its minimum value. These are the two main reasons why a resonating inverter is generally employed. Now, an inverter is said to be in the resonance condition when its switching frequency is equal to the resonant frequency of the load. For the construction of the inverter, four IGBTs are used and the resonance condition is produced by the load. This resonance condition can be summarized as follows,

$$X_L = X_C \text{ or } \omega L = \frac{1}{\omega C} \tag{1}$$

$$2\pi f_r L = \frac{1}{2\pi f_r C} \tag{2}$$

$$f_r = \frac{1}{2\pi \sqrt{LC}} \tag{3}$$

where X_L and X_c are the impedance of inductor and capacitor respectively and f_r is the resonant frequency.

2.1 Determining the Load Parameters and Calculation of Switching Frequency

For the induction heating coil to receive the maximum power, the resonating condition has to be precisely determined. This is necessary because any deviation from the ideal switching frequency can lead to low power output, improper temperature control, not attaining ZVS condition and low efficiency of the converter.

Therefore, mathematical equations are used to precisely determine the value of the resistor, capacitor and inductor. In this work, the type of resonance which has been employed is called parallel resonance. The parallel resonating circuit, also called a tank circuit, is composed of an inductor and resistor in parallel connection with a capacitor. When the circuit starts resonating, the admittance part of the equation is equal to zero. Admittance is given as,

$$Y = \frac{1}{R_L + jX_L} + \frac{1}{-jX_C} \tag{4}$$

$$Y = \frac{1}{R_L^2 + X_L^2} + j\left(\frac{1}{X_C} - \frac{1}{R_L^2 + X_L^2}\right) \tag{5}$$

When susceptance is zero,

$$\frac{1}{X_C} = \frac{1}{R_L^2 + X_L^2} \tag{6}$$

$$\omega C = \frac{\omega L}{R_L^2 + \omega^2 L^2} \tag{7}$$

From Eq. 7 we obtain,

$$\omega^2 L^2 = \frac{L}{C} - R_L^2 \tag{8}$$

$$\omega = \sqrt{\frac{1}{LC} - \frac{R_L^2}{L^2}} \tag{9}$$

The resonant frequency is thus given as,

$$f_r = \frac{1}{2\pi}\sqrt{\frac{1}{LC} - \frac{R_L^2}{L^2}} \tag{10}$$

In the present work, the load has been designed using the equations in the frequency domain which can be stated as Z(S) or Z(jw) [2]. Again, to create resonance, the imaginary part should be equal to zero.

$$-\omega C R^2 + \omega L - \omega^3 L^2 C = 0 \tag{11}$$

$$\omega = \sqrt{\frac{L - C R^2}{L^2 C}} \tag{12}$$

$$f = \frac{1}{2\pi} \sqrt{\frac{L - C R^2}{L^2 C}} \tag{13}$$

Using $R = 1 \, \Omega \, L = 57.1$ uH and $C = 0.2$ uf, the switching frequency is obtained as 47 kHz.

The full bridge resonant inverter should be operated at this switching frequency to optimize the operation. In practical situation, it is difficult to reach the exact resonating condition; therefore the switching frequency is kept near the theoretically determined value.

3 Diagrammatic Representation of the Method

As it has been stated above, the method which has been employed to compensate the voltage harmonic is a Series Active Power Filter [3, 8–11]. An SeAPF act as a series voltage source with the load. It is generally interfaced at the point of common coupling using a current transformer which for providing isolation between the filter circuit and the power grid. Firstly, the project consists of a high frequency resonant inverter as the main non-linear load which distorts supply utility signal. Secondly, the SeAPF which is in form of a voltage source inverter (VSI), the part of the filter which injects the phase shifted harmonics in the utility. The power filter stage is composed of three main parts, the VSI, the control circuit of the filter which generates the reference voltage and the hysteresis band controller. The VSI is composed of a three phase VSI with a capacitor and a ripple filter which controls the switching ripples. A diagrammatic representation has been shown in Fig. 1.

4 Control Strategy for the Filter

One of the most important steps in the implementation of the compensating strategy, the control of the filter plays a major role. With a properly devised control algorithm, the filter successfully is able to mitigate the harmonics. One of the most widely used algorithms for filter control is the instantaneous reactive power theory of commonly known as the PQ theory [4, 9]. But the calculations used in this control algorithm are susceptible to voltage distortions in the power grid. This is because the instantaneous reactive power, both imaginary and real power, is calculated based on the

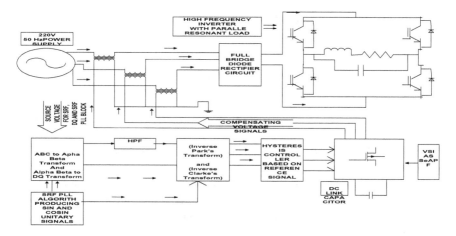

Fig. 1 Overall layout of the method used

source voltages and currents. Although an improvise PQ Theory based on synchronizing the supply voltage to the fundamental component, this algorithm is generally computationally heavy.

To overcome these disadvantages, another well know algorithm the Synchronous Reference Frame (SRF) or DQ theory had been used as the control algorithm. Thus, the model of the controller consists of the SRF block, the PLL block and the Hysteresis Controller for generating the switching pulses of the VSI. A detailed discussion of the theory has been provided below.

4.1 Generation of the Compensating Voltage Signals

Reference signal production for the control of the VSI is based on the DQ Theory. Among the control algorithms available for the SeAPF, the SRF theory [12] is one of the best strategy because it provides faster response to change in the system parameters, it is computationally less heavy and simpler to implement and also immune to system voltage changes as it makes use of a PLL for synchronization. A total of six voltage sensors are used three for sensing the source voltages and three to sense the load voltages, so that the filter works as per requirement. This algorithm for control is based on the transformation of source voltages which are in the *abc coordinate to dq coordinate.* The dq coordinates contains both the fundamental power component and the harmonic component, when the source is polluted by harmonics. To implement this algorithm first the voltages in *abc coordinates are converted to αβ coordinate* using the Clarke's Transformation.

$$
\begin{bmatrix} V_0 \\ V_\alpha \\ V_\beta \end{bmatrix} = \sqrt{\frac{2}{3}} \begin{bmatrix} \frac{1}{\sqrt{2}} & \frac{1}{\sqrt{2}} & \frac{1}{\sqrt{2}} \\ 1 & \frac{1}{2} & -\frac{1}{2} \\ 0 & \frac{\sqrt{3}}{2} & -\frac{\sqrt{3}}{2} \end{bmatrix} \begin{bmatrix} V_{sa} \\ V_{sb} \\ V_{sc} \end{bmatrix} \tag{14}
$$

Here, the V_0 part of the equation denotes the zero-sequence component. But, in this project this component has been excluded because the system is a three wire system.

After this $\alpha\beta$ coordinates generation, Park's transformation is applied to obtain the dq coordinates of the system voltages, this part of the voltages contains the fundamental voltage as well as harmonic voltages. The Park's Transformation is as follows,

$$
\begin{bmatrix} V_d \\ V_q \end{bmatrix} = \begin{bmatrix} \cos\theta & \sin\theta \\ -\sin\theta & \cos\theta \end{bmatrix} \tag{15}
$$

For extracting the high frequency harmonics from the V_d and V_q, an HPF is used. Now, we have the harmonic components in the dq frame. To convert these signals back to the three phase abc coordinates, first Inverse Park's Transformation is applied as follows obtain the $\alpha\beta\ coordinate$.

$$
\begin{bmatrix} V_\alpha \\ V_\beta \end{bmatrix} = \begin{bmatrix} \cos\theta & -\sin\theta \\ \sin\theta & \cos\theta \end{bmatrix} \tag{16}
$$

On applying Inverse Clarke's Transformation we get back the signals in the *abc coordinates,*

$$
\begin{bmatrix} V_{ca}^* \\ V_{cb}^* \\ V_{cc}^* \end{bmatrix} = \sqrt{\frac{2}{3}} \begin{bmatrix} 1 & 0 \\ \frac{-1}{2} & \frac{\sqrt{3}}{2} \\ \frac{-1}{2} & \frac{-\sqrt{3}}{2} \end{bmatrix} \begin{bmatrix} V_{s\alpha} \\ V_{s\beta} \end{bmatrix} \tag{17}
$$

The voltage components obtained in Eq. 17 are the reference voltages which will be required by the Hysteresis controller to produce the gating signals. Before applying these signals to the controller, the compensating signal V_{comp} is obtained by the difference of the reference signals and the source voltages.

As this topology is an SeAPF without an external power source, the voltage across the dc link capacitor has to be balanced using a controller for effective harmonic cancellation. A PI controller is employed for this purpose which takes in the error voltage. The error voltage is obtained by the difference of the fixed reference voltage and the actual dc link voltage measured across the capacitor. All the sin and cos components of the Park's and Inverse Park's Transformations are obtained using a PLL. The PLL is discussed in the following section.

SRF Based PLL

The PLL is an integral part of the SRF Control algorithm. A PLL [10], 13 circuit is basically implemented to track a particular system. It performs the task of synchronizing its output with the input signals (the reference signal) in terms of frequency and phase. Only when the PLL output signal phase synchronizes with the input signal, a locked stage in operation is achieved. After the lock is attained, even if the error tries to deviate from the actual value, the PLL will work automatically to synchronize the system. Here in the SRF theory, a similar concept is used to synchronize the output to the fundamental frequency of the supply voltage for producing the two sin and cos component required in the calculation. Phase detection, filtration and Voltage control oscillator are basic PLL parts.

In this present work, an SRF based PLL is used. Most of the PLL generally differ in the type of phase detector used. The SRF PLL, as the name suggests is based on the SRF theory. It mainly consists of the abc to dq transformation equations, followed by a PI controller which acts like an LPF an integrator and a VCO. The abc to dq transformation will act as the phase detector stage. Transformation of the grid voltages from abc to dq coordinate are shown in the SRF section above, which first uses the Clarke Transformation followed b Park Transformation. The direct and the quadrature components are given b Eq. 15.

$$\begin{bmatrix} V_d \\ V_q \end{bmatrix} = \begin{bmatrix} A\cos(\theta' - \theta) \\ A\sin(\theta' - \theta) \end{bmatrix} \quad (18)$$

$\theta' =$ output phase and $\theta =$ input phase.

Vd and Vq are the error and the amplitude component. Only the error component is required in this case (it contains the phase error) to obtain the phase lock condition. To put the PLL in locked state, the phase of the input and output signal as given in Eq. 18 should be brought to zero. Again in this block, a PI controller is employed for error reduction of the phase providing an output in angular frequency. This output is further fed into an integrator to obtain the phase angle. The phase angle acts as the input to a VCO block to produce a unitary PLL signal of sin and cos components.

Hysteresis Controller

The hysteresis controller is a variant of the PWM modulation technique for gating the VSI switches. It can be described as an OpAmp having a positive feedback. Among all the inverter modulation methods, hysteresis control is one of the most well known. It can be defined as a comparator with a positive feedback and the reference signal in fed into the negative terminal. The output versus input characteristic graph is in the form of a hysteresis curve, therefore the name. This comparator has two definite upper and lower levels called the upper threshold and the lower threshold. Every time the reference voltage signal tries to cross these upper and lower boundaries, the comparator output produces a significant change in the voltage levels, creating a PWM signal. The type of comparator employed here is a non-inverting Schmitt Trigger. To determine the values of the threshold levels, the following equation are used,

$$V_t = \frac{V_1}{R_1 + R_2} + \frac{V_o R_1}{R_1 + R_2} \tag{19}$$

V_I = Reference input signal and V_o = Output of comparator.
V_t = The positive feedback signal to comparator.
When a transition takes place, $V_t = 0$, therefore Eq. 19 can be modified as,

$$\frac{V_1}{R_1 + R_2} = -\frac{V_o R_1}{R_1 + R_2} \tag{20}$$

$$V_1 = \frac{-V_o R_1}{R_2} \tag{21}$$

When $V_o = -V_{saturation}$ value of V_1, the switch changes states from $-V_{sat}$ to $+V_{sat}$,

$$V_{UT} = \frac{-(-V_{sat})R1}{R2} \tag{22}$$

Similarly when $V_o = +V_{sat}$, the transition is from $+V_{sat}$ to $-V_{sat}$, giving the equation

$$V_{LT} = \frac{-(+V_{sat})R1}{R2} \tag{23}$$

Equations 22 and 23 give the upper and lower threshold levels respectively.

5 Simulation Diagrams and Results

To verify whether the proposed project is working as per standards, the project has been simulated in a simulation platform. All the simulations have been carried out in the PSIM environment. The simulation result consists of the simulation of the high frequency resonant inverter and the Series Active Power filter. The inverter has been simulated with and without the filter. The output signals and their FFT analysis have also been performed using the same PSIM tools.

Firstly, in Fig. 2, a high frequency resonant inverter has been simulated with a simple voltage balancing circuit installed before the diode bridge rectifier. The input to the inverter is a three phase 220 V and 50 Hz ac supply connected to a star delta transformer. Now to depict the load of the inverter, a resonant tank circuit has been chosen which consists of a series resistor and inductor branch in parallel to a capacitor. This entire configuration acts like an induction coil load to the inverter. An ammeter has been placed before the voltage balancing circuit to measure the input current. Figure 3 shows the input current to the load. It is quite clear from the graph that the current waveform is far from sinusoidal. To prove that the input current is

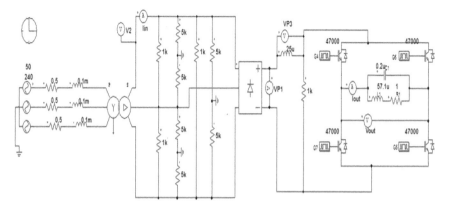

Fig. 2 Simulation diagram of the high frequency resonant inverter without an SeAPF

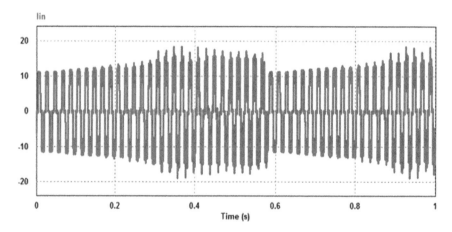

Fig. 3 Current graph of the input utility signal without the SeAPF

contaminated with harmonics, an FFT analysis of the signal is carried out (shown in Fig. 4), with the help of PSIM FFT tool. It clearly shows the presence of various high frequency harmonics. This part is responsible for the contamination of the input signal leading to a poor power factor. As has been mentioned before, these harmonics are removed using an SeAPF. Figure 5 shows the equivalent circuit diagram of the Full-bridge high frequency inverter along with an SeAPF configuration. The basic inverter parameters remain the same as shown in Fig. 2 but, with the addition of an active power filter. The filter consists for a voltage source inverter governed by a controller based on the DQ theory, the controller is further synchronized with the help of an SRF based PLL algorithm. The output of the SRF PLL is shown in Fig. 6, it clearly shows that the signals so produced are synchronized to the fundamental component of the utility supply, which is used in the control algorithm. Figure 7 depicts the input

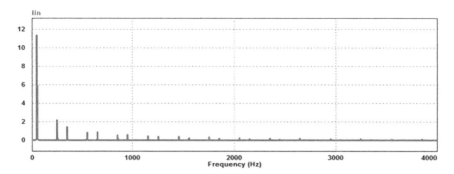

Fig. 4 FFT of the input voltage signal without SeAPF (from 0 to 4000 Hz)

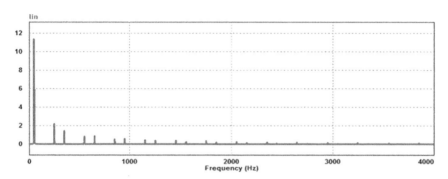

Fig. 5 Simulation of the high frequency resonant inverter along with an SeAPF

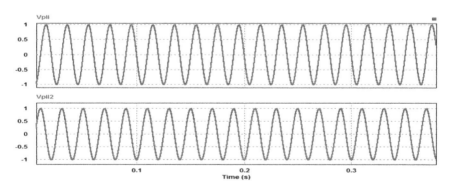

Fig. 6 SRF PLL output signals required in the SRF control algorithm. Vpll is the sinθ component, whereas Vpll2 is the cosθ component

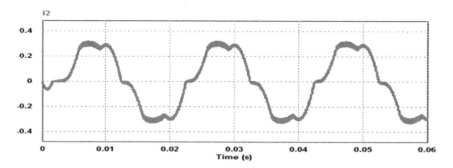

Fig. 7 Input current signal of the inverter with SeAPF

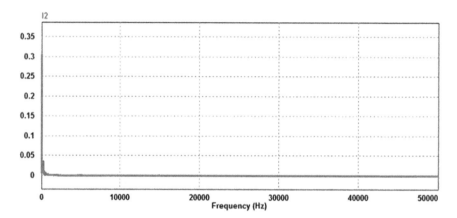

Fig. 8 FFT analysis of the signal with SeAPF

current to the inverter which has near sinusoidal profile when compared to Fig. 3. On doing the FFT analysis of the signal, we obtain the graph (Fig. 8) which shows that only the fundamental component is dominant; the remaining high frequency harmonics have almost been compensated. Now, the reason to show the current graphs before and after the installation of the active filter is as follows: it is a well know fact that the input voltage from the grid is quite immune to distortions, although there might be some harmonic contamination. This is attributed to the fact that our grid system is quite robust in nature. The main parameter in the circuit which is the matter of concern is the non-sinusoidal current waveform. This has been show in the above figures and the FFT analysis; which is highly non-sinusoidal. One thing which can be done in this condition is by controlling the voltage. Injecting an appropriate amount of voltage harmonics which is 180 phase shifted at the point of installation can actually reduce the current harmonics. Therefore, the main operation of this APF is to indirectly control the current harmonics by actually controlling the voltage parameters.

6 Harmonic Content Calculation

THD is the total count of the amount of harmonic which is present in the signal. A comparison between the FFT analyses shows the THD value of the signals with the filter and without the filter. For calculating the amount of distortion present in the signal the value of the fundamental voltage and the values of the corresponding harmonics are required. The calculation is down using the formula given below. The THD % has been calculated using the PSIM THD calculating tool.

$$THD = \frac{\sqrt{\sum_{n=2,3}^{\infty} I_{n\text{rms}}^2}}{I_{1\text{rms}}} \tag{24}$$

- **THD without active filter**:
  THD $= 35.25\%$
- **THD with active filter**:
  THD $= 1.46\%$

7 Conclusion

The above work deals with designing a high frequency resonant inverter along with a series active power filter for harmonic compensation. This resonant inverter being a non-linear load produces significant amount of harmonic leading to distortion of the grid supply. An active power filter is therefore used to compensate these harmonics so that these harmonics do not flow back to the utility supply. To validate whether the proposed scheme is working or not, all the circuits have been simulated using the PSIM software. The simulations with and without filter have been simulated along with their input signals respectively. A well designed controller along with the propose model is able to diminish the harmonics to a large extent. Further, the input signals are analyzed using the FFT calculation tool, this revels the various harmonics present in the signal. Without the filter, the number of harmonics is near about 14 and after using the filter, only the fundamental frequency is dominant. On performing the THD calculations, the value without filter is equal to 35.25% which is enough to cause major issues and after installation of the filter; the THD reduces to a value of 1.46%. These values clearly indicate that the SeAPF is working quite effectively and is able to mitigate the harmonics to a large extent. Moreover, without the filter a THD count of 35.25% shows that the power factor associated without a filter will be extremely poor. Therefore, it can be concluded that a well designed series active power filter composes an integral part of high frequency resonant inverter.

References

1. C. Gorain, P.K. Sadhu, R. Raman, Analysis of high-frequency class E resonant inverter and its application in an induction heater, in *Second International Conference on Power, energy and Environment: Towards Smart Technology* (IEEE, 2018). https://doi.org/10.1109/EPETSG. 2018.8659106

2. A. Kumar, P.K. Sadhu, R. Raman, J. Singh, Design analysis of full-bridge parallel resonant inverter for induction heating application using pulse density modulation technique, in *International Conference on Power Energy, Environment and Intelligent Control* (IEEE, 2018). https:// doi.org/10.1109/PEEIC.2018.8665571

3. J.L. Afonso, J.G. Pinto, H. Gonçalves, Active power conditioners to mitigate power quality problems in industrial facilities, in *Power Quality Issues* (INTECH)

4. H. Akagi, E.H. Watanabe, M. Aredes, *Instantaneous Power Theory and Applications to Power Conditioning* (Wiley-Interscience, Wiley, Inc, Publication, IEEE Press, 445 Hoes Lane, Picastaway, NJ 08854, United States of America, 2007).

5. R. Raman, P.K. Sadhu, A. Kumar, K. Sit, Design and analysis of RFI and EMI suppressor for high frequency induction heater using filters—a comparative study, in *4th international Conference on Recent Advances in Information Technology* (IEEE, 2018). https://doi.org/10. 1109/RAIT.2018.8389002

6. R. Raman, P. Sadhu, A. Kumar, P.K. Sadhu, Design and analysis of EMI and RFI suppressor for induction heating equipment using vienna rectifier, in *Second International Conference on Power, energy and Environment: Towards Smart Technology* (IEEE, 2018). https://doi.org/10. 1109/EPETSG.2018.8658871

7. P. Pal, P.K. Sadhu, N. Pal, S. Sanyal, An exclusive design of EMI-RFI suppressor for modified half bridge inverter fitted induction heating equipment. Int. J. Mechatron. Electri. Comput. Technol. (IJMEC) **15**, 2084–2100 (2015). 649123/10131

8. J.L. Torre, L.A.M. Barros, J.L. Afonso, J.G. Pinto (2019) Development of a propose single-phase series active power filter without external power sources, in *International Conference on Smart Energy Systems and Technologies (SEST)* (IEEE). https://doi.org/10.1109/SEST.2019. 8849010

9. A. Islam, M.A. Abeed, M.K.M. Rabby, M.H. Rahaman, A. Hossain, R. Al Nur, A.K.U. Mahbub, Series active power filter implementation using P-Q theory, in *International Conference on Informatics, Electronics and Vision* (IEEE, 2012). https://doi.org/10.1109/ICIEV.2012. 6317511

10. J.G. Pinto, H. Carneiro, B. Exposto, C. Couto, J.L. Afonso, Transformerless Series active power filter to compensate voltage disturbances, in *14th European Conference on Power Electronics and Applications, EPE 2011* (IEEE, 2011). pp. 1–6

11. A.J. Visser, J.H.R. Enslin, H. du T. Mouton, Transformerless series sag compensation with a cascaded multilevel inverter. Trans. Indus. Electron. IEEE **49**, 824−831 (2002). https://doi. org/10.1109/TIE.2002.801067

12. H. Usman, H. Hizam, M.A.M. Razdi, Simulation of single-phase shunt active power filter with fuzzy logic controller for power quality improvement, in *Conference on Clean Energy and Technology* (IEEE, 2013). https://doi.org/10.1109/CEAT.2013.6775655

13. S. Golestan, J.M. Guerrero, J.C. Vasquez, Three-phase PLLs: a review of recent advances. Trans. Power Electron. IEEE **32**(3), 1894–1907 (2016). https://doi.org/10.1109/TPEL.2016. 2565642

A New Electronically Controllable Active R-L Network Simulator

K. V. S. K. Dheeraj, A. R. Reddy, and Mayank Srivastava

1 Introduction

An inductor is a well-known passive element that has many applications in various areas of electrical engineering like power system, power electronics, instrumentation and measurement etc. But the use of inductors is limited in electronic circuits due to their large size, harmonic generation property, sensitivity to electromagnetic radiation etc.

Therefore, the implementation of inductor activity employing active circuit(s) has been a very attractive active research domain. In addition to an individual inductor, inductor based passive networks are also very useful. The applications of series/parallel R-L/L-C/LCR networks are well known. Hence active implementation of the inductor-based passive network is also a popular research field. In the last three decades, several research papers based on the active realization of grounded/floating, series/parallel, RL/LC/RLC networks employing numerous active building blocks (ABB) have been described by researchers.

The grounded parallel R-L network is one of the most useful passive impedance networks and finds many applications in filters and oscillators realization. Some active element based realizations for simulating the working of parallel R-L impedances have been reported in [1–10]. These presented active impedances exhibits more than one below given disadvantages:

K. V. S. K. Dheeraj (✉) · A. R. Reddy · M. Srivastava
National Institute of Technology, Jamshedpur, Jharkhand, India
e-mail: dheerajkotla6@gmail.com

A. R. Reddy
e-mail: arreddy.alluru@gmail.com

M. Srivastava
e-mail: mayank2780@gmail.com

© The Author(s), under exclusive license to Springer Nature Singapore Pte Ltd. 2021   479
J. Kumar and P. Jena (eds.), *Recent Advances in Power Electronics and Drives*, Lecture Notes in Electrical Engineering 707,
https://doi.org/10.1007/978-981-15-8586-9_42

(1) Circuits having more than one active building block (ABB).
(2) Realizations employing more than one resistance.
(3) Configurations having floating resistance(s).
(4) Configurations having floating capacitance(s).
(5) Circuits do not offer electronic control of both R_{eq} and L_{eq}.
(6) Circuits exhibiting independent control of L_{eq}.
(7) Circuits simulating parallel R-L behaviour with some matching constraints.

Therefore, the main aim of this research article is to develop an active simulator for ground parallel R-L impedance with the following advantages:

(1) Single active ABB based behaviour.
(2) Use of single resistance which is in a grounded state.
(3) A circuit using only a single grounded capacitance.
(4) Availability of non-interactive controllability of R_{eq} and L_{eq}.
(5) The facility of electronic tuning of both R_{eq} as well as L_{eq}.
(6) No matching requirement in resistance and VDCC transconductance.
(7) Unaffected behaviour with non-ideal VDCC model.

2   VDCC Circuit Idea

The VDCC is a popular active circuit idea proposed in [11]. The single block electrical representation of ideal VDCC has been illustrated in Fig. 1. It has six input/output terminals for applying and fetching current and voltages. The terminals P, N are voltage input terminals with very high impedance, terminal X is voltage output terminal with low impedance and the remaining three terminals Z, WP, WN are current output terminal with high impedance.

The relations among various ports of VDCC are described in Eq. 1.

Fig. 1 Single block electrical representation of ideal VDCC

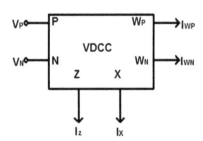

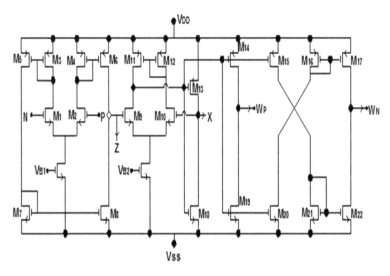

Fig. 2 MOS based realization of ideal VDCC

$$
\begin{bmatrix} I_N \\ I_P \\ I_Z \\ V_X \\ I_{W_P} \\ I_{W_N} \end{bmatrix} = \begin{bmatrix} 0 & 0 & 0 & 0 \\ 0 & 0 & 0 & 0 \\ g_m & -g_m & 0 & 0 \\ 0 & 0 & 1 & 0 \\ 0 & 0 & 0 & 1 \\ 0 & 0 & 0 & -1 \end{bmatrix} \begin{bmatrix} V_P \\ V_N \\ V_Z \\ I_X \end{bmatrix} \tag{1}
$$

The realization of ideal VDCC block using MOS transistors has been depicted in Fig. 2.

In the last one-decade various research papers describing the numerous analog circuits employing VDCC(s) have been reported in the open literature [12–18].

3   Proposed VDCC Based Grounded Parallel R-L Impedance Simulator

The proposed VDCC based grounded parallel R-L network simulator has been illustrated in Fig. 3. The presented configuration employs single VDCC, one grounded resistance and one grounded capacitance and found suitable for on-chip implementation.

On circuit analysis of configuration shown in Fig. 3 using Eq. 1, the input impedance of circuit can be calculated as:

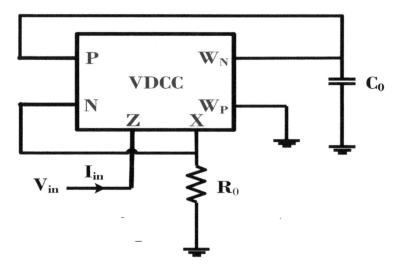

Fig. 3 Proposed configuration

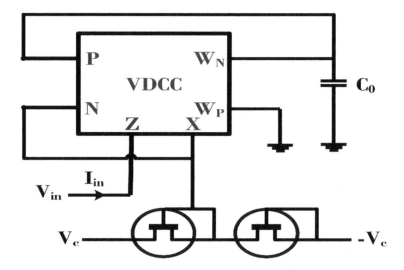

Fig. 4 Proposed configuration with independent electronic tunability of R_{eq} and L_{eq}

$$Z_{in} = \frac{V_{in}}{I_{in}} = \frac{R_0 s C_0}{g_m(1 + R_0 s C_0)} \tag{2}$$

Therefore, the realized admittance is

$$Y_{in} = \frac{I_{in}}{V_{in}} = \frac{g_m(1 + R_0 s C_0)}{R_0 s C_0} = \frac{g_m}{R_0 s C_0} + g_m = \frac{1}{s L_{eq}} + \frac{1}{R_{eq}} \tag{3}$$

From Eq. 3, it is clear that the presented circuit simulator the working of a parallel R-L network. The simulated equivalent inductance (L_{eq}) and resistance (R_{eq}) can be described as

$$L_{eq} = \frac{R_0 C_0}{g_m} \tag{4}$$

$$R_{eq} = \frac{1}{g_m} \tag{5}$$

From Eqs. 4 and 5 it can be noted that both the realized equivalent inductances as well as equivalent resistance exhibit the feature of electronic tuning through transconductance of VDCC (g_m). By slight variation in the proposed circuit, the independent electronic tuning of R_{eg} and L_{eq} can be achieved.

From Fig. 6 we can say that electronic tuning of L_{eq} can be achieved by biasing current of VDCC while the electronic control of R_{eq} can be obtained through control voltage (Vc) of MOSFET connected with X terminal.

4 Proposed Configuration with Non-ideal VDCC

The non-ideal mathematical model of VDCC can be defined by Eq. 6.

$$\begin{bmatrix} I_N \\ I_P \\ I_Z \\ V_X \\ I_{W_P} \\ I_{W_N} \end{bmatrix} = \begin{bmatrix} 0 & 0 & 0 & 0 \\ 0 & 0 & 0 & 0 \\ \alpha g_m & -\alpha g_m & 0 & 0 \\ 0 & 0 & 1 & 0 \\ 0 & 0 & 0 & \beta_P \\ 0 & 0 & 0 & -\beta_N \end{bmatrix} \begin{bmatrix} V_P \\ V_N \\ V_Z \\ I_X \end{bmatrix} \tag{6}$$

On mathematical analysis of presented design illustrated in Fig. 3 using the non-ideal model of VDCC the input impedance can be calculated as

$$Z_{in} = \frac{V_{in}}{I_{in}} = \frac{R_0 s C_0}{\alpha g_m (\beta_N + R_0 s C_0)} \tag{7}$$

On studying Eq. 7, it can be found that the proposed configuration realizes the working of a parallel R-L simulation circuit even using non-ideal VDCCs. In this case, the simulated resistance and inductance values can be evaluated as;

$$L_{eq} = \frac{\alpha R_0 C_0}{\beta_N g_m} \tag{8}$$

$$R_{eq} = \frac{1}{\alpha g_m} \tag{9}$$

5 High-Frequency Analysis Considering VDCC Terminal Parasitics

In order to find the high-frequency performance of the proposed design, the parasitic structure of VDCCs have been considered, the Fig. 7 shows the VDCC with its terminal parasitic impedances. Except for X terminal all the remaining terminals (P, N, Z, Wp, Wn) exhibits ground parasitic resistance and capacitance. The high-value resistances Rp, Rn, Pz, Rwp, Rwn are the parasitic resistances at P, N, Z, Wn and Wp terminals respectively. Similarly, Cp, Cn, Cwp and Cwn are the parasitic capacitances at the corresponding terminals. On using the parasitic model of VDCC, the circuit of the proposed configuration can be developed as shown in Fig. 5.

where R1 is the parallel combination of RP and RWP and C1 is a parallel combination of CP, C0 and CWP. The resistance R2 is Rz, C3 is CN and C2 is Cz.

The impedance of the circuit depicted in Fig. 5 can be calculated as:

$$Z_{in} = \frac{V_{in}}{I_{in}} = \frac{R_2 R_3 (1 + R_1 s C_0)}{R_3 (1 + R_1 s C_0)(1 + R_2 s C_2) + g_m R_2 [R_3 (1 + R_1 s C_0) + R_1 (1 + R_3 s C_3)]} \tag{10}$$

Hence the presented circuit does not simulate the behaviour of parallel R-L configuration under parasitic conditions.

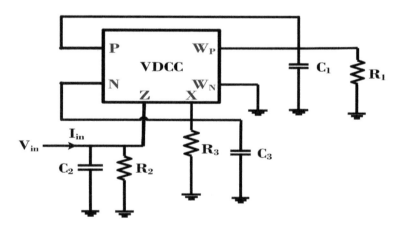

Fig. 5 Proposed configuration using a parasitic model of VDCC

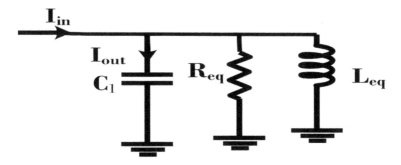

Fig. 6 Passive R-L-C current mode high pass filter

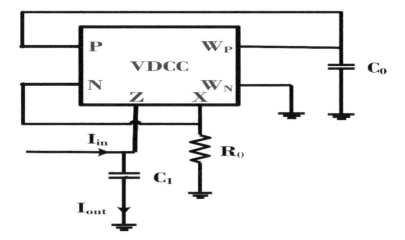

Fig. 7 Active structure of filtering circuit depicted in Fig. 6

6 Circuit Design Example of Presented Active R-L Circuit

To check the validity of the presented R-L simulation configuration a current mode high pass filter has been constructed by using the presented grounded R-L simulation configuration. The simple passive parallel R-L-C filter and its active realization using the presented configuration have been described in Fig. 6 and Fig. 7 respectively.

7 Simulation Results

In order to validate the working of the presented R-L simulation configuration shown in Fig. 3. The simulations in the PSPICE environment have been executed using CMOS VDCC. During the simulation the capacitance C_0 is taken as 0.1 nF. The biasing currents I_b and I_{b1} of CMOS VDCC are taken as 40 μA and 25 μA respectively. The PSPICE generated input impedance response with respect to frequency has been shown in Fig. 8. The corresponding phase response has been indicated in Fig. 9.

In order to prove the electronic tuning of the presented circuit, the simulations for different values of I_b have been executed and simulation results have been described in Fig. 10.

The CM high pass filter illustrated in Fig. 7 has been simulated for C0 = 0.1 nF, C1 = 0.1 nF and R0 = 1.5 kΩ. The simulation produced a high pass response is depicted in Fig. 11.

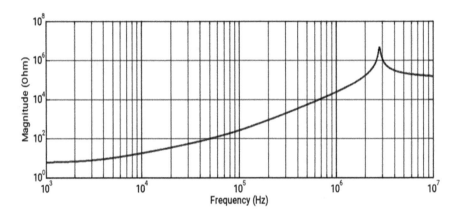

Fig. 8 Impedance response curve of the proposed circuit

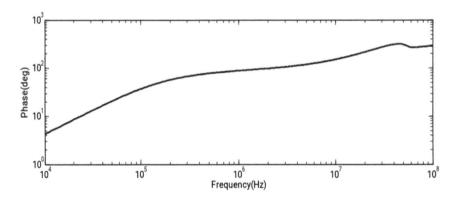

Fig. 9 Phase response curve for the impedance of the reported circuit

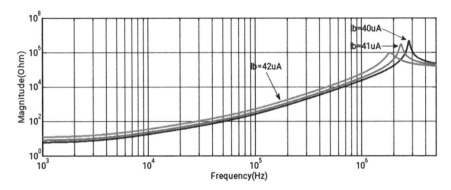

Fig. 10 Electronic tuning of the proposed circuit

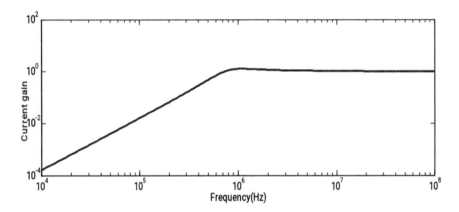

Fig. 11 High pass frequency response of the circuit shown in Fig. 7

8 Conclusion

A grounded parallel R-L network simulation circuit has been described. The circuit composed of single VDCC, one resistance and single capacitance. Both the resistance and capacitance are grounded hence the circuit is fit for on-chip realization. The simulated equivalent resistance and capacitance can be varied electronically through the bias current of VDCC. The working of the reported impedance simulator has been studied using non-ideal VDCC and high-frequency VDCC model. The behaviour of the constructed circuit configuration has been validated by designing a high pass current-mode filtering circuit. The simulations in the PSPICE environment have been executed to confirm the theoretical analysis.

References

1. A.M. Soliman, Grounded inductance simulation using the DVCCS/DVCVS. Proc. IEEE **66**(9), 334–336 (1979)
2. U. Cam, O. Cicekoglu, H. Kuntman, A. Kuntman, 2 novel two OTRA-based grounded immitance simulator topology. Analog Integr. Circuits Signal Process **39**(2), 169–175 (2004)
3. A.N. Paul, D. Patranabis, Active simulation of grounded inductors using a single current conveyor. IEEE Trans. Circuits Syst. **28**, 164–165 (1981)
4. R.L. Ford, F.E.G. Girling, Active filters and oscillators using simulated inductance. Electron. Lett. **2**(2), 481–482 (1966)
5. M. Incekaraoglu, U. Cam, Realization of series and parallel R-L and C-D impedances using single differential voltage current conveyor. Analog Integr. Circuits Signal Process. **43**, 101–104 (2006)
6. E. Yuce, Comment on "realization of series and parallel R-L and C-D impedances using single differential voltage current conveyor." Analog Integr. Circuits Signal Process. **49**, 91–92 (2006)
7. E. Yuce, S. Minaei, O. Cicekoglu, Limitations of the simulated inductors based on a single current conveyor. IEEE Trans. Circuits Syst. **53**(12), 2860–2867 (2006)
8. F. Kacar, H. Kuntman, CFOA-based lossless and lossy inductance simulators. Radioengineering **20**(3), 627–631 (2011)
9. B. Metin, Supplementary inductance simulator topologies employing single DXCCII. Radio-engineering **20**(3), 614–618 (2011)
10. P. Kumar, R. Senani, New grounded simulated inductance circuit using a single PFTFN. Analog Integr. Circuits Signal Process. **62**(1), 105–112 (2010)
11. D. Biolek, R. Senani, V. Biolkova, Z. Kolka, Active elements for analog signal processing; classification, review and new proposals. Radioengineering **17**, 15–32 (2008)
12. D. Prasad, D.R. Bhaskar, M. Srivastava, New single VDCC-based explicit current-mode SRCO employing all grounded passive components. Electron. J. **18**(2), 81–88 (2014)
13. M. Srivastava, D. Prasad, VDCC based dual-mode quadrature sinusoidal oscillator with current/voltage outputs at appropriate impedance level. Adv. Electr. Electron. Eng. (Czech Republic) **14**(2), 168–177 (2016)
14. M. Srivastava, New synthetic grounded FDNR with electronic controllability employing cascaded VDCCs and grounded passive elements. J. Telecommun. Electron. Comput. Eng. **9**(4), 97–102 (2017)
15. D. Prasad, M. Srivastava, J. Ahmad, New CM/VM 3rd-order quadrature oscillator using VDCCs, in *IEEE International Conference on Applied System Innovation*, Sapporo (Japan), May 2017, pp. 1328–1332
16. P. Gupta, M. Srivastava, A. Verma, A. Singh, Devyansi, A. Ali, A VDCC based grounded passive element simulator/scaling configuration with electronic control, in *Advances in Signal Processing and Communication*. Lecture Notes in Electrical Engineering, vol. 526 (Springer, 2019), pp. 429–441
17. M. Srivastava, P. Bhanja, S.F. Mir, A new configuration for simulating passive elements in floating state employing VDCCs and grounded passive elements, in *IEEE-International Conference on Power Electronics, Intelligent Control and Energy Systems (ICPEICES-2016)*, Delhi, India (2016), pp. 13–18
18. D. Prasad, M. Srivastava, A. Ahmad, A. Mukhopadhyay, B.B. Sharma, Novel VDCC based low-pass and high-pass ladder filters, in *IEEE-INDICON-2015*, New Delhi (2015), pp. 1–4

Improved Bridgeless Buck-Boost Converter Fed PMBLDC Motor Drive

Amitesh Prakash, Madhu Singh, Anumeha, and Niranjan Kumar

1 Introduction

A trapezoidal permanent magnet (PM) machine with non-salient poles and concentrated full-pitch windings in the stator gives performance closer to that of a DC motor and hence it is widely known as permanent magnet brushless dc motor (PMBLDM). These motors are gaining popularity in low as well as medium power applications such as automobiles, aircrafts, appliances, automation instruments, etc. [1]. BLDC motors with various power ratings are available in the market and hence have a wide range of applications [2]. It has many advantages such as high efficiency, low noise operation, fail-safe operation, better dynamic response, wide range of speed, low maintenance and improved torque-speed characteristics [3].

A PMBLDC motor is an electronically commutated synchronous motor that is usually energised by a dc current through an inverter. Permanent magnets are placed on the rotor and a three phase winding is present on its stator. As clear by the name, mechanical brushes and commutator arrangement for commutation are missing in PMBLDC motor and hence we employ electronic commutation using Hall Sensors and a three phase VSI so as to control direction of current on the stator unlike a

A. Prakash (✉)
Bhagalpur College of Engineering, Bhagalpur 813210, Bihar, India
e-mail: amitesh.nitjsr1@gmail.com

M. Singh · N. Kumar
National Institute of Technology Jamshedpur, Jamshedpur 831014, Jharkhand, India
e-mail: madhu_nitjsr@rediffmail.com

N. Kumar
e-mail: nkumar.ee@nitjsr.ac.in

Anumeha
Government Women's Polytechnic Jamshedpur, Jamshedpur 831014, Jharkhand, India
e-mail: amehanitjsr@gmail.com

© The Author(s), under exclusive license to Springer Nature Singapore Pte Ltd. 2021
J. Kumar and P. Jena (eds.), *Recent Advances in Power Electronics and Drives*, Lecture Notes in Electrical Engineering 707,
https://doi.org/10.1007/978-981-15-8586-9_43

conventional dc motor [4]. The absence of brushes and commutator assembly ensure that there are lesser wear and tear and sparking problems as compared to conventional DC machines [5]. The position sensing of the rotor is done using Hall-Effect Sensors that finally generates commutation logic for the VSI that contains MOSFET switches [6].

2 Performance of PMBLDM Drive in Absence of any Converter

The behaviour of the PMBLDM drive in the absence of any converter and quality of power on the input side of ac mains is observed by simulating the model in MATLAB/SIMULINK and the harmonic spectra of the input current is as shown below. A single phase AC supply along with a diode bridge rectifier is used to rectify the AC into DC. This DC power trough a capacitor is then fed to the inverter circuit which finally energises the PMBLDC motor [7]. The capacitor will undergo uncontrolled charging and hence the motor starts drawing current that is non-sinusoidal in nature from the AC supply. This simply means that the quality of power at the ac mains degrades which implies that the total harmonic distortion has increased significantly and hence a poor power factor which ultimately affects the behaviour of the PMBLDM drive [8]. Hence, Power Factor Correction (PFC) converters are used on the input AC side so as to enhance the performance of the drive. There are various PFC converter topologies [9], but here BL Buck-Boost converter has been discussed.

Simulation diagram and the harmonic spectra of input current for PMBLDM drive without using any converter is shown in Figs. 1 and 2. It is observed from the simulation results that in this case the THD is 65.92% which is quite high. The power

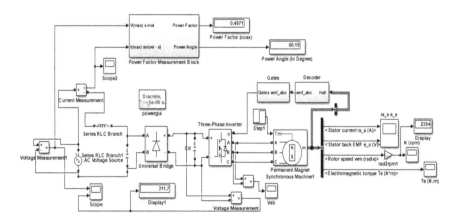

Fig. 1 Simulation diagram of PMBLDM drive without using PFC converters

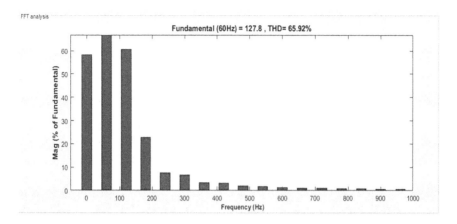

Fig. 2 Harmonic spectra of input current waveform without using any converter

factor is also very poor about 0.497. This shows that the PMBLDM drive has poor power quality being fed into in case we do not use any PFC converter. Hence is the need for PMBLDM drive fed with PFC converters.

3 Performance of PMBLDM Drive with PFC Converters

3.1 Behaviour of PMBLDM Drive with BL Buck-Boost Converter

PFC converters are employed at the input side of PMBLDM drives so as to reduce the THD and improve the power factor at the AC input mains [10]. There have been various PFC converter topologies discussed in literature like buck, boost, cuk, SEPIC, etc. [11]. Here, the rectifier circuit is replaced by a bridgeless Buck-Boost converter which leads to reduced conduction losses [12]. The control strategy used here is called the voltage follower approach. A dc link voltage (V^*_d) that is directly linked to the speed reference, is chosen as a reference voltage. This reference voltage V^*_d is compared with the actual voltage across the dc link (V_d) and the error signal generated (V_{er}) is then sent to a Proportional-Integral controller. The controlled output voltage (V_{cc}) from the PI controller is then processed through PWM generator which generates PWM signals for the switches of the converter. Any variation in the switching frequency will lead to a change in the duty ratio (D) and hence voltage across the dc link can be controlled effectively.

Mathematical modelling is a better explanation of the PFC control. A PI controller has been used in this case which converts the reference speed into an equivalent reference DC link voltage. Suppose for any nth time instant, V^*_d is the voltage reference across dc link and $V_d(n)$ the measured voltage across the dc link, then we find the error signal as [13]

$$V_{er}(n) = V_d^*(n) - V_d(n) \tag{1}$$

This is made to go through the PI controller and the controlled voltage at the nth instant is

$$V_{cc}(n) = V_{cc}(n-1) + k_p\{V_{er}(n) - V_{er}(n-1)\} + k_i V_{er}(n) \tag{2}$$

Here the integral and proportional gains of the PI controller are given by k_i and k_p.

This V_{cc} controlled output voltage is then sent to the PWM generator so as to generate PWM signals which are required to be sent to the MOSFETs of the PFC converter:

$$\text{For } V_s > 0 \begin{cases} \text{if } m_d < V_{cc} \text{ then } SW_1 = \text{'ON'} \\ \text{if } m_d \geq V_{cc} \text{ then } SW_1 = \text{'OFF'} \end{cases}$$

$$\text{For } V_s < 0 \begin{cases} \text{if } m_d < V_{cc} \text{ then } SW_2 = \text{'ON'} \\ \text{if } m_d \geq V_{cc} \text{ then } SW_2 = \text{'OFF'} \end{cases} \tag{3}$$

This way switch of the converter is made to work efficiently.

3.2 Design of PMBLCDM Drive for BL Buck-Boost Converter

One of the most important factors affecting the cost of the components and the rating of the device is the mode of operation of the converter that is employed in this case. Discontinuous Inductor Conduction Mode (DCM) is preferred over Continuous Inductor Conduction Mode (CCM) as it needs only one sensor in order to sense the DC link voltage. On the other hand, CCM requires three sensors that is never appreciable in low power applications owing to its high cost [14]. The discontinuous current mode (DCM) has an advantage that it provides an innate power factor correction at the ac input mains [15]. It is done in a manner that the inductor currents L_{i1} and L_{i2} happen to be discontinuous during a switching time period.

The purpose of designing this BL Buck-Boost converter fed PMBLDM drive is to ensure the proper control of speed of motor and enhancement in the quality of power at the ac input mains [16]

The dc link voltage in this case is given as

$$V_d = \frac{V_{in}D}{1-D} \tag{4}$$

The Buck Boost converter consists of input inductors L_{i1} and L_{i2}. The critical value of inductors is given by the equation

$$L_{ic} = R\frac{(1 - D^2)}{2f_s}$$ (5)

where, R is equivalent load resistance, D is the duty cycle and f_s is the frequency of switching.

For the converter to work in the DCM mode, L_{i1} and L_{i2} values should be very much smaller than the critical value.

A suitable input side low pass filter is made in order to get a ripple free input to the DBR by using R–C filter.

Similarly a DC link capacitor is modelled as per the equation given by

$$C_d = \frac{I_{dc}}{2\omega\Delta V_{dc}}$$ (6)

where, I_{dc} is the DC link current and ΔV_{dc} is maximum allowed ripple in the voltage across DC link (=3%).

The design parameter values obtained from the above equations are $L_{i1} = L_{i2} = 40\,\mu H$, $C_d = 2200\,\mu F$, $R_i = 1\,\Omega$ and Z_i consists of series R–C with $R = 1\,\Omega$ and $C = 330\,nF$.

3.3 Simulation Diagram and Results of Bridgeless Buck-Boost Converter Fed PMBLDM Drive

Figure 3 demonstrates complete simulation diagram of the PMBLDM drive fed by the Bridgeless Buck-Boost converter. A single phase supply voltage of amplitude 312 V is used along with bridgeless configuration of the Buck-Boost converter which simply means that there are two Buck-Boost converters, one for each half cycle. There is a DC link capacitor that is instrumental in giving supply to input side of the three phase inverter circuit. The outputs of the subsystem shown in Fig. 3 are given to the two MOSFET switches as the gate signals.

Figures 4, 5 and 6 shows the different waveforms that are the results obtained when the voltage follower approach is obtained to control speed of the PMBLDM drive discussed above.

Figure 4 shows waveform of input current at the AC mains side of the converter and its harmonic spectra. From this, it is evident that the current is nearly sinusoidal with THD as 1.73% which is very less whereas in contrast, the PMBLDM drive without using a PFC converter has very high THD of around 65.92%. Figure 5 shows the dc link voltage which is slowly increases and reaches its steady state with a value lower than the reference value. Figure 6 shows the stator back emf and stator phase

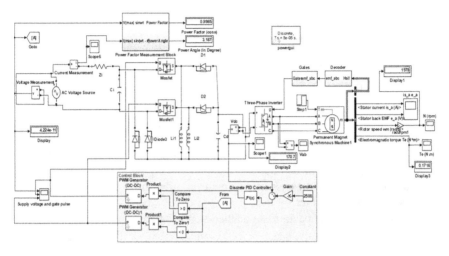

Fig. 3 Simulation diagram for Bridgeless Buck-Boost Converter fed BLDM drive

current waveform whereas Fig. 7 shows the simulation waveform for the rotor speed and electromagnetic torque.

It can be observed from these results that there is a direct relation between the motor speed and the voltage across the dc link as they are directly proportional. It is to be noted that the quality of power at the ac mains side is improved when BL Buck-Boost converter is employed in the PMBLDM Drive in contrast to that without any converter. This implies that the behavioural characteristics of the drive has got better by using PFC converter which simply means that the operating characteristics of the PMBLDC Drive has smoothened compared to its performance without using any converter. The value of THD can be made lesser than 1.73% by using some other advanced techniques for tuning of the PI controller which will be shown in the future works.

4   Results and Discussions

From the above simulation diagrams it is evident that speed of PMBLDM drive is dependent on supply voltage magnitude. The alteration in speed with the corresponding change in the supply voltage magnitude is as in Table 1.

It can be easily inferred from above Table 2 that speed of the PMBLDM drive is primarily dependent upon the voltage magnitude across the DC link.

Table 3 illustrates the comparative analysis of the result obtained without using any converter to that with using PFC BL Buck-Boost converter.

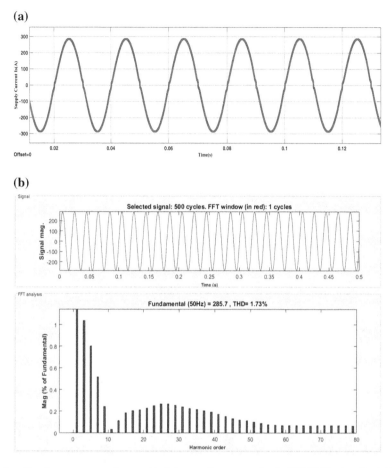

Fig. 4 **a** Supply current waveform and its **b** harmonic contents for the PFC Bridgeless Buck-Boost Converter fed PMBLDM drive

5 Conclusion and Future Scope

From the results obtained, finally it is deduced that PFC BL Buck-boost converter leads to a reduced supply current harmonic content of around 1.73% and a power factor near unity thereby providing a better power quality whereas in the absence of any converter, the THD is 65.92% and the power factor is very poor.

In future, further work can be done to include other converters in the comparative study and a better power quality can be achieved by proper tuning of the PI controller or by improving the filter design as here we have achieved a better power quality by using R–C filter instead of using L-C filter as discussed in various literatures. One more improvement can be done by designing a better Low-Pass filter as in this case the R–C filter will lead to more losses in the circuit.

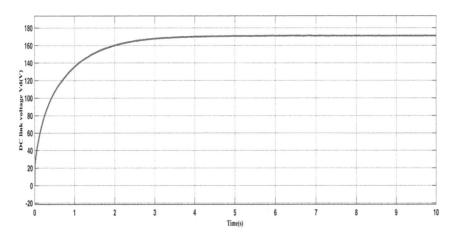

Fig. 5 DC link voltage waveform

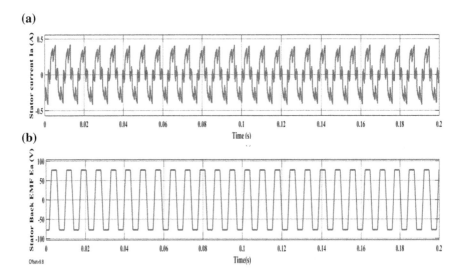

Fig. 6 **a** Stator phase current I_a **b** Emf WAVEFORM for BL Buck-Boost converter fed PMBLDM drive

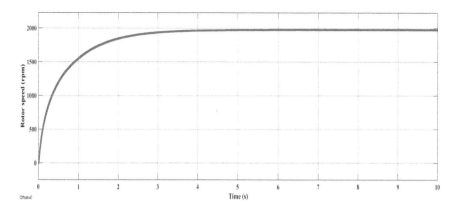

Fig. 7 Rotor speed waveform for the PMBLDM drive

Table 1 Performance of the drive with variation in the input voltage

Supply voltage (V_m)	Input current THD (%)	Power factor	Speed (rpm)
230	1.73	0.9985	1128
259	1.73	0.9984	1270
288	1.73	0.9985	1412
316	1.73	0.9985	1554
345	1.73	0.9985	1696

Table 2 Variation in speed corresponding to variation in V_d

Supply voltage (V_{sm})	DC link voltage (V_{dc})	Speed (rpm)
230	94.98	1128
259	107.4	1270
288	119.8	1412
316	132.4	1554
345	145	1696

Table 3 Comparison of Bridgeless Buck-Boost fed PMBLDM Drive to that without any PFC converter

Measurements	BL Buck-Boost converter	Without using any converter
THD (%)	1.73	65.92
Power factor	0.9985	0.497
Stator phase current, I_{sa} (A)	0.365	0.3279

References

1. P. Yedamale, *Microchip AN885, Brushless DC (BLDC) Motor Fundamentals* (Microchip Technology Inc., 2003), pp. 1–20
2. C.V. Pavithra, C. Vivekanandan, L. Amla Jenifer, Review of different types of converters for BLDC motors, in *Proceedings of International Conference on Energy, Environment and Engineering* (2016)
3. B. Singh, Recent advances in permanent magnet brushless DC motors. Sadhana J. **22**(6), 837–853 (1997)
4. R. Pireethi, R. Balamurugan, Power factor correction in BLDC motor drives using DC-DC converters, in *2016 International Conference on Computer Communication and Informatics (ICCCI)*, Coimbatore (2016), pp. 1–5
5. P. Pillay, R. Krishnan, Modeling, simulation, and analysis of permanent-magnet motor drives, Part II: The brushless DC motor drive. IEEE Trans. Ind. Appl. **25**, 274–279 (1989)
6. S. Jagadeeswari, N. Hemalatha, J. Baskaran, PFC Bridgeless buck-boost converter-fed BLDC motor drive, in *2015 International Conference on Computation of Power, Energy, Information and Communication (ICCPEIC)*, Chennai (2015), pp. 0171–0176
7. R. Chandran, A. Rajan, Mitigation of supply current harmonics in PMBLDM drive using Cuk converter, in *2014 International Conference on Computation of Power, Energy, Information and Communication (ICCPEIC)*, Chennai (2014), pp. 181–188
8. B. Das, S. Chakraborty, A. Chakraborti, P. Kasari, Performance analysis of BLDC motor using basic switching converters. Int. J. Innov. Technol. Explor. Eng. (IJITEE) (2012)
9. B. Singh, B.N. Singh, A. Chandra, K. Al-Haddad, A. Pandey, D.P. Kothari, A review of single-phase improved power quality AC-DC converters. IEEE Trans. Industr. Electron. **50**(5), 962–981 (2003)
10. A. Ramya, M. Balaji, C. Bharatiraja, Power quality improvement in BLDC motor drive using Bridgeless Modified Cuk converter, in *2018 IEEE International Conference on Power Electronics, Drives and Energy Systems (PEDES)*, Chennai (2018), pp. 1–6
11. S. Singh, B. Singh, Power quality improved PMBLDM drive for adjustable speed application with reduced sensor buck-boost PFC converter, in *2011 Fourth International Conference on Emerging Trends in Engineering & Technology*, Port Louis (2011), pp. 180–184
12. M. Kavitha, V. Sivachidambaranathan, Power factor correction in fuzzy based brushless DC motor fed by bridgeless buck boost converter, in *2017 International Conference on Computation of Power, Energy Information and Communication (ICCPEIC)*, Melmaruvathur (2017), pp. 549–553
13. S. Singh, B. Singh, A voltage-controlled PFC Cuk converter based PMBLDM drive for air-conditioners. IEEE Trans. Ind. Appl. **48**(2), 832–838 (2012)
14. K.U. Vinayaka, S. Sanjay, Adaptable speed control of bridgeless PFC buck-boost converter VSI fed BLDC motor drive, in *2016 IEEE 1st International Conference on Power Electronics, Intelligent Control and Energy Systems (ICPEICES)*, Delhi (2016), pp. 1–5
15. V. Bist, B. Singh, An adjustable-speed PFC bridgeless buck-boost converter fed BLDC motor drive. IEEE Trans. Industr. Electron. **61**(6), 2665–2677 (2014)
16. D.S.L. Simonetti, J. Sebastian, F.S. dos Reis, J. Uceda, Design criteria for SEPIC and Cuk converters as power factor pre-regulators in discontinuous conduction mode, in *Proceedings of the IEEE International Electronics and Motion Control Conference*, vol. 1 (1992), pp. 283–288

Design and Analysis of Series Resonant Inverter-Based Induction Heating Equipment Employing Power Factor Correction for Harmonic Attenuation

Rahul Raman, Mrigakshi Das, Priya Sarmah, Subrata Kumar Dutta, Amarjit Saikia, and Pradip Kumar Sadhu

1 Introduction

Modern world has a tremendous strong relationship with modern technology which is evolving each and every day with advance technologies paving our lifestyle. One such technology is induction heating which is widely used for household and industrial purposes. Induction heating equipments have several advantages over the conventional cooking methods and henceforth, dominated the market of domestic cooking [1–3]. It is highly efficient, safe and clean process of energy production. Induction heating offers a controllable and localised method of heat, generated by inducing an alternating magnetic field. However, problem with such kind of heating system from electrical point of view is the creation of considerable harmonic distortion. These harmonics have the tendency to flow back to the supply side and deteriorate the power quality of the system. With the advancement in the field of power semiconductor devices, passive and active circuit components, several theories have

R. Raman · M. Das (✉) · P. Sarmah · S. K. Dutta · A. Saikia
Department of Electrical Engineering, Jorhat Engineering College, Jorhat, India
e-mail: mailtomrigakshi@gmail.com

R. Raman
e-mail: rahulraman20@ee.ism.ac.in

P. Sarmah
e-mail: priya30sarada@gmail.com

S. K. Dutta
e-mail: sd6622484@gmail.com

A. Saikia
e-mail: saikiaamarjit2016@gmail.com

R. Raman · P. K. Sadhu
Department of Electrical Engineering, Indian Institute of Technology (ISM), Dhanbad, India
e-mail: pradip@iitism.ac.in

© The Author(s), under exclusive license to Springer Nature Singapore Pte Ltd. 2021
J. Kumar and P. Jena (eds.), *Recent Advances in Power Electronics
and Drives*, Lecture Notes in Electrical Engineering 707,
https://doi.org/10.1007/978-981-15-8586-9_44

been approached to improve the power quality, reduction of energy consumption and downsizing household electric appliances [4]. The key features of an efficient induction heating system are fast heating cycles, low noise interference and precise heating patterns. These characteristics are important in order to design high performance and highly efficient Induction Heating Equipment (IHE). Today, resonant inverters are widely used in IHE because of its qualities like lower losses by soft switching techniques and high-frequency operation. They can give rise to sinusoidal output waveform at high frequency [5, 6]. Different types of filter and boost circuits have been designed to enhance the induction circuit from Radio Frequency Interference (RFI) and Electromagnetic Interference (EMI), which may deteriorate the electrical circuit performance. In this paper, a boost PFC circuit is designed which are relatively simple and low-cost circuits. PFC circuits can reduce the Total Harmonic Distortion (THD) well below acceptable limit, increase the efficiency of power systems and reduce customers' utility bill [7].

1.1 RFI and EMI Sources

As the IHE is operated at very high frequency the voltage and current waveforms change very rapidly which increases the problem of electromagnetic interference and disturbs the operation of the system. The electromagnetic interference is propagated by two different ways known as conducted and radiated interferences [8]. Most electronic devices like the Induction Heating Equipment (IHE) operate in DC voltage and thus there is a need to convert the AC voltage from the grid to DC voltage. Linear power supplies and converter circuits introduce harmonics which flow back to the supply side. RFI problems are inherent in converter operation due to rapid rise and fall of current and voltage waveform during switching operation [9]. The switching transistor, rectifier, output diodes, protective diodes for the transistor as well as the control unit, all contribute towards switching noise [8]. The noise level radiated or conducted to the main input varies depending upon the design of the converter used. In some cases, it may vary drastically. These harmonics have a considerable effect on the power factor of the system. Because of the stringent EMI regulation, noises generated from high-frequency converters need to be analysed and suppressed. Use of passive or active filters is one way to eliminate these harmonics. But these filters have poor dynamic performance and low power factor. Henceforth there is a need to use power supplies with inbuilt Power Factor Correction Circuitry (PFC). The boost power factor correction converter is such a circuit which increases the power factor and reduces the harmonic currents to a considerable amount and hence improves the power quality [10].

1.2 Power Flow in the Proposed IHE

The input AC voltage is converted to pulsating DC by a rectifier circuit. The DC supply voltage derived from the rectifier output is controlled with a feedback loop control technique comprising the boost PFC circuit. The DC output of the rectifier is pulsating in nature so that it can be effectively operated in Zero Voltage Switching condition (ZVS) to obtain minimum switching losses. The controlled pulsating DC output is fed to the DC to AC inverter through a capacitor C_0 connected in parallel. This capacitor will charge when the boost PFC diode conducts and discharge when the switch conducts. The output of the DC to AC inverter is fed to the induction heating load at high frequency.

In the proposed paper, the work coil or the induction coil comprises a resistor and an inductor to generate the alternating magnetic field. A capacitor is externally connected in series with the work coil. A high-frequency AC source is fed to this coil. The switching frequency of the coil is chosen such that it is as close as possible to the resonant frequency of the RLC circuit, such that the current in the work coil increases and heating effect is enhanced. When the induction coil is subjected to an alternating magnetic field, power loss occurs due to some physical phenomena like hysteresis and eddy current. Heat is generated as eddy current flows against the resistivity of the workpiece, by the principle of Joule heating and hysteresis in the magnetic parts. Hysteresis loss is directly proportional to the switching frequency while eddy current loss is directly proportional to the square of the frequency. Thus, operation at higher frequency results in higher heating effect. Moreover, high-frequency application in induction heating equipment give rise to another phenomenon called skin effect. Due to skin effect the alternating current is forced to flow in a thin layer towards the outer surface of the workpiece. This results in efficient heating [1–3].

1.3 Use of Boost PFC Converter for Elimination of Harmonics

The Boost PFC converter is a digital converter which is used to enhance the power factor and thus reduce the harmonic distortion of power source [11]. An induction heating equipment requires an inductive load for the purpose of magnetisation. This gives rise to lagging reactive power at the load. The easiest way to eliminate this reactive power is to add a component in the circuit which can compensate this reactive power by using equal and opposite amount of the reactive power. But this method of power factor correction is not very accurate as it can be used only for linear loads and is neither cost effective. Therefore, for nonlinear loads, power factor cannot be corrected by such simple methods. If the input current can be controlled so that it tracks the input voltage waveform even if its magnitude is different, the power factor can be increased [12] (Fig. 1).

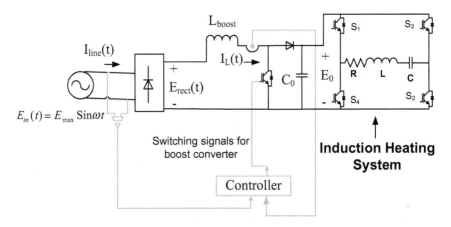

Fig. 1 Proposed structure of IH system with boost PFC

Figure 3a represents a high-frequency induction heating system with Boost PFC converter at the input side. A voltage sensor is used to calculate and monitor the amount of input voltage. Inductor current I_d is forced to follow a reference current which is in phase with the rectified voltage waveform. The PI controller calculates the error signal by taking the difference between the output and the input current. The boost PFC circuit is designed to pivot the MOSFET on and off with the help of the on–off controller S_1 at different intervals so that the input current and the input voltage V_{ac} are in phase and become sinusoidal in nature.

1.4 Mathematical Model of the Boost PFC

Power factor in the line may be expressed as follows:

$$pf = \frac{E_{1rms}(I_{line})_{1rms}\,Cos\varphi}{E_{rms}(I_{line})_{rms}} \tag{1}$$

We consider that the grid voltage is sinosoidal and will comprise only the fundamental voltage quantity.

Therefore, E1, rms = Erms

$$pf = \frac{(I_{line})_{1rms}\,Cos\varphi}{(I_{line})_{rms}} = R_H Cos\varphi \tag{2}$$

where R_H gives the harmonic content with respect to the fundamental and $Cos\varphi$ is the displacement factor.

$$\text{THD} = \frac{\sqrt{\sum\limits_{n=2}^{\infty} (I_{line})^2_{n,rms}}}{(I_{line})_{1,rms}} \tag{3}$$

$$R_H = \frac{(I_{line})_{1rms}}{(I_{line})_{rms}} = \frac{(I_{line})_{1rms}}{\sqrt{\sum\limits_{n=1}^{\infty} (I_{line})^2_{n,rms}}} \tag{4}$$

$$R_H = \frac{1}{\sqrt{1 + \frac{\sum\limits_{n=2}^{\infty} (I_{line})^2_{n,rms}}{(I_{line})^2_{1,rms}}}} = \frac{1}{\sqrt{1 + \text{THD}^2}} \tag{5}$$

$$pf = R_H \text{Cos}\varphi = \frac{\text{Cos}\varphi}{\sqrt{1 + \text{THD}^2}} \tag{6}$$

When $\text{Cos}\varphi = 1$, then pf can be expressed as follows:

$$pf = R_H = \frac{1}{\sqrt{1 + \text{THD}^2}} \tag{7}$$

The input voltage which is assumed to be purely sinusoidal and distortionless is defined as

$$E_{in}(t) = E_{\max} \text{Sin}\omega t \tag{8}$$

where $E_{\max}$ is the amplitude of input voltage and ω is the angular frequency.

The voltage after rectification can be expressed as

$$E_{rect}(t) = E_{\max} |\text{Sin}\omega t| \tag{9}$$

In discontinuous conduction mode, the peak value of inductor current $(I_L)_{peak}$ can be expressed as follows:

$$(I_L)_{peak} = \frac{E_{rect}(t)}{L_{boost}} \times D_{ON} \times T_{boost} \tag{10}$$

where D_{ON} and T_{boost} are the duty cycle and switching cycle of the switch in the boost converter while L_{boost} is the inductor in the boost converter.

Applying, the volt-second balance in the inductor:

$$E_{rect}(t) \times D_{ON} \times T_{boost} = \{E_0 - E_{rect}(t)\} \times D_{OFF} \times T_{boost} \tag{11}$$

E_0 is the output voltage of the boost converter.

$$D_{\text{OFF}} = \frac{E_{\text{rect}}(t) \times D_{\text{ON}}}{\{E_0 - E_{\text{rect}}(t)\}} \tag{12}$$

$$D_{OFF} = \frac{E_{\max} \text{Sin}\omega t \times D_{ON}}{\{E_0 - E_{\max} \text{Sin}\omega t\}} \tag{13}$$

Considering one switching cycle, the average current in the inductor may be expressed as follows:

$$I_L(t)_{\text{avg}} = \frac{1}{2} \times (I_L)_{\text{peak}} \frac{E_{\text{rect}}(t)}{L_{\text{boost}}} \times \{D_{\text{ON}} + D_{\text{OFF}}\} \tag{14}$$

$$I_L(t)_{\text{avg}} = \frac{1}{2} \times \frac{E_{\text{rect}}(t)}{L_{\text{boost}}} \times D_{ON} \times T_{\text{boost}} \times \frac{E_{\text{rect}}(t)}{L_{\text{boost}}} \times \{D_{\text{ON}} + D_{\text{OFF}}\} \tag{15}$$

$$I_L(t)_{\text{avg}} = \frac{E_{\max} |\text{Sin}\omega t| \times \{D_{ON}\}^2}{2 \times L_{\text{boost}} \times f_{\text{boost}} \times \left\{1 - \frac{E_{\max}}{E_0} |\text{Sin}\omega t|\right\}} \tag{16}$$

Thus, the line current can be expressed as follows:

$$I_{\text{line}}(t) = \frac{E_{\max} \text{Sin}\omega t \times \{D_{ON}\}^2}{2 \times L_{\text{boost}} \times f_{\text{boost}} \times \left\{1 - \frac{E_{\max}}{E_0} |\text{Sin}\omega t|\right\}} \tag{17}$$

It has been observed that the shape of peak current in the inductor is sinusoidal while that of the average current in the inductor is non-sinusoidal. The latter is distorted due to the presence of harmonics.

To get a simplified value in the expression of inductor current, its average value may be normalised on the following base: $\left[E_{\max} \times \{D_{ON}\}^2/2 \times L_{\text{boost}} \times f_{\text{boost}}\right] \times \left[1/\{1 - E_{\max}/E_0\}\right]$.

Therefore, the line current in the aforesaid expression may be expressed as follows:

$$I_{\text{line}}^*(t) = \left(1 - \frac{E_{\max}}{E_0}\right) \frac{\text{Sin}\omega t}{\left\{1 - \frac{E_{\max}}{E_0} |\text{Sin}\omega t|\right\}} \tag{18}$$

The aforementioned expression explains that the shape of average current in the inductor only depends upon the ratio of $E_{\max}$ and E_0. The decrease in this ratio decreases the distortion in the inductor current and makes it more and more sinusoidal. In each and every time period of line current, the duty cycle remains constant which in turn forces the peak value of current in the inductor to remain sinusoidal. Moreover, it also ensures sinusoidal behaviour in the average value of inductor current during the rising period. However, there is a problem associated with the inductor current during the time it is falling. The former depends upon $(E_0 - E_{\text{rect}})$ which in turn depends upon E_{rect} and that is variable in nature. This results in the non-sinusoidal nature of the current in the inductor during the time it is falling. This in turn adversely affects the desirable sinusoidal nature in the average value of inductor current.

To overcome the aforesaid problem, the falling time of current in the inductor is greatly reduced and made quite small by reducing the ratio of E_{max} and E_0. By using this strategy, the average value of current in the inductor achieves a shape which is very close to sinusoidal and almost free from high-frequency harmonic components.

The average power fed to the boost converter can be expressed as follows:

$$(p_{boost})_{in} = \frac{1}{T_{sys}/2} \int_0^{T_{sys}/2} E_{in}(t) I_{line}(t) dt \tag{19}$$

$$I_L(t)_{avg} = \frac{E_{max}^2 \times \{D_{ON}\}^2}{2 \times L_{boost} \times f_{boost}} \times \frac{1}{\pi} \int_0^{\pi} \frac{\operatorname{Sin}^2 \omega t}{1 - \frac{E_{max}}{E_0} |\operatorname{Sin} \omega t|} d\omega t \tag{20}$$

where T_{sys} refers to one line cycle.

If the boost converter under consideration is assumed to be ideal, then

$$(p_{boost})_{in} = (p_{boost})_{out} \tag{21}$$

Then, the duty cycle can be expressed as

$$D_{ON} = \frac{1}{E_{max}} \sqrt{\frac{2\pi \times L_{boost} \times f_{boost} \times (p_{boost})_{out}}{\frac{1}{\pi} \int_0^{\pi} \frac{\operatorname{Sin}^2 \omega t}{1 - \frac{E_{max}}{E_0}|\operatorname{Sin} \omega t|} d\omega t}} \tag{22}$$

Thus, the input power factor can be expressed as follows:

$$pf = \frac{(P_{boost})_{in}}{E_{rms}(I_{line})_{rms}} = \frac{(p_{boost})_{in}}{\frac{E_{max}}{\sqrt{2}} \sqrt{\frac{1}{\pi} \int_0^{\pi} [I_{line}(t)]^2 d\omega t}} \tag{23}$$

$$pf = \frac{\sqrt{\frac{2}{\pi} \int_0^{\pi} \frac{\operatorname{Sin}^2 \omega t}{1 - \frac{E_{max}}{E_0}|\operatorname{Sin} \omega t|} d\omega t}}{\sqrt{\left(\int_0^{\pi} \frac{\operatorname{Sin} \omega t}{1 - \frac{E_{max}}{E_0}|\operatorname{Sin} \omega t|}\right)^2 d\omega t}} \tag{24}$$

1.5 Simulation Diagram and Result

In the present paper PSIM software has been used to carry out simulation of the IHE. PSIM is a very fast and efficient tool designed to simulate any electronic circuit. Firstly, the simulation has been done without the Boost PFC circuit. Then simulation is performed when the Boost PFC circuit is installed between the input and the bridge converter. The simulation results thus obtained for the above two cases are studied.

Figure 2a is the simulation circuit diagram of an induction heating equipment without Boost PFC. A single-phase sinusoidal AC source, rated 240 V, 50 Hz is connected to the primary of a single-phase transformer. Secondary side of the transformer is connected to a single-phase uncontrolled bridge rectifier. The output of the bridge rectifier is connected to a high-frequency full bridge inverter. The bridge inverter is designed to operate at resonant condition such that the impedance offered by the load is minimum. A combination of inductor, resistor, and capacitor, comprising the load is connected in series and is used as the heating coil. An ammeter is connected to the secondary terminal of the transformer which measures the current drawn by the IH equipment. The input current measured by the ammeter is used to study the harmonics which are injected into the circuit because of high-frequency switching. Figure 2c gives the FFT analysis of the input current. It shows the various harmonic components that are present along with the fundamental component.

In Fig. 3a, a Boost PFC circuit has been introduced on the input side. Waveform of the input current after the installation of PFC is shown in Fig. 3b. The FFT analysis of it is shown in Fig. 3c, the measure option is then used to measure the magnitude of the frequency in PSIM.

2 THD Calculation from Simulation Results

1. Before Boost PFC installation:

THD calculation is done by performing FFT analysis of the input current waveform, fed to IHE. Upon analysis it is observed that along with the fundamental component 12 predominant harmonic components are present.

$$\text{THD} = \frac{\sqrt{\sum_{n=2}^{\infty} (I_{\text{line}})_{n,\text{rms}}^2}}{(I_{\text{line}})_{1,\text{rms}}}$$

$$\text{THD} = \frac{\sqrt{0.0513^2 + 0.0917^2 + 0.695^2 + 0.323^2 + 0.426^2 + 0.516^2 + 0.268^2 + 0.618^2 + 0.332^2 + 0.167^2 + 1.002^2 + 0.121^2}}{3.432}$$
$$\times 100 = 47.48\%$$

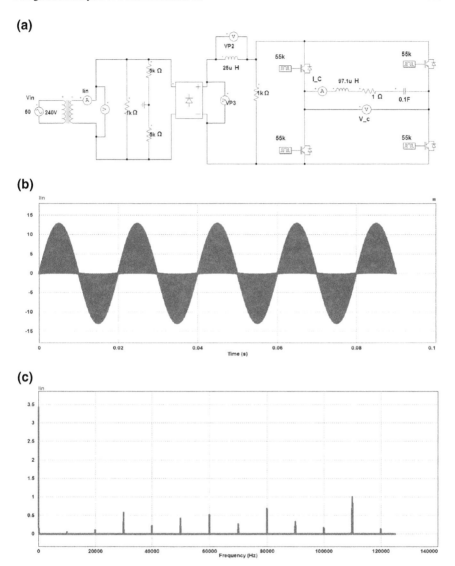

Fig. 2 a High-frequency IHE without Boost PFC configuration. **b** Input current waveform of high-frequency IHE without Boost PFC configuration. **c** FFT analysis of input current waveform of IHE without Boost PFC configuration

2. After the installation of Boost PFC:

After adding the Boost PFC circuit in the supply side of the IHE, the FFT analysis is again performed. The analysis shows the presence of four predominant harmonics besides the fundamental quantity (Table 1).

(a)

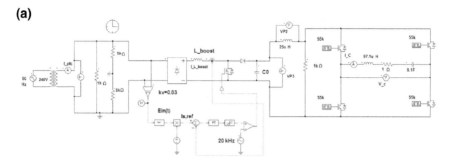

(b)

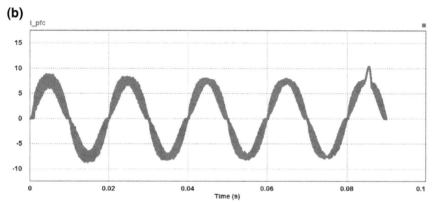

(c)

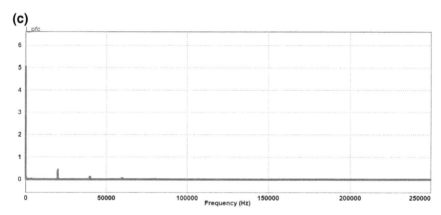

Fig. 3 a High frequency IHE with Boost PFC configuration. **b** Input current waveform of high-frequency IHE with Boost PFC configuration. **c** FFT analysis of input current waveform of IHE with Boost PFC configuration

Table 1 Comparison of THD with and without the application of boost PFC

	Without boost PFC	With boost PFC
No. of predominant harmonic components	12	4
THD	47.48%	3.445%

$$THD = \frac{\sqrt{\sum\limits_{n=2}^{\infty} (I_{line})^2_{n,rms}}}{(I_{line})_{1,rms}} = \frac{\sqrt{0.176^2 + 0.017^2 + 0.01^2 + 0.0001^2}}{5.12} \times 100 = 3.445\%$$

3 Conclusion

The present work illustrates the scheme that aims the design and analysis of high-frequency induction heating equipment using boost PFC converter with reduced total harmonic distortion. The boost PFC tries to reduce harmonics in the input side by turning on and off the MOSFET with varying duty cycle. It can efficiently reduce the THD, taking the power factor to nearly unity.

The THD in the input current waveform of the IHE with the installation of the boost PFC was found to be 3.445% and was free from distortions. However, the IHE without boost PFC circuit has THD 47.48% which gives very poor power factor quality. Thus, the boost PFC converter improves the power factor and efficiency of an IHE.

References

1. S. Chudjuarjeen, A. Sangswan, C. Koompai, An improved LLC resonant inverter for induction-heating applications with asymmetrical control. IEEE Trans. Ind. Electron. **58**(7), 2915–2925 (2011)
2. N.A. Ahmed, High-frequency soft-switching AC conversion circuit with dual-mode PWM/PDM control strategy for high-power IH applications. IEEE Trans. Ind. Electron. **58**(4), 1440–1448 (2011)
3. I. Millán, J.M. Burdío, J. Acero, O. Lucía, S. Llorente, Series resonant inverter with selective harmonic operation applied to all-metal domestic induction heating. IET Power Electron. **4**(5), 587–592 (2011)
4. M.A. Inayathullah, R. Anita, Single phase high frequency AC converter for induction heating application. Int. J. Eng. Sci. Technol. **2**(12) (2010)
5. K. Sayed, M. Nakaoka, Performance of induction heating power supply using dual control mode pulse width modulation, pulse density modulation, high frequency inverter. Electr. Power Compon. Syst. **13**(2) (2015)
6. M.H. Rashid, *Power Electronics: Circuits, Devices and Applications*, 3rd edn. (New Jersey, 2004)
7. E.E. EL-Kholy, S.A. Mahmoud, S.S. Shokralla, H.Z. Azazi, Review of passive and active circuits for power factor correction in single phase, low power Ac-Dc converter, in *14th International Middle East Power Systems Conference (MEPCON'10)* (Cairo University, Egypt, 2010), pp. 216–224
8. P. Pal, P.K. Sadhu, N. Pal, S. Sanyal, An exclusive design of EMIRFI suppressor for modified half bridge inverter fitted induction heating equipment. Int. J. Mechatron., Electr. Comput. Technol. (IJMEC) **5**(15), 2084–2100 (2015)
9. R. Raman, P.R. Sadhu, A. Kumar, K. Sit, Design and analysis of RFI and EMI suppressor for high frequency induction heater using filters—a comparative study, in *4th International Conference on Recent Advances in Information Technology, RAIT* (2018)

10. V. Vachak, A. Khare, A. Srivatava, Design of EMI filter for boost PFC circuit. IOSR J. Electr. Electron. Eng. (IOSR-JEEE) **9**(5) Ver. 1 (2014)
11. M. Asim, H. Parveen, M.A. Mallick, A. Siddiqui, Performance evaluation of PFC boost converters. Int. J. Innov. Res. Electr., Electron., Instrum. Control. Eng. **3**(11) (2015)
12. Y. Xiong, Y. Huang, Z. Zhong, T. Zhang, Research and design on digital PFC of 2kW On-Board charger. World Electr. Veh. J. **4** (2010). ISSN 2032–6653

Direct Torque Control of Asymmetrical Multiphase (6-Phase) Induction Motor Using Modified Space Vector Modulation

Vishal Rathore and K. B. Yadav

1 Introduction

The machine with high phase order, greater than three, increases the rating and power handling capability of a drives system. The drives with more number of phases in the stator of the machine are called multiphase drives. These multiphase drives possess various potential benefits over its three-phase counterparts, such as increase in frequency of torque pulsation by reducing the amplitude of pulsating torque, lowering per phase current without increase in voltage per phase, reduces the rotor and D.C link current harmonics, posses high fault tolerant capability. As an effect of which if a fault occurs in one or more phases the machine will continue to run without any interruption [1]. The Six-Phase Induction Machine (SPIM) is more commonly used multiphase machine and is also known as the dual star induction machine. Such type of machine is used where high power is required, as in case of aircraft applications, electric/hybrid vehicles, naval system, mine hoist, cement mills, rolling mills [2].

In the mid of 1980 I. Takahashi first proposed the direct torque control (DTC) of induction machine [3], this control strategy is highly efficient and more commonly used for AC machine drives system to provide quick flux and torque control. The DTC control strategy proposed here is based on the principle of rectifying error between the estimated values and the required values of corresponding torque and flux, the different states of two-level VSI are directly controlled to reduce the errors of flux and torque in between the band limits [4, 5].

V. Rathore (✉) · K. B. Yadav
Department of Electrical Engineering, NIT Jamshedpur, Jamshedpur 831014, Jharkhand, India
e-mail: vishalrathore01@gmail.com

K. B. Yadav
e-mail: kbyadav.ee@nitjsr.ac.in

© The Author(s), under exclusive license to Springer Nature Singapore Pte Ltd. 2021   511
J. Kumar and P. Jena (eds.), *Recent Advances in Power Electronics and Drives*, Lecture Notes in Electrical Engineering 707,
https://doi.org/10.1007/978-981-15-8586-9_45

Fig. 1 Generalized phasor
representation of a
multiphase machine

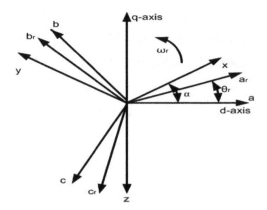

The work presented in this paper is discussed in five sections. The second section presents phase arrangement of A6PIM. Whereas the DTC based control strategy is implemented in section three. The results of simulation are discussed in section four. Finally, the summary of the work is made in conclusion.

2 A6PIM Phasor

The model implemented for the analysis, consists of two sets of six phases winding on the stator with a phase shift of $120°$ and displaced by an angle α. The rotor winding of 6-phase motor is same as that of its three-phase counterparts [6, 7]. The motor under consideration has no friction and windage losses and is also free from magnetic saturation of the core. The stator winding of a six-phase motor has two sets of winding, namely, 'abc and xyz', and their axes are phase displaced by $\alpha = 30°$ [8, 9]. The rotor windings a_r, b_r, c_r as shown in Fig. 1 has sinusoidal distribution with the phase displacement of $120°$.

3 Direct Torque Control

As the name indicate, DTC allows control of flux and electromagnetic torque directly and independently by selecting proper switching vector [10, 11]. Figure 2 represents the basic blocks of modified DTC method implemented to A6PIM. The reference values of torque (T_e^*) and flux (Φ_s^*) are analyzed with the obtained values and the resultant errors are fed into a PI controller which allows the control of the motor in both directions of rotation.

The idea of the DTC is to lower the value of errors in flux and torque between the specified hysteresis band of torque and flux by the proper selection of switching

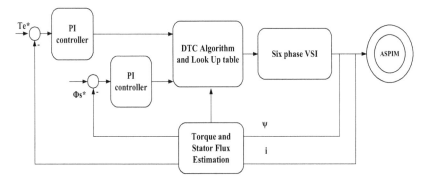

Fig. 2 Conventional DTC of SPIM

state of two levels inverter as shown in Table 1. In six-phase VSI, there are 64 states for switching that each state generates a voltage vector. In other words, there are 64 voltage vectors for the inverter. These voltage vectors can be shown in (x, y) and $(d-q)$ planes. Projecting the 64 voltage vectors in (x, y) plane, vectors with different lengths are appeared in this plane. In fact, the length of the switching vectors in (x, y) plane can be classified into four different sizes [12]. To obtain larger voltage amplitude, only switching vectors with largest size are selected in (x, y) plane [13, 14]. Figure 3 shows the projection of these 12 vectors in (d, q) and (x, y) planes.

Table 1 Switching states for DTC

Sector		*Dm*		
		$1\uparrow$	0	$-1\downarrow$
dψ	$1\uparrow$	V_{60}	V_0	V_{35}
	$0\downarrow$	V_{28}	V_0	V_3

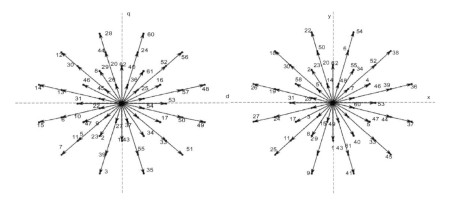

Fig. 3 **a** $d-q$ axis and **b** $x-y$ axis of voltage vectors [13]

The numbers written near the vectors represent the state of the six-phase VSI, when these numbers are converted to binary, six bits are obtained. Each zero/one in these binary numbers is corresponded to a switch in the related inverter leg.

The number zero means the lower switch is on and the number one means the upper switch is on. These vectors have the smallest lengths among vectors in (x, y) plane which decreases (d, q) voltage components; however, it is not enough a small amount of (x, y) voltage components results in large current of their axis. The DTC strategy should enforce (x, y) voltage components to become almost zero. If the flux and torque errors are increased, then the switching vector of the voltage is selected in such a way so that there is increase in stator magnitude and flux angle. Representation of space vectors of the voltage in $d–q$ and $x–y$ axis are illustrated in Fig. 3. All the space vectors of voltage that are active are categorized on the basis of their magnitude into four different groups (D_1, D_2, D_3, and D_4), where D_1 is classified as the group of largest magnitude vectors and D_4 is classified as the group of smallest magnitude vectors. Vectors of the same group in $d–q$ subspace fall in the same group of $x–y$ sub-space, for example, D_1 of $d–q$ falls in $D1$ of $x–y$ and vice versa. Similarly D_2 and D_3 in $d–q$ subspace fall in D_2 and D_3 of $x–y$ subspace, respectively.

4   Result of Simulation

The result of the proposed direct torque control is presented in this section where A6PIM operates at a speed of 1p.u and a load torque of 1.5 p.u under steady-state condition is illustrated in respective Fig. 4 and Fig. 5. Waveform of phase current (Ia) with THD of 35.46% with respect to reference voltage (Vr) is shown in Fig. 6. By observing the current waveform it has been found that the current harmonics of the order 5th and 7th are present in proposed method. Figure 7 shows the ripple contain in power of the A6PIM under steady-state condition.

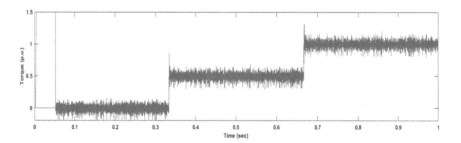

Fig. 4 Steady-state torque

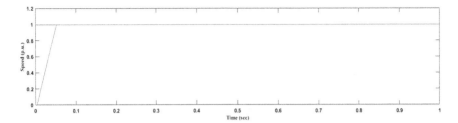

Fig. 5 Speed under steady state

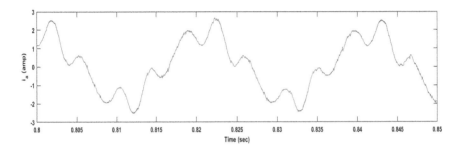

Fig. 6 Current waveforms of phases '*a*'

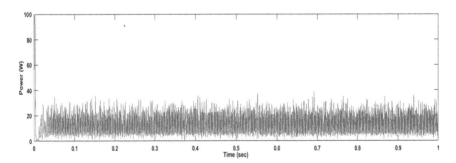

Fig. 7 Ripple contain in power

5 Conclusion

DTC of an asymmetrical multiphase 6-phase induction machine is proposed in this work. The proposed technique uses modified SVM to reduce the Total Harmonic Distortion around the acceptable limits, as obtained from commonly used conventional technique. The value of currents in x–y subspace is plotted. Vectors of groups D_1 and D_2 are formed by assigning proper times. The proposed method is free from switching sequence and no designed is required and dynamic behavior of the machine remains unchanged.

Appendix

Parameters and ratings of symmetrical six-phase induction motor under test

Poles (P)	2	Moment of inertia (J)	0.0030 Kg m^2
Stator and rotor resistance (r_s and r_r)	11.5 & 10.4 O	Power	1 HP
Frequency (f)	50 Hz	Mutual inductance (L_m)	550.7 mH
Speed (ω_{ref})	1450 rpm	Stator and rotor inductance (L_s, L_r)	579.17 mH

References

1. E. Levi, R. Bojoi, F. Profumo, H. Toliyat, S. Williamson, Multiphase induction motor drives—a technology status review. IET Electr. Power Appl. **1**(4), 489–516 (2007)
2. E. Levi, Multiphase electric machines for variable-speed applications. IEEE Trans. Industr. Electron. **55**(5), 1893–1909 (2008)
3. I. Takahashi, T. Noguchi, A new quick-response and high-efficiency control strategy of an induction motor. IEEE Trans. Ind. Appl. **IA-22**(5), 820–827 (1986)
4. E. Levi, Advances in converter control and innovative exploitation of additional degrees of freedom for multiphase machines. IEEE Trans. Ind. Electron. **63**(1), 433–448 (2016)
5. V. Rathore, K.B. Yadav, *Analytical Model Based Performance Characteristics Analysis of Six-Phase Induction Motor*. SSRN 3575381, April 2020
6. V. Rathore, M. Dubey, Speed control of asynchronous motor using space vector PWM technique international. J. Electr. Eng. Technol. (IJEET) **3**, 222–233 (2012)
7. K. Hatua, V. Ranganathan, Direct torque control schemes for split phase induction machine. IEEE Trans. Ind. Appl. **41**(5), 1243–1254 (2005)
8. K.D. Hoang, Y. Ren, Z.-Q. Zhu, M. Foster, Modified switching table strategy for reduction of current harmonics in direct torque controlled dual-three-phase permanent magnet synchronous machine drives. IET Electr. Power Appl. **9**(1), 10–19 (2015)
9. A.K. Mohanty, K.B. Yadav, Simulation based analytical investigation of multi-phase induction machine in motoring and generating mode. Int. J. Eng. Technol. (IJET) **8**(6), 2319–8613 (2017)
10. G. Boukhalfa, S. Belkacem, A. Chikhi, S. Benaggoune, Direct torque control of dual star induction motor using a fuzzy-PSO hybrid approach. Appl. Comput. Inf. (2018)
11. Y. Ren, Z. Zhu, Enhancement of steady-state performance in direct torque- controlled dual three-phase permanent-magnet synchronous machine drives with modified switching table. IEEE Trans. Industr. Electron. **62**(6), 3338–3350 (2015)
12. Y. Ren, Z. Zhu, Reduction of both harmonic current and torque ripple for dual three-phase permanent magnet synchronous machine using modified switching-table-based direct torque control. IEEE Trans. Industr. Electron. **62**(11), 6671–6683 (2015)
13. J.K. Pandit, M.V. Aware, R. Nemade, E. Levi, Direct torque control scheme for a six-phase induction motor with reduced torque ripple. IEEE Trans. Power Electron. **32**(9), 7118–7129 (2016)
14. J.K. Pandit, M.V. Aware, R. Nemade, Y. Tatte, Direct torque control of asymmetric six-phase induction motor with reduction in current harmonics, in *IEEE International Conference on Power Electronics, Drives and Energy Systems (PEDES)* (IEEE, 2016), pp. 1–6

DTC of Matrix Converter Fed Induction Machine Based on Fuzzy Logic

Raju Kumar Swami and Vinod Kumar

1 Introduction

The DC Motors are commonly used but need timely maintenance and replacement which makes these motors less reliable and also increases the running cost of the drive. The induction motor overcomes these drawbacks. Induction motors are used widely due to its rugged construction and low maintenance. Conventionally AC–DC–AC converters are used to produce the controlled AC output from the constant AC input for variable speed drive which requires higher maintenance cost, have poor efficiency due to two-stage power conversion and also the large size and high cost as a large size capacitor needed in the DC link. Matrix converter provides sinusoidal input/output current, able to flow the bidirectional power, and also it does not require any capacitor due to this it has compact size and less weight [1–3]. Initially the Matrix converter was introduced by Venturini [4]. The SVM is used in the control and analysis of AC to AC converters [5, 6] introduced. Direct torque control (DTC) initially proposed as direct self control by [7] and as DTC by [8] for the IM drive.

In DTC a comparative control of the flux and torque using the separate certain limits. The flux and torque control in DTC method is independent of each other [9]. DTC is popularly used in industrial world as it yields very fast response, simple structure, and easy to implement [10].

The drawback of DTC method is large ripples in electromagnetic torque, resulting distortion in stator current, and during starting the response is slow. These drawbacks

R. K. Swami (✉)
Pacific University, Udaipur, Rajasthan, India
e-mail: raju.swami0404@gmail.com

V. Kumar
College of Technology & Engineering, Udaipur, Rajasthan, India
e-mail: vinodctae@gmail.com

© The Author(s), under exclusive license to Springer Nature Singapore Pte Ltd. 2021   517
J. Kumar and P. Jena (eds.), *Recent Advances in Power Electronics and Drives*, Lecture Notes in Electrical Engineering 707,
https://doi.org/10.1007/978-981-15-8586-9_46

can be suppressed using artificial intelligence (AI) techniques. In various literatures the torque ripples minimization is done using the artificial neural [11, 12]. Using FLC the torque ripples are reduced in induction machine fed by MC [13, 14]. The FLC used direct torque control in three-level NPC fed IM [15]. The simulation study of inverter fed IM drive presented using the fixed, variable, and fuzzy hysteresis band [16].

In this paper the IM drive fed by MC controlled using fuzzy logic-based DTC (FLDTC). The controller controls the torque hysteresis limits resulting in the performance of the drive improve. The torque ripples in IM are minimized. The overall performance of the MC fed IM drive enhanced using the proposed controller.

2 Matrix Converter Control

The nine bidirectional switches yield the flexibility in matrix converter to allow all the output phases(three) can individually connected with the any of input phase. To avoid the short circuiting of the supply the input phase never be shorted and the output terminals never be open circuited [10]. Considering the above constraints

$$S_{ij} = \begin{cases} 1, & \text{switch } S_{ij} \quad \text{closed} \\ 0, & \text{switch } S_{ij} \quad \text{open} \end{cases} \tag{1}$$

where S_{ij} is the switching function, and

$$S_{ia} + S_{ib} + S_{ic} = 1 \tag{2}$$

The state of the MC switches represented by matrix T, which is represented as follows:

$$T = \begin{bmatrix} S_{Aa} & S_{Ab} & S_{Ac} \\ S_{Ba} & S_{Bb} & S_{Bc} \\ S_{Ca} & S_{Cb} & S_{Cc} \end{bmatrix} \tag{3}$$

The power transfer in matrix converter is instantaneous and thus the current and voltage at any side in the matrix converter can be reconstructed using the voltage and current of other side. The instantaneous input phase voltages and output phase voltages relation is given as below:

$$V_o = TV_i \tag{4}$$

$$
\begin{bmatrix} v_{Aph} \\ v_{Bph} \\ v_{Cph} \end{bmatrix} = \begin{bmatrix} S_{Aa} & S_{Ab} & S_{Ac} \\ S_{Ba} & S_{Bb} & S_{Bc} \\ S_{Ca} & S_{Cb} & S_{Cc} \end{bmatrix} \begin{bmatrix} v_{aph} \\ v_{bph} \\ v_{cph} \end{bmatrix} \tag{5}
$$

'An' denote for output and 'an' for input phase.

3 Direct Torque Control

The flux and torque of IM drive are controlled independently in the DTC method [10]. For any selected output vector of VSI, there are six switching configurations (SC). Since the DTC for the MC is derived using the conventional DTC for VSI. The six switching configurations like VSI are applied to the MC such that it generates the same voltage vector of output and current vector of input as shown in Fig. 1. There are twenty-seven maximum possible switching configurations (SCs) in 3×3 MC, twenty-one out of these twenty-seven SCs are useful. The 18 switching configurations among these 21 are represented as ± 1 to ± 9 and rest three switching configurations are for the zero output vectors.

The DTC for the MC fed IM drive is shown in Fig. 2. In the controller only the current of output and voltage of input are measured and rest are calculated. The use of a hysteresis controller (three-level) for electromagnetic torque error yields the flexibility to the IM drive to run in all the quadrants. $Sin\phi_i$ is the third controlling component where ϕ_i is the input voltage and the current.

Direction and magnitude of output vector of the MC depend upon the position of the input vector. Considering the flux vector position in sector I and error of flux at the lower end of the flux hysteresis, i.e., -1, then voltage vector V_2 and V_6 are selected which increases the magnitude of flux, and vector V_3 and V_5 are selected to

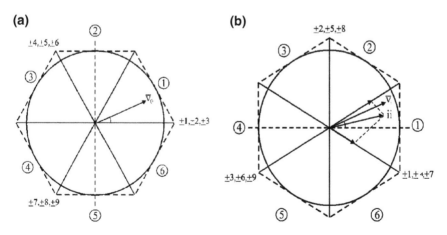

Fig. 1 **a** Voltage vector (output). **b** Current vector (input)

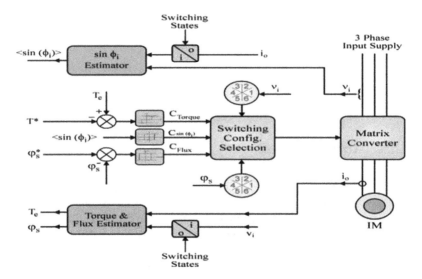

Fig. 2 DTC scheme for the MC fed IM drive

decrease the magnitude of the flux, and vector V_2 and V_3 increase the electromagnetic torque at the same time, and vector V_5 and V_6 decrease the electromagnetic torque. Irrespective of stator flux condition the zero vectors are selected in case of zero torque comparator output. Table 1 shows the basic DTC switching table.

For generating a same direction voltage vector and magnitude as V_1 the possible SCs are ± 1, ± 2, ± 3. Among these six vectors those having higher magnitude and having the same direction of V_1 are considered. Such vectors satisfy the flux and torque requirement. If the vector of input voltage lies in sector 1, the SCs $+1$ and -3 fulfill the above-mentioned conditions.

Table 1 Look-up Table for selecting output vector

Sector of stator flux		I	II	III	IV	V	VI
$C_{Flux} = -1$	$C_{Torque} = -1$	V_2	V_3	V_4	V_5	V_6	V_1
	$C_{Torque} = 0$	V_7	V_0	V_7	V_0	V_7	V_0
	$C_{Torque} = 1$	V_6	V_1	V_2	V_3	V_4	V_5
$C_{Flux} = 1$	$C_{Torque} = -1$	V_3	V_4	V_5	V_6	V_1	V_2
	$C_{Torque} = 0$	V_0	V_7	V_0	V_7	V_0	V_7
	$C_{Torque} = 1$	V_5	V_6	V_1	V_2	V_3	V_4

Table 2 Input variables fuzzy sets and the respective MFs

Fuzzy set or label	Set description (error)	Range	MFs
NL	Negative direction (large)	−1.0 to −1.0 −1.0 to −0.4	Gaussian
NS	Negative direction (small)	−0.7 to −0.4 −0.4 to 0.0	Gaussian
ZO	Around zero	−0.4 to 0.0 0.0 to 0.4	Gaussian
PS	Positive direction (small)	0.0 to 0.4 0.4 to 0.7	Gaussian
PL	Positive direction (large)	0.4 to 1 1 to 1	Gaussian

4 Fuzzy Logic Control

The DTC method is widely used due to its easy implementation and fast response. In DTC there are large ripples contents in the torque and this is the disadvantage of the DTC method. In order to overcome these drawbacks fuzzy logic is used in proposed system. The development of the fuzzy controller is done with the help of observations of the results of the conventional DTC method. The higher limit and lower limit of the torque hysteresis are controlled with FLC to minimize the ripples. The FLC has two inputs and both of the inputs have five membership functions. The fuzzy sets and respective membership function(MF) of the input variables are shown in Table 2.

5 Result and Discussion

Response of system is investigated at constant set speed of 100 rad/s and at zero load torque to test the proposed system in steady condition at no load. The results of the FLDTC-based system are compared with the conventional. The comparative speed responses are shown in Fig. 3a, the enlarge view is shown in Fig. 3b. The IM drive with FLC yields faster response. The IM drive using FLDTC meets the desired speed in 0.18 s while the conventional DTC-based system settle down in 0.21 s.

Figure 4 shows the comparative response of the stator flux. It is seen from Figs. 5a, b and 6a, b that the maximum ripple using proposed controller is 0.41 N-m whereas the ripples in torque of the IM using conventional DTC are 0.48 N-m. It is seen that the ripples in torque are lesser in the proposed controller-based induction motor drive then conventional controller.

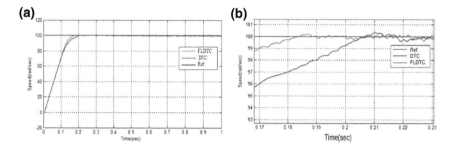

Fig. 3 **a** Comparative speed of IM, **b** enlarge view

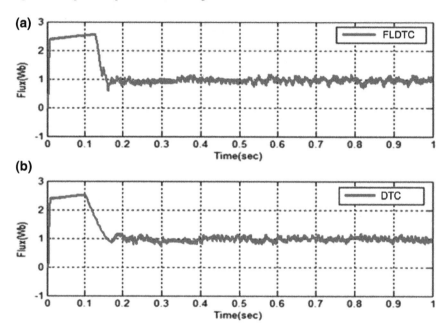

Fig. 4 Response of the flux **a** with and **b** without fuzzy logic controller-based IM drive

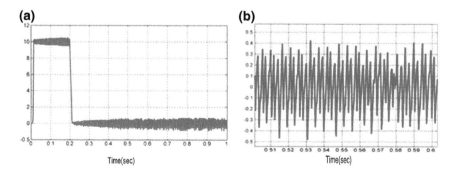

Fig. 5 **a** Torque response of IM using conventional DTC, **b** enlarge view

(a) **(b)**

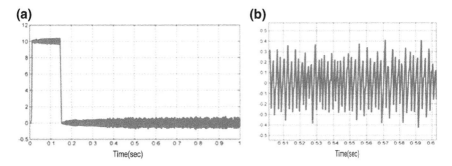

Time(sec)   Time(sec)

Fig. 6 **a** Torque response of IM using proposed controller, **b** enlarge view

6 Conclusion

In this paper the FLDTC for IM fed by matrix converter (MC) is proposed. The DTC yields fast response and the matrix converter has less weight and compact in size. The drawback of the large ripples of torque in conventional DTC method is overcome using fuzzy logic. Ripple contents are reduced by using the proposed controller. Also the starting performance of the IM is improved. The comparison of the results of IM drive using FLDTC compared with the conventional DTC is shown. The result shows the enhanced performance using the proposed controller. The comparative results in steady state and at starting conditions are shown and analyzed.

References

1. E. Yamamoto, T. Kume, H. Hara, T. Uchino, J. Kang, H. Krug, Development of matrix converter ans its applications in industry, in *35th Annual Conference of the IEEE Industrial Electronics Society IECON, Porto, Portugal*, (2009)
2. J. Rodriguez, M. Rivera, J. Kolar, P. Wheeler, A review of control and modulation methods for matrix converters. IEEE Trans. Indus. Electron. **59**, 8–70 (2012)
3. J. Rodriguez, A new control technique for AC–AC converters, in *Proceedings of the IFAC Control in Power Electronics and Electrical Drives Conference* (Lausanne, Switzerland, 1983), pp. 203–208
4. M. Venturini, A new sine wave in sine wave out, conversion technique which eliminates reactive elements, in *Proceedings of the POWERCON*, vol. 7 (1980), pp. E3_1–E3_15
5. M. Braun, K. Hasse, A direct frequency changer with control of input reactive power, in *Proceedings of the IFAC Control in Power Electronics and Electrical Drives Conference* (Lausanne, Switzerland, 1983), pp. 187–194
6. G. Kastner, J. Rodriguez, A Forced Commutated Cycloconverter With Control Of The Source And Load Currents, in *Proceedings of the EPE'85* (1985), pp. 1141–1146
7. M. Depenbrok, Direct self-control (DSC) of inverter fed induction machine. IEEE Trans. Power Electron. **3**, 420–429 (1988)
8. I. Takahashi, T. Noguchi, A new quick-response and high efficiency control strategy of an induction motor. IEEE Trans. Ind. Appl. **22**, 820–827 (1986)

9. D. Xiao, F. Rahman, A modified DTC for matrix converter drives using two switching configurations, in *Power Electronics and Applications. EPE'09. 13th European Conference on* (2009), pp. 1–10
10. C. Ortega, A. Arias, C.J. Balcells, G.M. Asher, Improved waveform quality in the direct torque control of matrix-converter-fed PMSM drive. IEEE Trans. Indus. Electron. **57**(6), 2101–2110 (2010)
11. V.M.V. Rao, A. Anand Kumar, Artificial neural network and adaptive neuro fuzzy control of direct torque control of induction motor for speed and torque ripple control, in *2nd International Conference on Trends in Electronics and Informatics (ICOEI)* (Tirunelveli, 2018), pp. 1416–1422
12. K. Bouhoune, K. Yazid, M.S. Boucherit, B. Nahid-Mobarakeh, Simple and efficient direct torque control of induction motor based on artificial neural networks, in *IEEE International Conference on Electrical Systems for Aircraft, Railway, Ship Propulsion and Road Vehicles & International Transportation Electrification Conference (ESARS-ITEC)* (Nottingham, 2018), pp. 1–7
13. C. Venugopal, Fuzzy logic based DTC for speed control of matrix converter fed induction motor, in *IEEE International Conference on Power and Energy* (2010)
14. R.K. Swami. V. Kumar, R.R. Joshi, FPGA-based implementation of fuzzy logic DTC for induction motor drive fed by matrix converter. IETE J. Res. (2019)
15. O. Aissa, S. Moulahoum, N. Kabache, H. Houassine, DTC of induction motor drive fed by NPC three level inverter based on fuzzy logic, in *IEEE 16th International Conference on Harmonics and Quality of Power (ICHQP)* (2014), pp. 214–218
16. N. Farah, M.H.N. Talib, Z. Ibrahim, S.N. Mat Isa, J.M. Lazi, Variable hysteresis current controller with fuzzy logic controller based induction motor drives, in *2017 7th IEEE International Conference on System Engineering and Technology, (ICSET)* (2017), pp. 122–127

CPSIA information can be obtained
at www.ICGtesting.com
Printed in the USA
LVHW022325281220
675234LV00001B/25